ORGANIC LIGHT EMITTING DEVICES

Edited by **Jai Singh**

Organic Light Emitting Devices
http://dx.doi.org/10.5772/2589
Edited by Jai Singh

Contributors

Jai Singh, Meiso Yokoyama, Byoungchoo Park, Soon Moon Jeong, Hideo Takezoe, S. Ayachi, A. Mabrouk, M. Bouachrine, K. Alimi, Dashan Qin, Jidong Zhang, M.G.Kaplunov, S.N. Nikitenko, S.S. Krasnikova, Elmars Zarins, Aivars Vembris, Valdis Kokars, Inta Muzikante

Published by InTech

Janeza Trdine 9, 51000 Rijeka, Croatia

Publishing Process Manager Mirna Cvijic
Technical Editor InTech DTP team
Cover InTech Design team

Additional hard copies can be obtained from orders@intechopen.com

Organic Light Emitting Devices, Edited by Jai Singh
p. cm.
ISBN 978-953-51-0850-4

For EU product safety concerns:
IN TECH d.o.o., Prolaz Marije Krucifikse Kozulić 3, 51000 Rijeka, Croatia,
info@intechopen.com or visit our website at intechopen.com.

Meet the editor

Jai Singh is a Professor of Physics at the School of Engineering and IT, Charles Darwin University, Australia. He is an internationally recognised theoretician in the field of excitonic processes in condensed matter, both bulk and nanostructures, optoelectronic properties of thin film organic and inorganic solar cells, organic light emitting devices and scintillators. He has made significant contributions to the literature in these fields with 200 research articles and two research monographs: (i) Excitation Energy Transfer Processes in Condensed Matter (Plenum, 1994) and (ii) Advances in Amorphous Semiconductors with K. Shimakawa (Taylor and Francis, 2003). He has also edited a book Optical Properties of Condensed Matter and Applications (John Wiley and Sons, 2006). He has held numerous visiting posts around the world, including Germany (Max-Planck Institute and Hahn Meitner Institute Berlin), Japan (as a JSPS Fellow) and USA (UC-San Diego and Lawrence Berkeley National Laboratory, Berkeley).

Contents

Preface

The world has seen a huge research development in the field of organic light emitting devices (OLEDs) (commonly known as the organic light emitting diodes) in the last two decades. This is because such devices exhibit a number of extra technical features in flat panel displays that conventional display devices do not. These advantageous features include high brightness and contrast, high luminous efficiency, fast response time, wide viewing angle, low power consumption, flexible structures, light weight and low cost manufacturing chemical technology. Furthermore, the advancement in developing the concept of rendering white light from OLEDs for domestic, commercial and street lighting has boosted the research activity in this field. The organic white light emitting devices (OWLEDs), traditionally known as the white organic light emitting devices (WOLEDs) show promise to have a major share in future ambient lighting due to their favourable properties mentioned above. In addition, as similar chemical technology can be used to fabricate both, WOLEDs and organic solar cells (OSCs), it opens new ways in lighting design such as light ceilings or luminous objects of almost any shape. A common flexible substrate can be used to fabricate WOLEDs on one side facing a room and OSCs on the other side facing the sun. The device thus fabricated can generate power in the day time and light a room/house at night. This book covers developments on OLEDs, WOLEDs and briefly on OSCs as well.

Several other books have recently appeared on OLEDs and related topics but the field is too active to cover up to date development in one volume for very long. Although the existing resources can provide valuable fundamental information on organic electroluminescence, this book provides the state-of-the-art advancement in the field of organic electroluminescence contributed by many established experts in the field.

This volume contains eight chapters contributed by many expert researchers worldwide. Chapter 1 introduces the mechanism of light extraction from the excited singlet and triplet exciton states. The concept of capturing the triplet exciton emission has been fully understood only recently through the invention of the new time-dependent exciton-spin-orbit-photon interaction operator and its role is very critical in fabricating WOLEDs. Chapter 2 deals with the development of field emission organic light emitting diodes (FEOLEDs) which have a better balance (equality) in the number of injected electrons and holes and hence have higher luminous efficiency. Both fabrication and the performance of the resulting devices are discussed.

Chapters 3 and 4 focus on the extraction of polarized light from OLEDs. In chapter 3 the fabrication of OLEDs for generating highly polarized electroluminescent (EL) through the innovative approach of using photonic reflective polarizers is presented and their electroluminescent properties are fully characterized. In chapter 4, the minimisation of trapping of light within OLEDs is described with a view to maximise the light extraction by incorporating nanostructures. The polarization of electroluminescence through the use of innovative reflectors is also covered.

Chapter 5 presents a very comprehensive study of the photophysical properties of two new donor-acceptor conjugated copolymers and their model compounds. These properties are measured as well as calculated and results are used to analyse possible applications in fabricating polymer light emitting diodes (PLEDs) and polymer photovoltaic cells (PPCs). Chapter 6 describes state-of-the-art advancement of charge carrier injection techniques that can be used for fabricating high-power OLEDs. Different designs of OLEDs have been fabricated and tested in order to optimise the injection.

Chapter 7 describes the role of exciplex electroluminescence of new organic materials which may be used for fabrication of WOLEDs. Many structures are explored and their electroluminescent properties are characterized. Finally in chapter 8 are described the synthesis and physical properties of pyranylidene and isophorene fragments containing donor-π-acceptor type chromphores which have the ability to emit light in the visible part of the spectrum. The chapter presents a very comprehensive study of many organic structures and discusses their potential applications in the fabrication of OLEDs.

This book has resulted from the sole effort of InTech from inviting the contributors to publication. I would like to express my gratitude to all the contributors for their enthusiasm and dedication in preparing their contributions. Without their efforts the realisation of this book would not have been possible. It is my great pleasure to acknowledge the support of Ms Mirna Cvijic and Ms Viktorija Zgela from InTech in completing my editorial job successfully and efficiently.

Prof. Jai Singh
Charles Darwin University, Darwin
Australia

Chapter 1

Harvesting Emission in White Organic Light Emitting Devices

Jai Singh

Additional information is available at the end of the chapter

http://dx.doi.org/10.5772/53130

1. Introduction

The current use of lighting in buildings and streets accounts for a significant percentage of the electricity consumed in the world at present and nearly 40% of that is consumed by inefficient thermoluscent incandescent lamps, only about 15 lm/W. This has created interest in investigating more efficient electroluminescent sources of white light for use in domestic, industrial and street lighting. The total light output efficieny η_{out} efficiency of an electroluminescent lighting device depends on the internal qantum efficiency η_{int} and the photon out-coupling efficiency η_{ph} as [1]:

$$\eta_{out} = \eta_{\text{int}}\eta_{ph}, \tag{1}$$

where η_{int} is the ratio of number of radiative recombinations to the number of electrically injected electrons and holes from opposite electrodes of the device and it is given by:

$$\eta_{\text{int}} = \gamma\eta_{ex}. \tag{2}$$

Here γ is the ratio of number of electrons to that of holes, or vice versa, injected from the opposite electrodes of a device so that $\gamma \le 1$ is maintained. η_{ex} is the fraction of the injected electron (e) and hole (h) pairs that recombine radiatively due to their Coulomb interaction. $\eta_{ph} \approx \dfrac{1}{2n^2}$, where n is the index of refraction of the substrate through which the light comes out. In the case of a glass substrate with $n = 1.5$, $\eta_{ph} \approx 20\%$.

The schematic of a very simple electroluminescent device can be envisaged as a single thin film of an electroluminescent layer sandwiched between anode and cathode electrodes, as shown in Fig. 1. In this case the anode is made of a transparent conducting oxide (usually indium tin oxide (ITO) and cathode is a metal, usually Al, Ca, Ag, etc). If the electroluminescent layer is of any direct band gap inorganic semiconductor, for example, based on GaAs and InP, then the injected electrons and holes from the cathode and anode, respectively, remain free electron and hole pairs and recombine radiatively by emitting light. In inorganic semiconductors, the static dielectric constant is relatively high (12.9 for GaAs and 12.5 for InP) which to a relative extent prevents the injected free charge carriers from forming bound hydrogenic excited states, called excitons. This is easy to understand as the atractive Coulomb potential energy between e and h is given by:

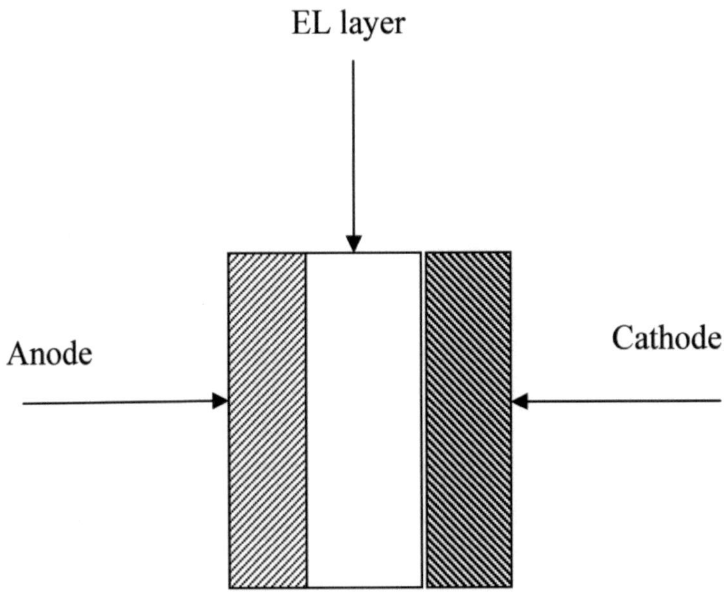

Figure 1. Schematic design of a single layered electroluminescent (EL) device sandwiched between anode and cathode electrodes.

$$E_p = -\frac{\kappa e^2}{\varepsilon r}, \tag{3}$$

where $\kappa = (4\pi\varepsilon_0)^{-1} = 8..9877 \times 10^9$, e is the electronic charge, ε is the static dielectric constant and r is the average separation between the injected electrons and holes. According to Eq. (3), materials with larger ε will have reduced binding energy(E_B) between the injected electrons and holes and hence they remain free charge carriers. The binding energy is equal to the magnitude of E_p ($E_B = |E_p|$) in Eq. (3).

In contrast organic semiconductors, both of small molecules and polymers, have lower dielectric constant ($\varepsilon \approx 3$) which enhances the binding energy about four times larger than that in inorganic materials. Such a large binding energy between electrons and holes enables them to form excitons immediately after their injection from the opposite electrodes. On one hand, the formation of excitons due to their Coulomb interaction assists their radiative recombination leading to electroluminescence. On the other hand, excitons can be formed in two spin configurations, singlet and triplet and this complicates the mechanism of radiative recombination because the recombination of singlet excitons is spin allowed but that of triplet excitons is spin forbidden. The singlet and triplet exciton configurations are shown in Fig. 2 and accordingly the probability of forming singlet and triplet excitons may be in the ratio of one to three (1:3). If the triplet excitons cannot recombine due to forbidden spin configuration, then the light emission can occur only through singlet excitons and that means internal quantum effciency η_{int} can be only about 25%, and 75% of the injected electron-hole pairs will be lost through the non-radiative recombination due to the formation of triplet excitons. This limits the light-out efficiency $\eta_{out} = 0.25x0.2 = 0.05 = 5\%$ according to Eq. (1).

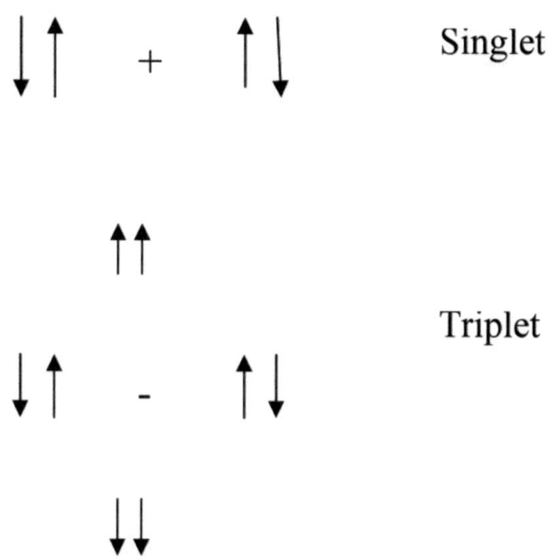

Figure 2. Spin configurations of electron and hole pairs in forming an exciton. Pairs of arrows represent pairs of electron and hole. The upper combination of spin configurations represents the single possibility for formation of a singlet exciton and lower three spin configurations represent the three posibilities for formation of a triplet exciton.

However, Cao *et al.* have found that the ratio of quantum efficiencies of EL with respect to PL in a substituted PPV-based LED can reach as high as 50% [2]. This higher quantum internal efficiency is attributed to larger cross section for an electron-hole pair to form a singlet exciton than that to form a triplet exciton [3] as explained below. If one denotes the cross section of the formation of a singlet exciton by σ_S and that of a triplet by σ_T then by assuming that all pairs of injected e and h form excitons, the internal quantum efficiency can be expressed in terms of cross sections as $\eta_{ex} = \dfrac{\sigma_S}{\sigma_S + 3\sigma_T}$. Thus, for $\sigma_S = \sigma_T$ one gets

$\eta_{ex} = 0.25$ (or 25%), for $\sigma_S = 3\sigma_T$, $\eta_{ex} = 0.5$ (50%) and for $\sigma_T = 0$, $\eta_{ex} = 1$ (100%). This suggests that if one can minimise the cross section of the formation of triplet excitons one can maximise the internal quantum efficiency in OLEDs. However, for modifying the cross sections one has to know the material parameters on which these cross sections depend and then one has to manipulate those parameters to minimise the triplet cross section. This approach has not been applied yet probably because the dependence of cross sections on the material parameters has not been well studied. The other approach of increasing η_{ex} to 100% is by harvesting the radiative emissions from all triplet excitons as well as has been achieved by Adachi et al. [1]. The mechanism of this approach and process will be presented in detail here.

Thus, as the formation of triplet excitons is more probable than singlet, it is very desirable to capture the full emission from triplet excitons in OLEDs. It may be noted that the mechanisms of singlet and triplet emissions are different because of their different spin configurations and therefore the emission from singlet excitons is known as electrofluorecence and that from triplet excitons as electrophosphorescence in analogy with the terms used in photoluminescence. The description presented above may raise a question in your mind why then one should make any effort in organic solids/polymers for fabricating light emitting devices if the emission from triplet excitons cannpot be harvested. This is because OLEDs have the potential of being produced by one of the very cost effective chemical technolgies.

In additon, by harvesting emissions from both singlet and triplet excitons not only the 100% internal quantum efficiency (η_{int}) can be achieved but also the white light emission can be achieved by incorporating fluorescent blue emitters (emission from singlet excitons) combined with phosphorescent green and red emitters (emission from triplet excitons) in the electroluminescent layer of OLEDs. Materials from which singlet emission can be harvested are called fluorescent or electro-fluorescent materials and those from which triplet emission is availed are called phosphorescent or electr- phosphorescent materials. An OLED that can emit white light is called white OLED (WOLED) actually it is an organic white light emitting device (OWLED). A successful cost effective technological development of WOLEDs is going to provide a huge socio economic benefit to mankind by providing brighter and cheaper lighting. WOLEDs show promise to have a major share in the future ambient lighting due to their very favourable properties such as homogenous large-area emission, good colour rendering, and potential realization on flexible substrates. This is expected to open new ways in lighting design such as light emitting ceilings, curtains or luminous objects of almost any shape [4-5]. Therefore, much research efforts are continued in developing more cost effective and efficient white organic light emitting devices (WOLEDs) [4,6-7].

The performance of a WOLED can be optimised by finding optimum emitting materials, manipulating the charge carrier balances and location of the recombination zone and energy transfer. The first WOLED fabricated [6] had a single poly (N-vinylcarbazole) emission layer doped with three fluorescent dyes. To achieve higher power efficiency, a combined use of blue fluorescent and green and red phosphorescent emitters in WOLEDs has been made recently [4,7]. This concept is based on the coincidence of a physical phenomenon of formation a singlet spin configuration with probability 25% and triplet with 75% between an electron and hole injected from the opposite electrodes of a device with that of a natural phenomenon that white light consists 25% of blue light and 75% of red and green lights. Thus, the combination of fluorescent (blue singlet emission) and phosphorescent (red and green or orange triplet emission) emitters is capable of reaching 100% internal quantum efficiency of white light emission by harvesting 25% singlet emission and 75% triplet emission. Although by trial and error experimental techniques on WOLEDs the triplet radiative recombination is activated by a heavy metal atom compound (phosphor) that enhances the spin-orbit interaction and hence triplet radiative recombination, the mechanism has not been fully understood theoretically until recently [8]. This is because the well known spin-orbit interaction is a stationary operator that cannot cause transitions[8-9].

In this chapter, the radiative recombination of both singlet and triplet excitons in organic solids/polymers is reviewed. Rates of spontaneous emission from both singlet and triplet excitons are calculated in several phosphorescent materials by using the recently invented new time-dependent exciton-spin-orbit-photon interaction operator [8] and found to agree quite well with the experimental results.

2. Emission from singlet excitons

Let us consider an excited pair of electron and hole created such that the electron (e) is excited in the lowest unoccupied molecular orbital (LUMO) and hole (h) in the highest occupied molecular orbital (HOMO) of organic layer sandwiched between two electrodes, and then they recombine radiatively by emitting a photon. The interaction operator between a pair of excited e and h and radiation can be written as:

$$\hat{H}_{xp} = -(\frac{e}{m_e^*}\mathbf{p}_e - \frac{e}{m_h^*}\mathbf{p}_h)\cdot\mathbf{A}, \tag{4}$$

where m_e^* and $\mathbf{p}_e$ and m_h^* and $\mathbf{p}_h$ are the effective masses and linear momenta of the excited electron and hole, respectively, and $\mathbf{A}$ is the vector potential given by:

$$\mathbf{A} = \sum_{\lambda}\left(\frac{\hbar}{2\varepsilon_0 n^2 V\omega_\lambda}\right)^{1/2}\left[c_\lambda^+\hat{\varepsilon}_\lambda e^{-i\omega_\lambda t} + c.c.\right], \tag{5}$$

where n is the refractive index, V is the illuminated volume of the material, ω_λ is the frequency and c_λ^+ is the creation operator of a photon in a mode λ, $\hat{\varepsilon}_\lambda$ is the unit polarization vector of photons and *c.c.* denotes complex conjugate of the first term. The second term of **A**, which is the complex conjugate of the first term, corresponds to the absorption and will not be considered here onward. It may be noted that in organic solids and polymers the effective masses of charge carriers are approximated by the free electron mass m_e, i.e., $m_e^* = m_h^* = m_e$.

Using the centre of mass, $\mathbf{R}_x = \dfrac{m_e^*\mathbf{r}_e + m_h^*\mathbf{r}_h}{M}$ and relative $\mathbf{r} = \mathbf{r}_e - \mathbf{r}_h$ coordinate transformations, the interaction operator $\hat{H}_{xp}$ [Eq. (4)] can be transformed into [10-11]:

$$\hat{H}_{xp} = -\frac{e}{\mu_x}\mathbf{A}\cdot\mathbf{p}, \tag{6}$$

where $\mathbf{p} = -i\hbar\nabla_r$ is the linear momentum associated with the relative motion between e and h and μ_x is their reduced mass ($\mu_x^{-1} = m_e^{*-1} + m_h^{*-1} = 2m_e^{-1} \Rightarrow \mu_x = 0.5m_e$ in organics). The operator in Eq. (6) does not depend on the centre of mass motion of e and h. Therefore, this operator [Eq. (6)] is the same for the exciton-photon interaction or a pair of e and h and photon interaction.

The field operator of an electron in LUMO can be written as:

$$|\,\mathrm{LUMO}> = \sum_{\sigma_e} |\,\Psi_{\mathrm{LUMO}} > a_{LUMO}(\sigma_e), \tag{7}$$

where $\Psi_{LUMO}(\mathbf{r}_e)$ is the molecular orbital wave function of an electron excited in the LUMO, $\mathbf{r}_e$ is the position coordinate of the electron and $a_{LUMO}(\sigma)$ is the annihilation operator of an electron with spin σ_e.

Likewise the field operator of a hole excited in HOMO can be written as:

$$|\,\mathrm{HOMO}> = \sum_{\sigma_h} |\,\Psi_{\mathrm{HOMO}} > d_{HOMO}(\sigma_h), \qquad d_{HOMO}(\sigma_h) = a^{+}_{HOMO}(-\sigma_h), \tag{8}$$

Using Eqs. (5), (7) and (8), the operator $\hat{H}_{xp}$ [Eq. (6)] of interaction between an excited e-h pair and a photon can be written in the second quantized form as:

$$\hat{H}_{xp} = -\frac{e}{\mu_x}\sum_{\lambda,\sigma_e,\sigma_h}\left(\frac{\hbar}{2\varepsilon_0 n^2 V\omega_\lambda}\right)^{1/2} Q_{LUMO,HOMO}\, c_\lambda^{+}, \tag{9}$$

where

$$Q_{LUMO,HOMO} = <\Psi_{LUMO}\,|\,\hat{\varepsilon}_\lambda.\mathbf{p}\,|\,\Psi_{HOMO} > a_{LUMO}(\sigma_e) d_{HOMO}(\sigma_h), \tag{10}$$

We now consider a transition from an initial state $|\,i>$ to a final state $|\,f>$. The initial state is assumed to have one singlet exciton created by exciting an electron in LUMO and a hole in HOMO. The spin configurations for singlet and triplet excitons used here are given [8,12] as:

$$\frac{1}{\sqrt{2}}[a_e(+1/2)d_g(-1/2)+a_e(-1/2)d_g(+1/2)]= \\ \frac{1}{\sqrt{2}}[a_e(+1/2)a_g^+(+1/2)+a_e(-1/2)a_g^+(-1/2)] \tag{11}$$

for singlets and

$$\begin{aligned} &a_e(+1/2)d_g(+1/2)=a_e(+1/2)a_g^+(-1/2) && \text{(a)} \\ &\frac{1}{\sqrt{2}}[a_e(+1/2)d_g(-1/2)-a_e(-1/2)d_g(+1/2)]= && \text{(b)} \\ &\frac{1}{\sqrt{2}}[a_e(+1/2)a_g^+(+1/2)-a_e(-1/2)a_g^+(-1/2)] \\ &a_e(-1/2)d_g(-1/2)=a_e(-1/2)a_g^+(+1/2) && \text{(c)} \end{aligned} \tag{12}$$

for triplets. We assume that there are no photons in the initial state and the final state has no excitons but only a photon in a λ mode. The transition matrix element is then obtained for singlet excitons as [10,11]:

$$<f|\hat{H}_{xp}|i>=-\frac{e}{\mu_x}\sum_{\lambda}\left(\frac{\hbar}{2\varepsilon_0 n^2 V\omega_\lambda}\right)^{1/2} p_{LUMO,HOMO}, \tag{13}$$

where

$$p_{LUMO,HOMO}=<\Psi_{LUMO}|\hat{\varepsilon}_\lambda\cdot\mathbf{p}|\Psi_{HOMO}>=i\omega\mu_x|r_{e-h}|. \tag{14}$$

Here the energy difference between the LUMO and HOMO levels is given by $\hbar\omega = E_{LUMO} - E_{HOMO}$ and $|r_{e-h}|$ is the mean separation between the excited electron and hole. It may be noted that for triplet excitons the transition matrix element vanishes. This can be easily verified using Eqs. (9) and (12). Using Fermi's golden rule for such a two level system and the transition matrix element [Eq. (13)], the rate of spontaneous emission, R_{sp12}, is obtained as [11]:

$$R_{sp12}=\frac{4\kappa e^2\sqrt{\varepsilon}\omega^3|r_{e-h}|^2}{3\hbar c^3}, \tag{15}$$

where $\varepsilon = n^2$ is the static dielectric constant and hole and $\kappa = 1/(4\pi\varepsilon_0)$. For a quantitative evaluation $|r_{e-h}|$ one should evaluate the integral in Eq. (14) using the LUMO and HOMO molecular orbitals. However, for excitons $|r_{e-h}|$ can be replaced by their excitonic Bohr radius as $|r_{e-h}| = a_x^S/\varepsilon$, a_x^S being the singlet excitonic Bohr radius and given by [10,12]:

$$a_x^S=\frac{\alpha^2}{(\alpha-1)^2}a_x^T \quad \text{where} \quad a_x^T=\frac{\mu\varepsilon}{\mu_x}a_0, \tag{16}$$

where a_x^T is the excitonic Bohr radius of a triplet exciton, $a_0 = 0.0529$ nm is the Bohr radius, and μ reduced mass of electron in hydrogen atom which is used here as equal to the free electron mass. The parameter α depends on the energy difference, ΔE_x, between singlet and triplet exciton states as [12]:

$$\alpha = \left[1 - \sqrt{1 - \frac{\Delta E_x}{C_M}}\right]^{-1} \quad \text{and} \quad C_M = \frac{\mu_x e^4 \kappa^2}{2\hbar^2 \varepsilon^2} \tag{17}$$

In organic solids, ΔE_x is estimated to be 0.7 eV [9,13] which gives α =1.38 with $\varepsilon = 3$ and the triplet exciton Bohr radius as $a_x^T = 6a_0$. According to Eq. (16) then we get the singlet exciton Bohr radius as $a_x^S \approx 79a_0$ and $|r_{e\text{-}h}| = 26.3\ a_0$. As an example, 4,40-bis(9-ethyl-3 carbazovinylene)-1,10-biphenyl [BCzVBi] used as a fluorophor in WOLEDs [4,7] has a singlet energy of 2.75 eV and corresponds to $\omega = 4.21 \times 10^{15}$ Hz. Using these in Eq. (15), one finds the rate of spontaneous emission from singlet excitons in BCzVBi is $R_{sp12} = 2.7\text{x}10^{10}\text{s}^{-1}$ and the radiative lifetime $\tau_R = R_{sp12}{}^{-1} = 3.7 \times 10^{-11}$ s. This radiative lifetime may be considered to be much shorter than the singlet lifetime usually found in the ns range. The discrepancy may be attributed to the approximations involved and to the fact that the rate depends on third power of the frequency of emitted light (ω^3), which is quite high in this case.

3. Emission from triplet excitons

As recombination of a triplet exciton state to the ground state is spin forbidden, it cannot occur unless either the triplet goes through an intersystem crossing to a singlet or a source of flipping the spin is introduced to make such a radiative recombination possible. Unlike inorganic solids, most organic solids and polymers have significant exchange energy between singlet and triplet excitons states. Therefore the mechanism of intersystem crossing may not be very efficient without doping the solids with another material of lower singlet energy state. This is possible and usually the host material is doped with a fluorescent material but some loss of energy is inevitable due to the difference in energy [4]. A more efficient way of harvesting triplet is to dope the host material with phosphorescent compounds containing heavy metal atoms, like platinum (Pt), palladium (Pd) or iridium (Ir) [1]. Here again the energy matching needs to be carefully examined otherwise an energy loss will occur. Thus, in the fabrication of a WOLED, the host polymer is doped with a fluorophore to emit the blue emission from singlet excitons and two phosphorescent compounds to emit green and red from the triplet radiative recombination [4,7]. A most efficient such combination is the host polymer being doped with a blue fluorophore 4,4'-bis(9-ethyl-3-carbazovinylene)-1,1'-biphenyl (BCzVBi) 12 in a region separate from the phosphorescent dopants, which are fac-tris (2-phenylpyridine) iridium(Ir(ppy)3) for emitting green and iridium(III) bis(2-phenyl quinolyl-N,C20) acetylacetonate (PDIr) for emitting red [4]. In some cases an orange phosphorescent dopant is used in place of red and green. It is commonly well established that the transfer of singlet excitons to blue fluorophore occurs efficiently due to the Förster transfer and that of triplet excitons to phosphorescent dopants due to Dexter or diffusive transfer. However, after that how the radiative recombination occurs by the enhanced spin-orbit interaction due to the introduction of heavy metal atoms is not thoroughly explored. The problem is that the well known expression for an electron spin-orbit interaction in an atom is given by:

$$\hat{H}_{so} = \frac{Ze^2}{2m_e^2c^2r^3}\mathbf{s.L}, \tag{18}$$

where Z is the atomic number and r is the distance of an electron from the nucleus. s and L are the spin and orbital angular momentum of the electron, respectively. It is obvious that the spin-orbit interaction, $\hat{H}_{so}$ in eq. (18) is zero for $\mathbf{s} = \mathbf{L} = \mathbf{0}$, i.e. for all s-state orbitals with l = 0 and also for singlet excitations ($\mathbf{s} = \mathbf{0}$). It is only non-zero for p- type or higher state orbitals. As the interaction in Eq. (18) is derived for a single electron in an atom, it cannot be applied for excitons which consist of a pair of electron and hole. Therefore, it cannot contribute to the radiative recombination of a triplet exciton in a semiconductor where both the singlet and triplet spin configurations arise from the first excited s-state with $n = 1$ and $l =$ 0. However, the photoluminescence spectra from both singlet and triplet excitons in the first excited state have been observed in amorphous semiconductors [14-15] as well as in WOLEDs [1].

Furthermore, the interaction operator given in Eq. (18) is a stationary interaction operator, i.e., **s** and **L** are intrinsic properties of charge carriers (electrons and holes) and are always with them. These are present in all atoms all the time like the Coulomb interaction between electrons and nucleus. Such an interaction can give rise only to the stationary effects, like splitting the degeneracy of a triplet state but it cannot cause any transitions. As the splitting depends on the strength of the spin-orbit interaction, which increases with Z, the splitting usually increases with the atomic number of the constituting atoms. However, in solids its magnitude can usually be estimated only from the experimental data (see, e.g., [16]). To the author's knowledge any such splitting in semiconductors has not been calculated theoretically.

We have recently addressed the problem [8-9] of finding a new time-dependent exciton-spin-orbit-photon interaction operator as described below.

3.1. Electron-spin-orbit-photon interaction

We consider the case of an atom of atomic number Z excited to a triplet state. Instead of using the interaction operator given in Eq. (18), we start from the interaction of an electron of spin angular monentum **s** , linear momentum **p** moving under the influence of the electric field **E** created by the nucleus as [17]:

$$\hat{H}_{so}^{at} = -\frac{eg}{2m_e^2c^2}\mathbf{s}\cdot\mathbf{p}\times\mathbf{E} \tag{19}$$

where g is the gyromagnetic ratio (g = 2), **s** and **p** are the spin angular and orbital momenta of the electron, respectively, and **E** is the electric field experieced by the electron due to the nucleus. If we now shine light on the atom then the interaction operator in Eq. (19) changes to:

$$\hat{H}_{so}^{at} = -\frac{eg}{2m_e^2c^2}\mathbf{s}\cdot(\mathbf{p}+\frac{e}{c}\mathbf{A})\times(-\frac{1}{c}\frac{\partial\mathbf{A}}{\partial\mathbf{t}}-\nabla V_n)+\frac{eg}{m_ec}\mathbf{s}\cdot\mathbf{H}, \tag{20}$$

where **A** is the vector potential of photons as used in Eq. (5) but expressed in a different form here (see Eq. (21)), V_n is the scalar potential of the nucleus and $\mathbf{H} = \nabla\times\mathbf{A}$ is the magnetic field of the electromagnetic radiation. The interaction operator in Eq. (19) gets modified in Eq. (20) due to the interaction with the electromagnetic radiation, which changes the electron orbital momentum as well as the electric field and introduces interaction between the spin of electron and magnetic field of radiation.

Within the dipole approximation ($e^{i\mathbf{k}_\lambda\cdot\mathbf{r}} \approx 1$), the vector potential is given by:

$$\mathbf{A} = \sum_\lambda A_0\hat{\varepsilon}_\lambda c_\lambda^+ e^{-i\omega_\lambda t} + c.c., \tag{21}$$

where $A_0 = \left[\frac{2\pi c^2\hbar}{\varepsilon_0\omega_\lambda V}\right]^{1/2}$. The nuclear electric field $\mathbf{E} = -\nabla Vn$, where the scalar nuclear potential V_n is given by:

$$V_n = \frac{Ze\kappa}{r_e}, \quad \text{and} \quad \nabla V_n = -\frac{Ze\kappa}{r_e^3}\mathbf{r}_e \tag{22}$$

where $\mathbf{r}_e$ is the position vector of the electron from the nucleus and $|\mathbf{r}_e| = r_e$. For $Z > 1$, the interaction between the excited electron and other valence electrons in the atom is considered to be negligible [18].

The interaction operator in Eq. (20) can be further simplified by noting that within the dipole approximation we get $\nabla\times\mathbf{A} = \mathbf{0}$, which makes the magnetic contribution vanish and also two other terms vanish because of the following:

$$\frac{e}{c^2}\mathbf{s}\cdot(\mathbf{A}\times\frac{\partial\mathbf{A}}{\partial t}) = \mathbf{0} \quad \text{(a)}$$

and

$$\frac{1}{c}\mathbf{s}\cdot\mathbf{p}\times\frac{\partial\mathbf{A}}{\partial\mathbf{t}} = -\frac{i\hbar}{c}\mathbf{s}\cdot\nabla\times\frac{\partial\mathbf{A}}{\partial\mathbf{t}} = -\frac{i\hbar}{c}\mathbf{s}\cdot\frac{\partial}{\partial t}(\nabla\times\mathbf{A}) = 0 \quad \text{(b)} \tag{23}$$

Even otherwise, the contribution of the term in (23b) is expected to be small and therefore will not be considered here.

Substituting Eqs. (22) and (23) in Eq. (20) the interaction operator contains only the following two non-zero terms:

$$\hat{H}_{so}^{at} = -\frac{eg}{2m_e^2c^2}\left(-\frac{Ze\kappa\mathbf{s}\cdot\mathbf{L}}{r_e^3} - \frac{e}{c}\mathbf{s}\cdot(\mathbf{A}\times\nabla V)\right), \quad (24)$$

where $\mathbf{L} = \mathbf{r}_e \times \mathbf{p}$ is the orbital angular momentum of electron. The first term of Eq. (24) is the usual stationary spin-orbit interaction operator as given in Eq. (18) and it is obtained in the absence of radiation. Its inclusion in the Hamiltonian as a perturbation can only split the degeneracy of a triplet state. As explained above, this term is a stationary operator and hence it cannot cause a transition. Only the last term, which depends on spin, radiation and time can be considered as the time-dependent perturbation operator and hence can cause transitions. Using Eqs. (21) and (22), the last term of Eq. (24), denoted by $\hat{H}_{so}^{at(t)}$, can be written for an atom or a two level system as:

$$\hat{H}_{so}^{at(t)} = -\frac{e^3gZ\kappa}{2m_e^2c^2r_e^2}\sum_{\lambda}\left(\frac{2\pi\hbar}{\varepsilon_0\omega_\lambda V}\right)^{1/2}e^{-i\omega t}\,\mathbf{s}\cdot(\hat{\varepsilon}_\lambda\times\hat{\mathbf{r}}_e)c_\lambda^+, \quad (25)$$

where $\hat{\mathbf{r}}_e = \frac{\mathbf{r}_e}{r_e}$ is a unit vector. For evaluating the triple scalar product of three vectors, without the loss of any generality we may assume that vectors $\hat{\varepsilon}_\lambda$ and $\hat{\mathbf{r}}_e$ are in the xy-plane at an angle φ_λ, then we get $\hat{\varepsilon}_\lambda\times\hat{\mathbf{r}}_e = \sin\phi_\lambda\,\hat{\eta}$, $\hat{\eta}$ being a unit vector perpendicular to the xy-plane. This gives $\mathbf{s}\cdot(\hat{\varepsilon}_\lambda\times\hat{\mathbf{r}}_e) = \mathbf{s}\cdot\hat{\eta}\sin\phi_\lambda = s_z\sin\phi_\lambda$, which simplifies Eq. (25) as:

$$\hat{H}_{so}^{at(t)} = -\frac{e^3gZ\kappa}{2m_e^2c^2r_e^2}\sum_{\lambda}\left(\frac{2\pi\hbar}{\varepsilon_0\omega_\lambda V}\right)^{1/2}e^{-i\omega t}\sin\varphi_\lambda\, s_z\, c_\lambda^+. \quad (26)$$

For an atom, the field operator for an electron in the excited state and a hole in the ground state can be respectively written as:

$$|\hat{\psi}_e(r_e)\rangle = \sum_{\sigma_e}\varphi_e(r_e,\sigma_e)a_e(\sigma_e), \quad \text{(a)}$$

and (27)

$$|\hat{\psi}_h(r_e)\rangle = \sum_{\sigma_h}\varphi_g{}^*(r_e,\sigma_h)d_g(\sigma_h), \quad d_g(\sigma) = a_g^+(-\sigma) \quad \text{(b)}$$

where $\varphi(r_e,\sigma)$ is the electron or hole wave functions as a product of orbital and spin functions corresponding to spin σ = ½ or – ½, and $a_e(\sigma)$ and $d_g(\sigma)$ are the annihilation operators of an electron in the excited state and hole in the ground state, respectively. It may be noted that in an atom it is the same electron that is excited from the ground to the excited state therefore the same coordinate r_e is used for both the electron and hole in Eq. (27).

Using Eq. (27), the interaction operator in Eq. (26) can be expressed in second quantization as:

$$\hat{H}_{so}^{at(t)} = -\frac{e^3 gZ\kappa}{2m_e^2 c^2 r_e^2}\sum_\lambda \left(\frac{2\pi\hbar}{\varepsilon_0 \omega_\lambda V}\right)^{1/2} e^{-i\omega t} \sin\phi_\lambda < \varphi_h(r_e) | r^{-2} | \varphi_e(r_e) > \\ \times \delta_{\sigma_e, -\sigma_h} s_z a_e(\sigma_e) d_g(\sigma_h) c_\lambda^+ \tag{28}$$

Using the property of the spin operator $s_z a_e(\pm\frac{1}{2}) = \pm\frac{1}{2}\hbar a_e(\pm\frac{1}{2})$ we find that only the integral from Eq. (12b) is non-zero and then the operator in Eq. (28) becomes:

$$\hat{H}_{so}^{at(t)} = -\frac{\hbar e^3 gZ\kappa}{4m_e^2 c^2}\sum_\lambda \left(\frac{2\pi\hbar}{\varepsilon_0 \omega_\lambda V}\right)^{1/2} e^{-i\omega t} \sin\phi_\lambda [< \varphi_h(r_e) | r_e^{-2} | \varphi_e(r_e) > \\ \times [\frac{1}{\sqrt{2}}(a_e(+1/2)d_g(-1/2) + a_e(-1/2)d_g(+1/2))]c_\lambda^+ \tag{29}$$

It may be noted that the operator s_z has flipped the triplet spin configuration to a singlet configuration and hence the recombination can now occur.

We now consider a transition from an initial state with a triplet excitation whose spin has been flipped by the spin-orbit interaction but it has no photons to a final state with no excitation (ground state) and one photon created in a mode λ. Within the occupation number representation, such initial $|i>$ and final $|f>$ states can be respectively written as:

$$|i> = |\frac{1}{\sqrt{2}}[a_e^+(+1/2)d_g^+(-1/2) + a_e^+(-1/2)d_g^+(+1/2)]|0>|0_p>, \tag{30}$$

$$|f> = c_\lambda^+ |0>|0_p>, \tag{31}$$

where $|0>$ and $|0_p>$ represent the vacuum states of electrons (no excitations) and photons (no photons), respectively. Using Eqs. (29) - (31) and the usual anti-commutation rules for fermion and commutation rules for boson operators, the transition matrix element is obtained as:

$$< f | \hat{H}_{so}^{at(t)} | i > = -\frac{\hbar e^3 gZ\kappa}{4m_e^2 c^2}\left(\frac{2\pi\hbar}{\varepsilon_0 \omega_\lambda V}\right)^{1/2} e^{-i\omega_\lambda t} \sin\iota_\lambda < \varphi_h(r_e) | r_e^{-2} | \varphi_e(r_e) >, \tag{32}$$

Using Fermi's golden rule and Eq. (32), the rate of spontaneous emission of a photon from the radiative recombination of a triplet exitation in an atom denoted by R_{sp}^{atom} (s^{-1}), is obtained as:

$$R_{sp}^{atom} = \frac{2\pi}{\hbar}\sum_\lambda |< f | \hat{H}_{so}^{at(t)} | i >|^2 \, \delta(E_e - E_g - \hbar\omega_\lambda), \tag{33}$$

where the sum over λ represents summing over all photon modes and E_e and E_g are the energies of the excited and ground states. This can be evaluated as follows: Considering that a wave vector $\mathbf{k}$ can be associated with every photon mode, one can write: $\sum_{\lambda} = \frac{2V}{(2\pi)^3}\int d^3\mathbf{k}$, with $k = \omega_\lambda / c$, $k^2 dk = \frac{\omega_\lambda^2}{c^3\hbar} d(\hbar\omega_\lambda)$ and then $\sum_{\lambda} = \frac{2V}{(2\pi)^3}\int d^3\mathbf{k} = \frac{2V}{(2\pi)^3}\int_0^{\hbar\omega_{12}}\int_0^{\pi}\int_0^{2\pi} \frac{\omega_\lambda^2}{c^2\hbar}\sin\theta \sin^2\phi_\lambda d(\hbar\omega_\lambda) d\theta\, d\phi_\lambda$. Using this we can replace the sum in Eq. (33) by a triple integration and then substituting g = 2 we get:

$$R_{sp}^{atom} = \frac{e^6 Z^2 \kappa^2 \hbar\omega_{12}}{2m_e^4 c^7 \varepsilon_0 |r|^4}, \tag{34}$$

where $\hbar\omega_{12} = E_e - E_g$ and $|<\varphi_h(r_e)|r_e^{-2}|\varphi_e(r_e)>|^2 = |r|^{-2}$ with $|r|$ being the average distance of an electron in the triplet excited state from the nucleus. It is to be noted that the rate of the spontaneous emission derived in Eq. (34) is very sensitive to the separation between the excited electron and nucleus, $|r|$, and the electronic mass but not so sensitive to the emitted photon energy. These properties are different from the rate of spontaneous emission from a singlet state derived in Eq. (15). The inverse of the rate of spontaneous emission gives the radiative lifetime τ_R [$(R_{sp}^{atom})^{-1} = \tau_R$], which can easily be calculated provided ω_{12} and r are known.

The rate of spontaneous emission obtained in Eq. (34) is derived within the two level approximation may be applied to organic solids and polymers [9] where excitation gets confined on individual molecules/monomers as Frenkel excitons and also referred to as molecular excitons [19]. Until the late seventies excitons in organic solids, like naphthalene, anthracene, etc., were regarded in this category. Furthermore, the concept that an exciton consists of an excited electron and hole pair was considered to be applicable only for excitons created in onorganic solids, known as Wannier excitons or Wannier-Mott excitons. These were also known as the large radii orbital excitons because of the small binding energy the separation between electron and hole is relatively larger than that in Frenkel excitons in organic solids. However, this distinction has blurred since the development of OLEDs where electrons and holes are injected from the opposite electrodes, as described above, and form Frenkel excitons. This proves the point that Frenkel excitons also consist of the excited electron and hole pairs but they indeed form a molecular excitations because of the small overlap betwen the intermolecular electronic wavefunctions.

Assuming that the Frenkel excitons are molecular excitons in organic solids/polymers, the above theory has been extended to organic solids [9] and the rate of spontaneous emission is obtained as:

$$R_{sp}^{mol} = \frac{e^6 Z^2 \kappa^2 \hbar \omega_{12}}{2 m_e^4 c^7 \varepsilon^3 \varepsilon_0 |r|^4} \tag{35}$$

where ε is the static dielectric constant of the solid $|r|$ is the average separation between the electron and hole and $|r| = a_x^T / \varepsilon$, a_x^T being the excitonic Bohr radius of a triplet exciton given by $a_x^T = \frac{\mu\varepsilon}{\mu_x} a_0$ [8,12], $a_0 = 0.0529$ nm is the Bohr radius. Substituting this in equation (35), the rate of spontaneous emission from a triplet excitation in molecular semiconductors and polymers is obtained as:

$$R_{sp} = \frac{e^6 Z^2 \kappa^2 \varepsilon \hbar \omega_{12}}{2 \mu_x^4 c^7 \varepsilon_0 a_x^4}. \tag{36}$$

As the rate of spontaneous emission is proportional to Z^2, it becomes very clear why the presence of heavy atoms enhances the rate of radiative emission of triplet excitons. The radiative lifetime of triplet excitons is calculated from the inverse of the rate in equation (36), $\tau_R = 1 / R_{sp}$.

The rate of spontaneous emission in equation (36) is used to calculate the triplet radiative rates in several organic molecular complexes, conjugated polymers containing platinum in the polymer chain and some organic crystals [9]. For all polymers considered from ref.[20], where the effective mass of charge carriers and excitonic Bohr radius are not known, it is assumed that $m_e^* = m_h^* = m_e$ giving $\mu_x = 0.5 m_e$ and $\varepsilon = 3$, which give the triplet excitonic Bohr radius $a_x = \frac{m_e \varepsilon}{\mu_x} a_0 = 6 a_0$. The first three polymers P_1, P_2 and P_3 are chosen from ref. [20], where the rates of radiative recombination have been measured in many polymers containing platinum atoms. We can calculate the radiative rates for all the polymers studied in [20] but as they are all found to be of the same order of magnitude only the rates for the first three polymers are listed here. The triplet emission energy used in the calculation, and the calculated rate and the corresponding radiative lifetime are listed in table 1 along with the observed experimental rates and radiative lifetimes. For conjugated polymers incorporated with platinum atoms, the rates of radiative recombination in P_1, P_2 and P_3 are found to be of the order of 10^3 s^{-1} , which agrees very well with the experimental results [20]. In table 1 are also included the rate of spontaneous emission and radiative lifetime calculated for platinum porphyrin (PtOEP) used as a phosphorescent dye in organic electroluminescent devices [21] and phenyl-substituted poly (phenylene-vinylene) (PhPPV)[9]. From table 1, it is quite clear that the rate in equation (3) can be applied to most organic semiconductors and polymers because the calculated rates and radiative lifetimes agree very well with the experimental results.

Material	$\hbar\omega_{12}$ (eV)	R_{sp} (s^{-1}) Eq. (3)	R^{exp} (s^{-1})	$\tau_R = 1/R_{sp}$ (s)	τ^{exp} (s)
Benzene	3.66 [22]	0.63	-	1.6	4-7 [22]
Naphthalene	2.61 [22]	0.45	-	2.2	2.5 [22]
Anthracene	1.83 [22]	0.31	-	3.19	0.1 [22]
P1	2.40 [20]	5.5x10^3	(6 ± 4)x10^3 [23]	1.82x10^{-4}	
P2	2.25 [20]	5.1x10^3	(1.8 ± 0.9)x10^3[23]	1.96x10^{-4}	
P3	2.05 [20]	4.6x10^3	(1 ± 1)x10^3 [23]	2.17x10^{-4}	
Pt(OEP)	1.91[24]	4.9 x10^3		2. 03x10^{-4}	7.00x10^{-4}

Table 1. Assuming $m_e^* = m_h^* = m_e$, which gives $\mu_x = 0.5m_e$ and taking $\varepsilon = 3$, rates of spontaneous emission are calculated from equation (36) for a few molecular crystals, conjugated polymers and platinum porphyrin [Pt(OEP)]. Using these the triplet excitonic Bohr radius becomes $a_x = 6a_0$.

3.2. Exciton -spin-orbt-photon interaction

In the above section, it is shown that a time-dependent electron-photon-spin-orbit interaction operator does exist and it can be applied for triplet state transitions. The theory is also extended to Frenkel excitons or molecular excitons without considering them as consisting of electron and hole pairs. However, the formalism presented above is relevant to an excited electron in an atom/molecule which is not consistent with the situation occurring in a WOLED, where electrons and holes are injected from the opposite electrodes and they form excitons before their radiative recombination. Thus, for WOLEDs we need a time-dependent exciton-photon-spin-orbit interaction operator. For a pair of injected carriers in a solid with N atoms, an operator analogous to Eq. (19) and denoted by H_{so}^{sol} can be written as [9]:

$$\hat{H}_{so}^{sol} = -\frac{eg}{2\mu_x^2 c^2}\mathbf{s}_e \cdot (\mathbf{p}_e \times \sum_{n=1}^{N} \mathbf{E}_{ne}) + \frac{eg}{2\mu_x^2 c^2}\mathbf{s}_h \cdot (\mathbf{p}_h \times \sum_{n=1}^{N} \mathbf{E}_{nh}) \tag{37}$$

where μ_x is the reduced mass of exciton as described above. Other quantities with subscript e represent the electron and with subscript h represent the hole. In the presence of radiation, Eq. (37) becomes:

$$\begin{aligned} \hat{H}_{so}^{a-sem} &= -\frac{eg}{2\mu_x^2 c^2}\mathbf{s}_e \cdot (\mathbf{p}_e + \frac{e}{c}\mathbf{A}_e) \times (-\frac{1}{c}\frac{\partial \mathbf{A}_e}{\partial t} - \sum_{n=1}^{N} \nabla V_{ne}) \\ &+ \frac{eg}{2\mu_x^2 c^2}\mathbf{s}_h \cdot (\mathbf{p}_h - \frac{e}{c}\mathbf{A}_h) \times (-\frac{1}{c}\frac{\partial \mathbf{A}_h}{\partial t} - \sum_{n=1}^{N} \nabla V_{nh}) \end{aligned} \tag{38}$$

where the zero magnetic contribution is neglected. In analogous with Eq. (24), one gets two non-zero terms for the electron and two for the hole as:

$$\hat{H}_{so}^{a-sem} = -\frac{eg}{2\mu_x^2 c^2}\left(-\sum_{n=1}^{N}\frac{Z_n e\kappa \mathbf{s}_e \cdot \mathbf{L}_{en}}{\varepsilon r_{en}^3} - \frac{e}{c}\mathbf{s}_e \cdot (\mathbf{A}_e \times \sum_{n=1}^{N} \nabla V_{ne})\right)$$
$$+\frac{eg}{2\mu_x^2 c^2}\left(-\sum_{n=1}^{N}\frac{Z_n e\kappa \mathbf{s}_h \cdot \mathbf{L}_{hn}}{\varepsilon r_{en}^3} + \frac{e}{c}\mathbf{s}_h \cdot (\mathbf{A}_h \times \sum_{n=1}^{N} \nabla V_{nh})\right) \quad (39)$$

Here Z_n is the atomic number of nth atom and r_{en} and r_{hn} are, respectively, the electron and hole distances from their nuclear site n. s_{ez} and s_{hz} are the spin projections along the z-axis of the electron and hole, respectively, in an exciton. Other symbols have their usual meanings [9]. It may be pointed out here that the interaction operator as obtained in Eq. (26) is the same for a triplet exciton and an excited pair of electron and hole in a triplet spin configuration. Following the procedures applied in deriving Eq. (26) for a single electron, we get the time-dependent exciton-photon-spin-orbit interaction in a solid as [9]:

$$\hat{H}_{so}^{a-sem(t)} = -\frac{e^3 g\kappa}{2\mu_x^2 \varepsilon c^2}\sum_{\lambda,n}\left[\frac{Z_n}{r_{en}^2}\left(\frac{2\pi\hbar}{\varepsilon_0 \varepsilon \omega_\lambda V}\right)^{1/2}\sin\phi_{\lambda en}\, s_{ez}\right.$$
$$\left.+\sum_{\lambda,n}\frac{Z_n}{r_{hn}^2}\left(\frac{2\pi\hbar}{\varepsilon_0 \varepsilon \omega_\lambda V}\right)^{1/2}\sin\phi_{\lambda hn}\, s_{hz}\right] e^{-i\omega_\lambda t} c_\lambda^{+} \quad (40)$$

Using the field operators in Eqs. (7) and (8) and Eq. (40), the time-dependent operator of exciton-photon-spin-orbit interaction is obtained in second quantisation as [9]:

$$\hat{H}_{so}^{(t)} \approx -\frac{e^3 g Z\kappa}{\mu_x^2 c^2 \varepsilon r^2}\left(\frac{2\pi\hbar}{\varepsilon_0 \varepsilon V}\right)^{1/2}\sum_{\lambda}\sum_{\sigma_e,\sigma_h}\frac{\sin\lambda}{\sqrt{\omega_\lambda}} e^{-i\omega t}(s_{ez}+s_{hz})a_e(\sigma_e)d_g(\sigma_h)\delta_{\sigma_e,\sigma_h} c_\lambda^{+} \quad (41)$$

where r is the average separation between electron and hole in an exciton and it is approximated by:

$$< \mathrm{HOMO} | \mathrm{r}_{en}^{-2} | \mathrm{LUMO} > \approx < \mathrm{HOMO} | \mathrm{r}_{hn}^{-2} | \mathrm{LUMO} > \approx (r/2)^{-2}. \quad (42)$$

The other important approximation made in Eq. (41) is that the sum over sites n has disappeared. This is briefly because of the fact that the interaction operator depends on the atomic number Z_n and the inverse square of the distance between an electron and nucleus and hole and nucleus. Therefore only the heaviest and rearest atom will contribute most and the contribution of other atomic sites will be negligible. Using this approximation the summation over n is removed.

Using the triplet spin configuration in Eq. (12) and the property of s_{ez} and s_{hz} operators as $s_{ez}a_e(\pm\frac{1}{2}) = \pm\frac{1}{2}\hbar a_e(\pm\frac{1}{2})$ and $s_{hz}d_g(\pm\frac{1}{2}) = \mp\frac{1}{2}\hbar d_g(\pm\frac{1}{2})$, here again we find that only the contribution of Eq. (12b) is non-zero and then the operator in Eq. (41) becomes:

$$\hat{H}_{so}^{at(t)} \approx -\frac{2\hbar e^3 gZ\kappa}{\mu_x^2 c^2 \varepsilon r^2}\left(\frac{2\pi\hbar}{\varepsilon_0 \varepsilon V}\right)^{1/2} \sum_{\lambda} \frac{\sin\lambda}{\sqrt{\omega_\lambda}} e^{-i\omega_\lambda t}$$
$$\times[\frac{1}{\sqrt{2}}(a_e(+1/2)d_g(-1/2)+a_e(-1/2)d_g(+1/2))]c_\lambda^+ \qquad (43)$$

It is to be noted here also that the operator ($s_{ez} + s_{hz}$) in Eq. (41) has flipped the spin of triplet sconfiguration (compare with Eq. (12b)) to a singlet configuration and hence the recombination can occur. Thus the mechanism of the occurrence of radiative recombination of triplet excitons through the new transition operator can be described in the following two steps:

1. The new operator is attractive for excitons so it attracts a triplet exciton to the heaviest atom as it is proportional to the atomic number. As the magnitude of attraction in inversely proportional to the square of the average distance between an electron and nucleaus, only the nearest heavy nucleus will play the dominant role.
2. As soon as a triplet exciton interacts with such a spin-orbit-exciton-photon interaction, the spin gets flipped to a singlet configuration and exciton recombines radiatively by emitting a photon.

3.3. Rate of spontaneous emission from triplet excitons

We now consider a transition from an initial state $|i>$ with a triplet exciton whose spin has been flipped by the spin-orbit interaction but it has no photons to a final state $|f>$ with no excitation (ground state) and one photon created in a mode λ. These states in the second quantization are analogous to Eqs. (30) and (31), respectively. Using Eqs. (30) - (31) and the interaction operator in Eq. (43), the transition matrix element is obtained as:

$$< f \mid \hat{H}_{so}^{(t)} \mid i >= -\frac{2\hbar e^3 gZ\kappa}{\mu_x^2 c^2 \varepsilon |r|^2}(\frac{2\pi\hbar}{\varepsilon_0 \varepsilon V})^{1/2} e^{-i\omega_\lambda t} \frac{\sin\phi_\lambda}{\sqrt{\omega_\lambda}}, \qquad (44)$$

Using Fermi's golden rule and the transition matrix element in Eq. (44), the rate of spontaneous emission of a photon from the radiative recombination of a triplet exiton in an organic solid/polymer denoted by R_{sp} (s^{-1}), can be written as:

$$R_{sp} = \frac{2\pi}{\hbar}\sum_{\lambda} |< f \mid \hat{H}_{so}^{(t)} \mid i >|^2 \, \delta(E_{LUMO} - E_{HOMO} - \hbar\omega_\lambda), \qquad (45)$$

where the sum over λ represents summing over all photon modes and E_{LUMO} and E_{HOMO} are the energies of the LUMO and HOMO energy levels. This can be evaluated in a way analogous to Eq. (33) and then we obtain:

$$R_{sp} = \frac{8e^6 g^2 Z^2 \kappa^2 \hbar\omega_{12}}{\mu_x^4 c^7 \varepsilon_0 \varepsilon^3 |r|^4}, \qquad (46)$$

For triplet excitons using $|r| = a_x^T / \varepsilon$ and g = 2, the rate in Eq. (46) becomes [20]:

$$R_{sp} = \frac{32e^6 Z^2 \kappa^2 \varepsilon \hbar\omega_{12}}{\mu_x^4 c^7 \varepsilon_0 a_{ex}^4} \qquad \mathrm{s}^{-1} \tag{47}$$

For different phosphorescent materials only the atomic number of the heavy metal atom and the emitted energy will be different so the rate of spontaneous emission in Eq. (37) can be simplified as follows: Using $\varepsilon \approx 3$ which gives the triplet exciton Bohr radius $a_{ex} = 6a_0$ ($\mu = m_e$ and $\mu_x = m_e / 2$; m_e being the free electron mass), the rate in Eq. (47) can be expressed as:

$$R_{sp} \approx 25.3Z^2(\hbar\omega_{12}) \qquad \mathrm{s}^{-1} \qquad (\hbar\omega_{12} \text{in eV}) \tag{48}$$

For phosphorescent materials like fac-tris (2-phenylpyridine) iridium (Ir(ppy)3) and iridium(III) bis(2-phenyl quinolyl-N,C20) acetylacetonate (PDIr), where Ir has the largest atomic number $Z = 77$, other atomic numbers can be neglected being mainly of carbon. The rate in Eq. (48) depends linearly on the emission energy $\hbar\omega_{12}$ and other quantities are the same for all iridium doped materials. Thus, for iridium complexes doped in organic polymers the rate is obtained as: $R = 1.5 \times 10^5 \hbar\omega_{12}$ s^{-1} ($\hbar\omega_{12}$ in eV). For green phosphor $Ir(ppy)_3$ has been doped for emission energy of 2.4 eV, for orange phosphor Ir(MMQ) [25] and FIrpic [4] have been doped for emission at 2.00 eV. In all these films the rate of spontaneous would be of the same order of magnitude (3 - 4 $\times 10^5$ s^{-1}). This agrees quite well with the measured rate for Ir complexes [26].

Both rates of spontaneous emission derived in Eq. (37) on the basis of single electron excitation (atomic case) and that obtained in Eq. (46) for an electron-hole pair excitation have been applied to calculate it in organic solids and polymers [9, 27]. Apparently for platinum complexes Eq. (37) gives rates that agree better with experimental results but for iridium complexes Eq. (46) produces more favourable results.

In addition to developing the introduction of the phosphorescent materials to enhance the radiative recombination of triplet excitons, a step progression of HOMO and LUMO of the organic materials to confine the injected carriers within the emission layer has been applied [25]. This enables the injected e and h confined in a thinner space that enhances their recombination. This scheme has apparently proven to be most efficient so far.

Another approach for meeting the requirement of availing different energy levels for singlet and triplet emissions within the same layer of a WOLED is to incorporate nanostructures, particularly quantum dots (QDs), in the host polymers [28]. As the size of QDs controls their energy band gap, the emission energy can be manipulated by the QD sizes. It is found that the energy band gap of a QD depends on its size as [29]:

$$Eg(\mathrm{eV}) = Eg_{\mathrm{bulk}} + C / d^2 \tag{49}$$

where C is a confinement parameter and d is the size of a QD. Such a hybrid structures of organic host and inorganic QDs have been tried successfully [29-30]. It would be interesting if in future organic QDs could be grown on polymers and then the fabrication would be very cost effective.

This chapter is expected to present up to date review of the state-of-the art development in the theory of capturing emissions from triplet excitons in WOLEDs.

Author details

Jai Singh
School of Engineering and IT, B-purple-12, Faculty of EHSE, Charles Darwin University, Darwin, NT, Australia

4. References

[1] C. Adachi, M. A. Baldo, M.E. Thompson, and S. E Forrest, J. Appl. Phys. 90, 5048 (2001).

[2] Y. Cao, I.D. Parker, G. Yu, C. Zhang and A.J. Heeger, Nature 394, 414 (1999).

[3] Z. Shuai, D. Beljonne, R.J. Silbey and J. L. Bredas, Phys. Rev. Lett. 84, 131 (2000).

[4] G. Schwartz, S. Reineke, T. C. Rosenow, K. Walzer and K. Leo, Adv. Funct. Mat. 19, 1319 (2009).

[5] J. Singh, Phys. Status Solidi C8, 189 (2011).

[6] J. Kido, K. Hongawa, K.Okuyama, K Nagai, Appl. Phys. Lett. 64, 815 (1994).

[7] Y. Sun, N. C. Giebink, H. Kannao, B. Ma, M. E. Thompson and S. R. Forest, Nature, 440, 908 (2006).

[8] J. Singh, Phys. Rev. B76, 085205(2007)

[9] J. Singh, H. Baessler and S. Kugler, J. Chem. Phys. 129, 041103 (2008).

[10] J. Singh, Photoluminescence and photoinduced changes in noncrystalline condensed Matter in *Optical Properties of Condensed Matter and Applications*, J. Singh (Ed.) (John-Wiley, Chichester, 2006), Ch.6.

[11] J. Singh and I.-K. Oh, J. Appl. Phys. 97, 063516 (2005).

[12] J. Singh and K. Shimakawa, *Advances in Amorphous Semiconductors* (Taylor & Francis, London, 2003).

[13] A. Köhler, J. S. Wilson, R. H. Friend, M. K. Al-Suti, M. S. Khan, A. Gerhard, and H. Baessler, J. Chem. Phys. 116, 9457 (2002).

[14] T. Aoki, in *Optical Properties of Condensed Matter and Applications*, J. Singh (Eds.) (John Wiley and Sons, Chichester, 2006), Ch.5, pp 75 and references therein.

[15] T. Aoki, T. Shimizu, S. Komedoori, S. Kobayashi and K. Shimakawa, *J. Non-Cryst. Solids* 338-340, 456 (2004) and T. Aoki, *J. Non-Cryst. Solids* 352, 1138 (2006).

[16] J. Singh, *Physics of Semiconductors and their Hetrostructures* (McGraw-Hill, Singapore, 1993).

[17] S. Gasiorowicz, *Quantum Physics*, 2nd Edition (John Wiley & Sons, N.Y., 1996).
[18] H. F. Hameka in *The Triplet State* (Cambridge University Press, Cambridge, 1967), p. 1.

[18] J. Singh, *Excitation Energy Transfer Processes in Condensed Matter* (Plenum, N.Y., 1994).

[19] J. S. Wilson, N. Chaudhury, R.A. Al-Mandhary, M. Younus, M.S. Khan, P.R. Raithby, A. Köhler and R.H. Friend, J. Am. Chem. Soc. 123, 9412 (2001).

[20] M.A. Baldo, D.F. O'Brien, Y. You, A. Shoustikov, S. Sibley, M.E. Thompson, and S.R. Forrest, Nature 395, 151 (1998).

[21] J. B. Birks, *Photophysics of Aromatic Molecules* (John Wiley and Sons, London, 1970).

[22] D. Beljonne, H.F. Wittman, A. Köhler, S. Graham, M. Younus, J. Lewis, P.R. Raithby, M.S. Khan, R.H. Friend and J.L. Bredas, J. Chem Phys. 105, 3868 (1996).

[23] F. Laquai, C. Im, A. Kadashchuk and H. Baessler, Chem. Phys.Lett. 375, 286 (2003).

[24] S.-J. Su, E. Gonmori, H. Sasabe and. Kido, Adv. Mater. 20, 4189(2008).

[25] N. R. Evans, L. S. Devi, C. S. K. Mak, S. E. Watkins, S. I. Pascu, A. Köhler, R. H. Friend, C. K. Williams, and A. B. Holmes, J. Am. Chem. Soc. 128, 6647 (2006).

[26] J. Singh, Phys. Status Solidi A 208, 1809 (2011).

[27] Park, et al., Appl. Phys. Lett. 78, 2575 (2001).

[28] Hsueh Shih Chena! and Shian Jy Jassy Wang, Appl. Phys. Lett. 86, 131905 (2005)

[29] H.-S. Chen, C.-K. Hsu, and H.-Y. Hong, IEEE Phot. Tech. Lett. 18, 193(2006).

Chapter 2

Field Emission Organic Light Emitting Diode

Meiso Yokoyama

Additional information is available at the end of the chapter

http://dx.doi.org/10.5772/52487

1. Introduction

Several flat panel displays (FPDs) technologies, such as liquid crystal displays (LCDs), plasma display panels (PDPs), light-emitting diodes (LEDs), organic light-emitting devices (OLEDs) and field emission displays (FEDs), have been developed. They coexist because each technology has its own unique properties and applications.

In recent years, the developments in the OLEDs have gradually reached very advantageous of existence. These advantageous characteristics include self-luminous, wide viewing angle and low power consumption, etc. which make OLEDs very useful for numerous display applications and lighting devices. To effectively improve the characteristics of an OLEDs, there are many ways to be adopted. Such as: (a) structure: using the quantum well structure or multilayer structure to enhance efficiency by promoting the radiative recombination capability; (b) material: use of a low work function metal as the cathode or a high carrier mobility material to allow efficient carrier injection into OLEDs structure; (c) doping: by doping the guest material into the host material to increase the efficiency of recombination, such as phosphorescent sensitizer.

However, the performance of OLEDs using above methods will be limited. Therefore, in this chapter, we propose new FEOLEDs with electron multiplier.[1,2-5, 6]

An effective enhancement in the lighting efficiency is achieved by using the external electron source supplement into the OLEDs. The FEOLEDs can simply be divided into two types: FEOLED (original diode type) and triode type. The structure of FEOLEDs is similar to that of the field emission diodes (FEDs), but formers utilize an organic EL light emitting layer instead of an inorganic phosphor thin film used in FEDs. The mechanism of operation of FEOLEDs is the same as OLEDs.

In FEOLEDs also a hole blocking layer is used to confine the electron-hole pairs to enhance recombination in the organic light emitting layer. Besides, to avoid damage to the organic material by electron beam bombardment, an aluminum (Al) thin film is coated on the organic light emitting layer facing to the carbon nanotubes (CNTs) template to protect the organic light emitting materials and thereby enhance the luminous efficiency of the FEOLEDs. In a triode FEOLED an electron multiplier layer is inserted between the anode and cathode. This layer amplifies the field emission electrons and then injects them into the organic light emitting layer.

In this chapter, as stated above, we have proposed and discussed two kinds of electron multipliers: 1) a dynode and 2) a strip electron multiplier. The dynode is formed between the cathode and organic light emitting layer to provide electron amplification capability as well as it makes a FEOLED more stable. The electrons emitted from the cathode move towards the dynode as they are attracted by the applied electric field. The primary electrons impact the secondary electron material of dynode to produce the secondary electrons. Finally, both primary and secondary electrons are directly injected into the organic light emitting layer and increase the current density of FEOLEDs. Due to the presence of a dynode it is difficult to fabricate very thin FEOLEDs. Therefore, the strip electron multiplier has been proposed as it can easily be incorporated in FEOLEDs. The strip electron multiplier is formed on the organic light emitting layer facing to the CNTs cathode, and made by strip Al coated with secondary electron material (Cesium iodide, CsI). CsI does not completely cover the strip Al.

The mechanism of electron amplifier ability of the strip electron multiplier is the same that of a dynode, but the process of fabrication is easier. The organic light emitting layer of FEOLEDs integrated with the strip electron multiplier forms an OLEDs, which can operate independently. Accordingly, applying an electric field to the CNTs template and strip electron multiplier one can attract electrons to impact the strip electron multiplier to generate the secondary electrons. Therefore, the current density of OLEDs is increased by supplementing the electrons into the multilayer of the organic light emitting layer. In this way, the luminance efficiency of FEOLEDs with strip electron multiplier can further be enhanced by one and a half times more than of OLEDs.

The organization of this chapter is as follows. First the concept and mechanism of operation of FEDs and OLEDs are introduced. Then are illustrated the basic concepts and luminescent mechanisms of FEOLEDs and experimental procedure of fabricating them, FEOLEDs of diode structure and triode structure and their characteristics of electron multiplier are discussed next. Finally, we discuss the advantages and disadvantages of the conventional OLEDs and novel FEOLEDs, including some suggestions for future work..

2. Field emission light emitting diodes (FEDs)

A FED is a vacuum electron device, sharing many common features with the vacuum fluorescent displays (VFDs) and cathode ray tubes (CRTs) [7]. Like in a VFDs or CRTs, the image in a FED is created by impacting electrons from a cathode onto a phosphor coated screen. In a CRTs the electron source is made up of up to three thermionic cathodes [8]. A set of electromagnetic deflection coils raster the electron beam across a phosphor screen, which is typically held at a potential of 15-30 kV [9]. In a FED the electron source consists of a matrix-addressed array of millions of cold emitters. This is field emission arrays (FEAs), which is placed in closed proximity (0.2mm) to a phosphor faceplate and is aligned such that each phosphor pixel has a dedicated set of field emitters [10].

The idea of a FED dates back to the 1960s, when Ken Shoulders of the Stanford Research Institute (SRI) proposed electron beam micro devices based on FEAs [11]. The first operating FEAs were demonstrated by Capp Spindt, also of SRI, in 1968 [12]. Despite many advantages of the spindt-type FEA fabrication technique, scaling this method to large area substrates (>400 mm on the side) is still a major challenge. Another difficulty associated with the scale up of spindt process is the large size of the evaporator required to deposit the spindt tips. Most phosphors have low luminous efficiency at voltages below 3 kV because of the low electron penetration depth and high, non-radiative recombination rates at the surface. While raising the emission current density increase the brightness, high current density leads to faster aging of the phosphor, thus further decreasing the brightness [13].

Among various kinds of emitters in field emission devices, carbon nanotubes (CNTs) have been attracting a considerable attention due to their excellent field emission characteristics of high field emission current density and low turn-on electric field [14]. In order to enhance the field emission electron ability and emission uniformity in large area CNT-FED panels, additional methods are required to improve uniformity by inserting the gate design for electron multiplier and focusing. A gate coated with the secondary electron emission (SEE) materials for obtaining electron amplification is called a dynode [15]. In general, any insulator with low work function is suitable for SEE application [16]. The mechanism of dynode can be simplified by the following processes: (i) the primary electrons penetrate into a certain depth of an insulating layer; (ii) through collision, the energy of the primary electrons is transferred to the bound electrons of the insulator, leading to a release of electrons; (iii) the released electrons migrate to the surface and escape into the vacuum as secondary electrons.

Therefore, the field emission involves the extraction of electrons from a solid by tunneling through the surface potential barrier. The emitted current depends directly on the local electric field E at the emitting surface and metal's work function (Φ) as shown in Fig.1 [17]. The field-emission properties of wide band gap materials (WBGMs) is favorable for the emission, as it is considered a property unique to the surface of emitter [18]. The role of the WBGMs in CNTs field emission is to decrease the effective work function of emitters, which increases emissivity. We can assure that the carbon nanotubes are excellent electron sources, providing a stable current at very low fields and capable of operating in moderate vacuum.

In summary, the light emitting principle of FEDs is that the electrons are excited and accelerated by the high electric field under vacuum, so as to become sufficiently energized to bombard the inorganic phosphor to emit light. Although CNT-FED was very successful in achieving the result in different low voltage phosphors such as $ZnGa_2O_4$+ In_2O_3, ZnO:Zn low voltage phosphors research, the applied voltage was about 300 V [19]. Thus, it does not yet meet the requirement for low voltage flat panel display usage.

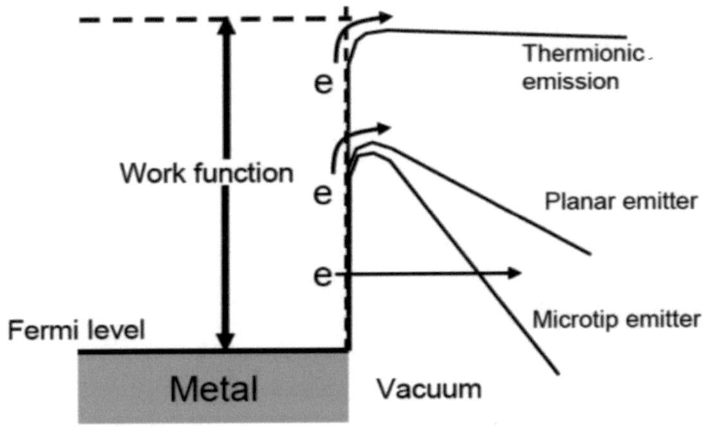

Figure 1. Schematic diagram of the field emission barriers for a planar and a micro-tip emitter.

3. Organic light emitting diodes (OLEDs)

Using organic material for light emitting diodes (LEDs) is fascinating due to their vast variety and relative ease of controlling their composition to tune their properties by chemical means. For example, by applying an electric field to an anthrancene single crystal, Pope *et al.* in 1965 observed blue electroluminescence (EL) [20]. Soon after alternating current EL was also achieved using an emissive polymer [21].The observation of efficient bright EL, defined as the number of photons emitted from the face of the device per injected electron or hole, the investigation on the organic optoelectronic device commenced to investigate and developed slowly until Tang and Vanslyke demonstrated efficient green electroluminescence [22] from a vapor deposited organic compound in 1987. Till now, OLEDs are the best flat light emitting source.

3.1. Structure of organic light emitting diodes (OLEDs)

A typical OLED is composed of an emissive layer, a conductive layer, a substrate, and both anode and cathode terminals. The basic structure of a typical OLED is shown in Fig. 2. The first layer above the glass substrate is a transparent conducting anode, typically indium tin oxide (ITO). In Flexible OLEDs the anode is made of a transparent conductance plastic substrate. There are two different types of OLEDs. Traditional OLEDs use small organic molecules deposited on glass to produce light. The other type of OLEDs uses large plastic molecules called polymers. The single or multilayer small organic molecular or polymer film is deposited on the transparent anode. Appropriate multilayer structures are used to enhance the performance of the device by lowering the barrier for hole injection from the anode and by controlling the electron and hole recombination region. The injected holes move from the interface of the organic/electrode into the organic light emitting layer, where the defect density is high. Therefore, the organic layer deposited on the anode should generally be a good hole transport layer (HTL). Similarly, the organic layer in contact with the cathode should be an optimized electron transporting layer (ETL). Generally, the anode of OLEDs is an ITO film, the cathode is typically a low-to-medium work function (Φ) metal such as Ca (Φ= 2.87 eV), Al (Φ= 4.3 eV) or $Mg_{0.9}Ag_{0.1}$ (Mg, Φ= 3.66 eV) deposited either by e-beam or thermal evaporation [23].

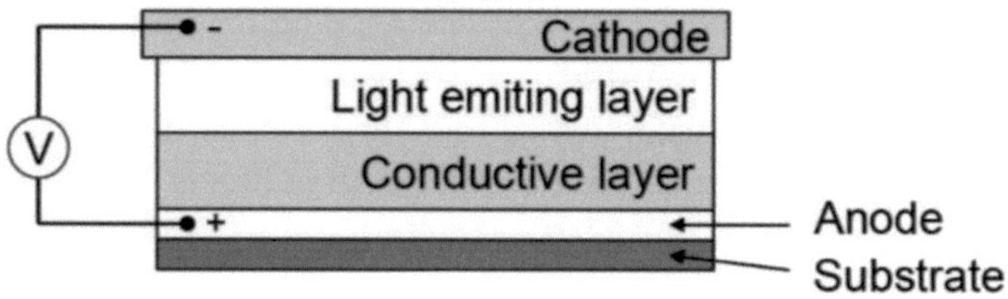

Figure 2. Schematic structure of organic light emitting diodes.

3.2. Mechanism of the operation of organic light emitting diodes (OLEDs)

The light emission from OLEDs is through electroluminescence (EL), which can be described in three steps (see Fig. 3) as follows: step 1: when a forward bias voltage is applied to an OLED, holes and electrons are injected. These injected charge carriers have to overcome their respective interface barriers and then holes occupy into the highest occupied molecular orbital (HOMO) energy level of the hole transport layer (HTL) and electrons into the lowest unoccupied molecular orbital (LUMO) energy level of the electron transporting layer (ETL). The HOMO of HTL is similar to the valence band in bulk semiconductors, and LUMO of ETL is similar to the conduction band. Step 2: The externally applied field on OLED drives the injected holes and electrons to the interface of HTL and ETL, where they are accumulated; holes in HOMO of HTL and electrons in LUMO of ETL. Step 3: Due to organic solids have low dielectric constant and strong binding energy both carriers (holes and electrons) move toward the interface between the two transport layers (HTL and ETL) and recombine in the light emitting layer (EML) to form excitons. Then these excitons emit light through the transparent electrode (ITO coated on glass substrate). In general environment, the exctions exist in an unstable state and their radiative move to recombination releases energy in the form of the light and heat. According to the above three steps, illustrated in Fig. 3, the light emission from an OLED is current driven and hence called electroluminescence (EL) .

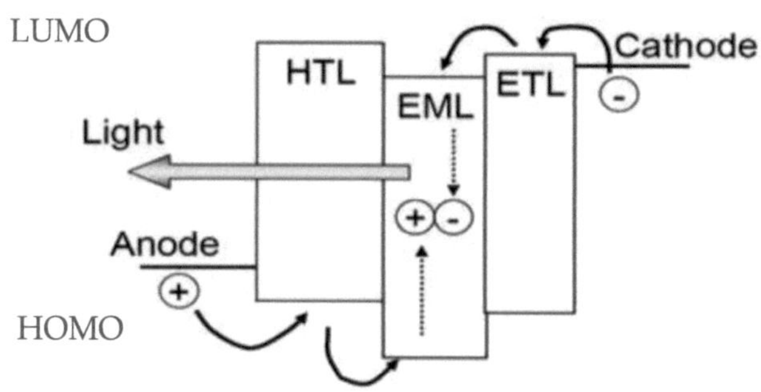

Figure 3. Schematic sketch of the active energy levels of organic light emitting diodes (OLEDs)

The carrier transmission in organic molecules is different from that in inorganic semiconductors or crystalline materials. There are no continuous energy bands in organic semiconductors. In organic semiconductors, consisting of organic molecules, there are delocalized π electrons, which are relatively free but confined on individual molecules due to weak intermolecular interactions. Therefore, the hopping theory is the most commonly used to describe the phenomenon of carrier transfer in organic solids. Driven by the electric field, electrons are excited or injected into the LUMO of one molecule in ETL and hop to the LUMO of neigh-boring molecule and thus electron transport occurs. Likewise, injected holes get transported by hopping from the HOMO of one molecule to another in HTL. In fact, as the charge carriers are injected externally and do not exist before the application of the electric field, the location of holes in HOMO and electrons in LUMO deforms the associated bond length and structure. Therefore, the movement of an injected electron or hole is coupled with the local deformation zone to form a unit, this unit is called polaron. Hence, in organic semiconductors the movement of the electrons or holes is often accompanied by the deformation of its structure, which is called self-trapped electron or hole. As such self-trapped charge carriers move slower the carrier mobility in organic semiconductors is in general lower than that in inorganic semiconductors or metals. The hole mobility in organic materials is typically 10^{-7}-10^{-3} cm^2/Vs and the electron mobility is typically lower by a factor of 10-100 [24].

The organic materials are usually insulators (such as plastic). Generally only a very small amount of current can be injected into the organic material by applying certain electric field, and EL occurs from the recombination of these injected electrons and holes. Therefore, if the current is less then injected carriers will be less and number of excitons to recombine will be limited. Therefore, to a large extent EL depends on improving the carrier injection efficiency from both electrodes, and on obtaining balanced and controlled electron–hole recombination within a well-defined zone.

3.3. Full Color of organic light emitting diodes (OLEDs)

There are five potential methods to make an OLED emit in red (R), green (G) and blue (B) color spectral regions [35]: 1). side by side patterning of red (R), blue (B) and green (G) OLED's, 2). absorptive filtering of white OLED, 3). fluorescent down-conversion of blue OLED's, 4). microcavity filtered OLED's and 5). color-tunable OLED's.

Method (1) This method is employed a precisely positioned shadow mask to selectively deposit the R、G and B OLEDs with individual pixels of R、G and B emission.

Method (2) A white OLED device can be made using materials with very broad emission spectra, or using two or more sequentially deposited light emitting layers and then color filters on white light OLEDs are used to change the emission into R, G and B colors.

Method (3) The full color pixel can be using a single blue OLED to pump wavelength down-converters, which efficiently absorb blue light and re-emit the energy as green or red light.

Method (4) The emission from a white OLED is filtered by a microcavity, which is composed of a dielectric quarter wavelength stack as the bottom mirror, the metal contact as the top mirror and an inactive material as a filler layer to adjust the cavity thickness [26]. However, the microcavity resonance causes strong viewing angle dependence of emitted colors, limiting this method to applications which need small viewing angle. In this method about $\pm 15^0$ viewing angle can be achieved [27].

Method (5) The color variation is achieved by voltage and/or polarity tuning. Only molecular OLEDs are capable of three color tuning. This method shows low efficiency and/ or requires high driving voltage. Hence, the color variable devices based on the polarity and/ or voltage-tuning are still far from applications. White light emission OLEDs can also serve as backlight panels of LCDs. White is the most important color in the lighting industry. A number of device structure concepts have been proposed to achieve white emission. These include the mixing of three primary colors from respective layers in a multilayer structure [28], the doping of appropriate amount of red, green, and blue dopants in the same host [29], the microcavity effect of one emission layer [30], use of exciplex formation, etc.

OLEDs have become viable now for flat panel displays after intensive research and progress in the past decade. Through proper material design/choice and device fabrication, various OLEDs with colors of high brightness have been developed for use in single- or full-color applications. As the operation of an OLED depends on the carrier transport in HTL and ETL, hole and electron confinement in EML and then their recombination to emission light. In most cases the number of injected holes in an OLED is more than electrons. Therefore, improving efficient electron injection is essential for efficient and stable OLEDs.

4. The field emission organic light emitting diodes (FEOLEDs)

4.1. Structure of FEOLEDs

As shown in Fig. 4, the basic structure of a FEOLED is to utilize an organic EL light-emitting material instead of inorganic phosphor thin film in FEDs [31]. The anode of the FEOLEDs can be a multi-layered organic solid or an OLED . But both have the same structure, which includes a hole injection layer (HIL), a hole transport layer (HTL) and a light emitting layer (EML). The cathode of FEOLEDs is made of CNTs template as electron source. In such a structure, it is not only difficult to protect the light emitting layer (EML) from high-energy electron bombardment [32], but also not easy to control highly efficient emission.

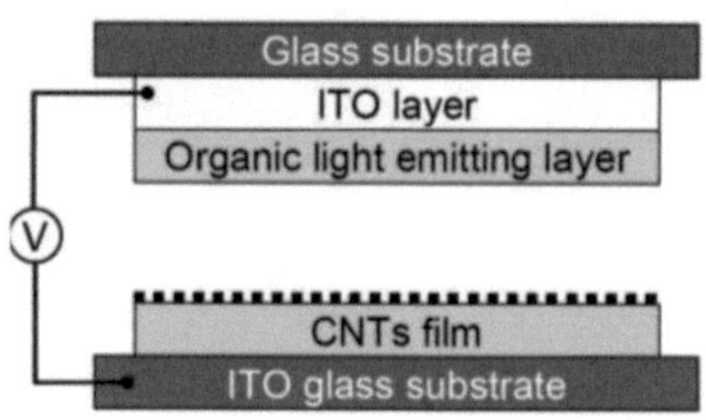

Figure 4. The schematic diagram of a FEOLED.

Since the basic structure of a FEOLED uses a direct current (DC), a protection layer is needed. Additionally, it can be deposited on EML. Such a protection layer in FEOLEDs increases the operating lifetime since this layer protects against electron bombardment .To improve this, a few different structures of FEOLEDs are presented in Figs.5. and 6.

Fig. 5 (a) shows the device structure of a FEOLED with the protection layer, which is made of a secondary electron material used as an electron multiplier. By apply operating voltage to FEOLEDs, the electron and hole recombination occurs in the EML by mechanisms similar to OLEDs. As OLEDs are current injection devices, electron injection must be improved to ensure their efficiency and stability. Therefore, as shown in Fig. 5(b), a field emission electrons layer is introduced in OLEDs to increase the electron density, which exhibits a higher luminous efficiency in FEOLEDs than conventional OLEDs [33]

Fig.6 shows the schematic diagrams of FEOLEDs with a dynode in (6a) and a strip electron multiplier in 6(b). According to the above described mechanism of FEOLEDs, the luminance intensity increases with the increase in electron injection. A common way to solve this problem is to introduce a dynode or an electron multiplier into the FEOLEDs . As shown in Fig.6 (a), dynode has holes whose whole inner surface is coated with a secondary electron material such as Be, Mg, or Ca oxide to increase the electron amplification factors. Fig.6 (b) shows a FEOLED device with strip electron multiplier. The strip electron multiplier was first proposed here by the author, which consisted of a strip of Al coated with another striped secondary electron material [3]. Both of the dynode and strip electron multiplier used in FEOLEDs are made for increasing the number of electrons and allow carriers to achieve a more balanced state in OLEDs and then enhance the luminance efficiency of OLEDs.

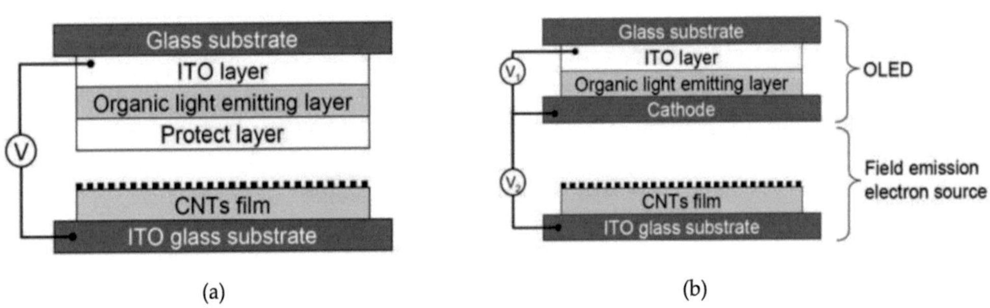

Figure 5. Two configurations of FEOLEDs (a) with protection layer and (b).with Al metal cathode.

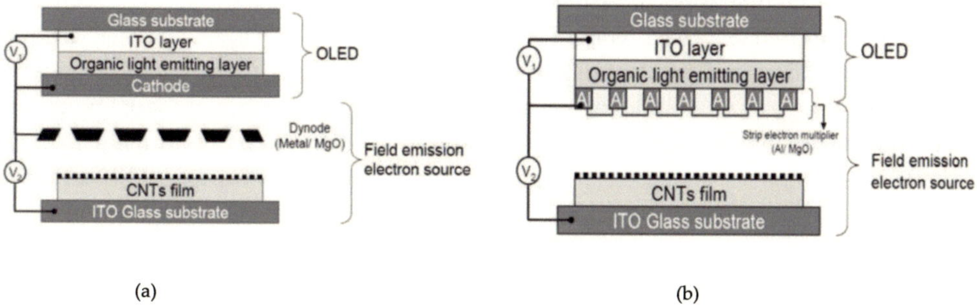

Figure 6. Two configurations of FEOLEDs with (a).Dynode and (b).strip electron multiplier

4.2. Fabrication of FEOLEDs

A FEOLED is that an organic emission layer (organic EL) is utilized instead of inorganic phosphor thin film in field emission display (FED). The organic EL in a FEOLED consists of a hole injection layer (HIL), a hole transport layer (HTL) and light emitting layer (EML). A FEOLED is able to attain higher luminance and low power consumption than conventional OLED.

Fig.7 (a) shows the structure of a basic FEOLED. The anode of FEOLED is ITO on which are coated the organic multi layers. The cathode consists of a field emission electron, which provides the electron injection into the organic material layer. Fig.7 (b) shows the FEOLED structure with a dynode, which acts as an electron multiplier to increase the number of electrons injected into the organic layer.

Fig. 8 (a) shows the structure of FEOLED with the strip electron multiplier. Due to the dynode is difficult to set in the narrow space of FEOLEDs. Therefore, the strip electron multiplier was to be proposed. Refer to Fig. 7(b), the dynode can be of the metal channel, box, line focus, or MPC type [1]. The secondary electron material in the dynode can be Cu-Be or Ag-Mg alloys. However, the strip electron multiplier was form with the secondary electron material (MgO or CsI) and Al, as shown in Fig. 8 (b).

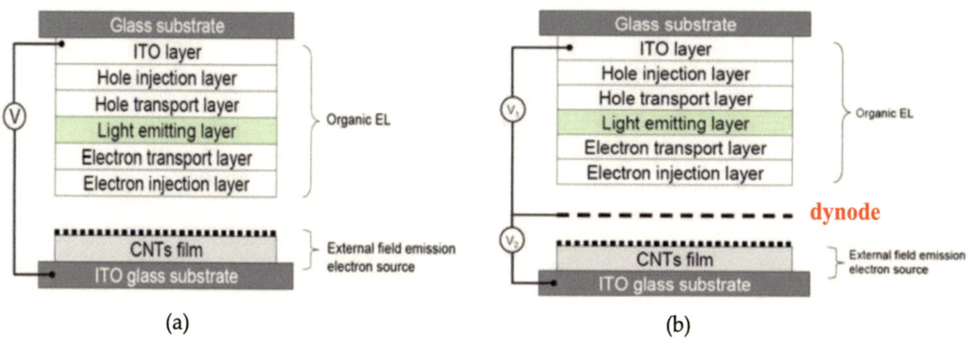

Figure 7. The components of FEOLED with (a) the basic device and (b) dynode structure.

Fig. 8 (c) shows the operating principle of a dynode, which is a cross-sectional schematic view of the metal channel type dynode. The secondary electron emission generated from the dynode can be understood by the following processes: (a) the primary electrons penetrate into a certain depth of an insulating layer (the secondary electron material); (b) through collision the energy of primary electrons is transferred to bound electrons of the insulator to release them and (c) released electrons migrate to the surface and escape into the vacuum as secondary electrons.

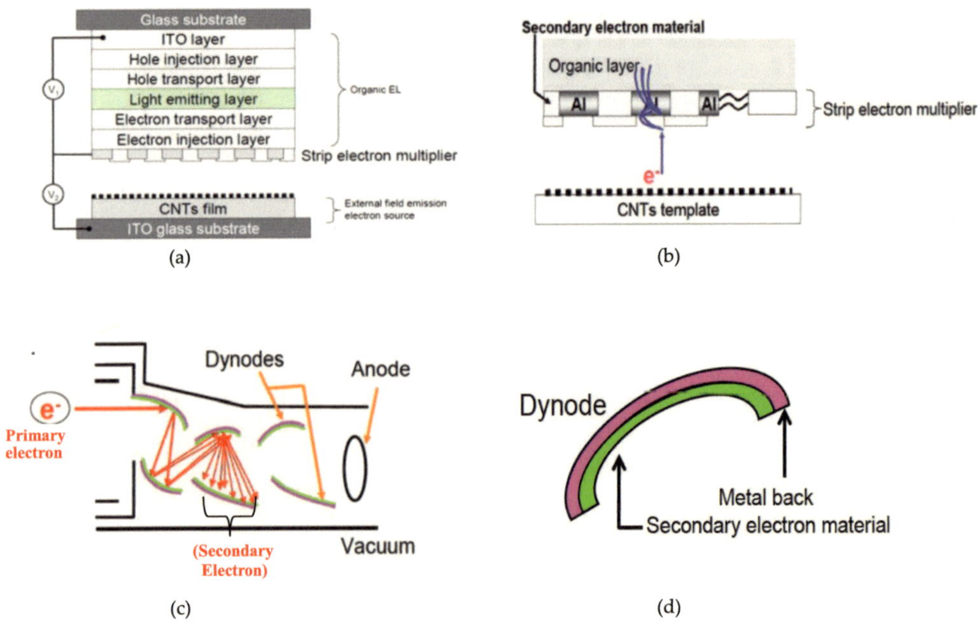

Figure 8. Schematic structure of a FEOLED with (a) strip electron multiplier and (b) strip Al. (c) the operating principle of a dynode and (d) A sheet of a dynode

4.3. Organic light emitting diode in FEOLED

In Fig. 8 (a) is shown a FEOLED with the strip electron multiplier formed on an OLED as a part of anode. The organic multilayer structure of FEOLED is the same as that of OLED, which works with the same mechanism.

One of the most important issues related with the characteristics of OLEDs is the number of injected electrons and holes should be balanced. It is well know that the direct electron-hole recombination in the light emitting layer occurs due to OLEDs. Therefore, an effective cathode structure for efficient electron injection is critical to optimal performances of OLEDs. A nanometer-size interfacial layer between the metal cathode and organic material in OLEDs plays the critical role in the carrier injection efficiency. In order to improve the injection efficiency of electrons, the low work function metal or alloys such as LiF are usually used to form low energy barriers for electron injection from the cathode to the organic material [34]. It has been shown that LiF is very effective in terms of facilitating electron injection. However, recently alkali metal carbonates (Cs_2CO_3) have also been reported to be efficient electron injection material [35, 44-45]. We believe that if the electrons are efficiently injected from the cathode to organic layer, then this will improve the charge carrier balance in OLEDs and hence improve the device efficiency. This idea will work the same way in FEOLEDs.

4.4. Carbon nanotubes field emission template in FEOLEDs

In this section, FEOLEDs are discussed for their optoelectronic characteristics in terms of external electron supplement into the organic light emitting layer. Hence, FEOLEDs must have an excellent field emission cathode to emit electrons. The carbon nanotubes (CNTs) template is the best choice to be adopted in FEOLEDs as the field emission cathode. CNTs can work in less stringent vacuum conditions ($<1x10^{-5}$ torr) and have higher emission currents than metal and semiconductor micro-tip field-emission sources. Iijima discovered carbon nanotubes (CNTs) in 1991[36]. CNTs have a superior mechanical strength, good heat conductance, and ability to emit cold electrons at relatively low voltages because of their high aspect ratios and nanometer-scale tips.

The traditional method of fabricating field emitters is based on the use of multi-needle field emission cathodes and precision technological processes based on electron lithography techniques. Metal and semiconductors are usually used as cathode materials, which, unfortunately, have rather high work functions (4-5eV) [37]. The application of CNTs in field emission template is very extensive. For example, CNTs can be directly synthesized on a substrate by CVD on an anodic aluminum oxide (AAO) template, and by screen print. However, the fabrication of a CNTs template is time consuming. There is a new and more convenient method to fabricate the CNTs templates by the spray method [38-39].

CNTs thin films are usually fabricated by two methods, such as: drop-drying from solvent [41] and filtration and spin-coating [42], but these methods have severe limits in the film quality, like in uniformity, homogeneity, and production efficiency. CNTs thin films consisting of multi-walled CNTs (MWCNTs) are fabricated by the spray method, which is an easy and convenient method to deposit CNTs and can achieve large area deposition [43]. The procedures of the fabrication of CNTs template are described as below: First CNTs are suspended in 1, 2- dichloroethylene (DCE) and second, sonication of 30 mg CNTs in 50 ml DCE solvent for 2 hours. To obtain good adhesion between CNTs and ITO glass substrate, an indium (In) metal layer is deposited onto the ITO glass substrate by thermal evaporation. After annealing at 300 °C for 15 min in N2 atmosphere, CNTs are firmly adhered to the In layer and produce good field emission characteristics.

4.5. Operation principle of FEOLEDs

According to the above mentioned FEOLED structure, FEOLEDs can be divided into two models: 1) Original type FEOLEDs and 2) triode devices.

4.5.1. Original type FEOLEDS

As shown in Fig. 7 (a), FEOLEDs (original type) are similar to FEDs, but they use organic light emitting material instead of a phosphor. In these FEOLEDs an ITO film is used as anode and CNTs template as cathode. By applying the driving voltage to both the electrodes (ITO and CNTs template), then electrons and holes move toward the interface between the two transport layers (HTL, and ETL) and recombine to form excitons. Finally, these excitons emit light through the ITO substrate..

4.5.2. Triode FEOLEDs with a dynode structure (strip electron multiplier)

As shown in Fig, 7 (b), FEOLEDs with a dynode structure are classified as triode devices. It comprises the dynode and an organic EL light emitting layer and CNTs template. The dynode is formed between the cathode and the organic EL lighting layers.

As shown in Fig. 8 (a), there is another kind of triode FEOLED, which is a FEOLED with a strip electron multiplier. In this case a strip electron multiplier is used instead of a dynode, but it is attached directly to the organic EL light emitting layer for protecting the electron injection layer from high-energy electron bombardment and allows the electrons to disperse evenly in the EML.

4.5.3. Luminescence mechanisms in FEOLEDs

To further confirm that the mechanism of operation of FEOLEDs is the same as that of OLEDs, the following experiment was conducted. A hole blocking layer (BCP) is inserted between the hole transport layer (NPB) and the emission layer (Alq_3) of the organic formation, as shown in Fig. 9(a). If an OLED is applied a voltage, hole carriers injected from the ITO anode electrode to the hole transport layer would be blocked at the interface of the NPB layer and the BCP layer. The electrons (emitted from the cathode and passing through Alq_3) would then recombine with the holes accumulated in the NPB layer. Refer to Fig. 3, this NPB excitation generated by the recombination, according to the active energy levels of OLEDs would give rise to blue light. The Alq_3 layer in such case would generate no light. If a cathode luminescence mechanism device is applied with the BCP layer, on the other hand, electron bombardment on the organic material would generate light in the emission layer Alq_3, which should have green color. As such, when a BCP layer is inserted in FEOLEDs, if blue light is observed, the luminescent mechanism of the FEOLED must be similar to that of the conventional OLED: if green light is observed, the luminescent mechanism must be similar to that of cathode luminescence. The experiment results showed that blue light was observed, as shown in Fig. 9(b), which clearly illustrates that both FEOLEDs and OLEDs operate on similar mechanism of emission.

Thus, the light emission in FEOLEDs also occurs via the following five processes as in OLEDs: 1) both electrons and holes injected from anode and cathode into organic layers, 2) these injected charge carriers are transported towards each other across the organic layer, 3) formation of singlet excitons due to the Coulomb interaction between the injected electrons and holes, 4) migration singlet excitons to organic emitting layer and 5) radiative recombination of single excitons.

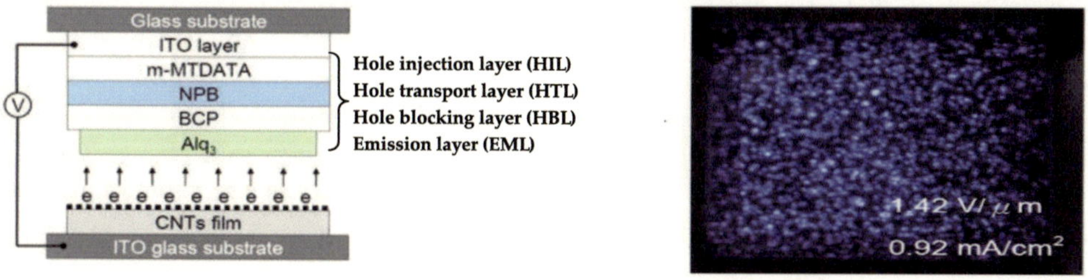

(a) (b)

Figure 9. (a) Structure of a FEOLED with Hole blocking layer and (b) emission of blue light.

4.6. A FEOLED with strip electron multiplier

As shown in Fig. 10, the typical structure of FEOLEDs in this work comprises a CNTs template cathode and the strip electron multiplier formed on an OLED as a part of anode. It can be assembled in a vacuum chamber.

4.6.1. *Experimental*

In this section, the current density (J)-applied voltage (V) characteristics and the optical performances of a FEOLED with the strip electron multiplier are studied experimentally.

Fig.10 shows the configuration of FEOLED in this work, where the upper portion is an OLED lower part is the external electron source of CNTs cathode. The S_{oled} switch is use to control OLED and S_{ext} switch is used to control the external electron source of CNTs template.

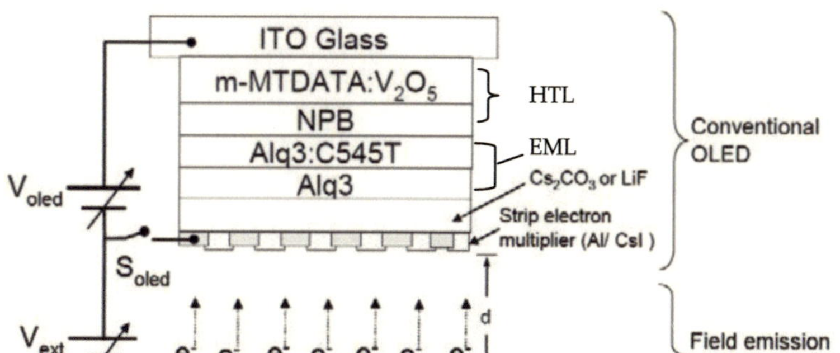

Figure 10. Schematic presentation of an apparatus for characterizing a FEOLED with strip electron multiplier

The structure of an OLED is ITO glass/ m-MTDATA:V_2O_5(10 nm, 10 %)/ NPB (30 nm)/ Alq3:C545T (30nm, 3 %)/ Alq3 (10 nm)/ Cs_2CO_3(1 nm)/ Al, where, the V_2O_5 doped m-MTDATA was chosen as the hole transport layer (HTL) and it has a high conductance, intrinsically leading to the formation of many intrinsic carriers between ITO and the organic surface significantly enhancing the hole injection and transport [46]. Additionally, the emission layer (EML) of OLED uses Alq_3 as host and C545T dye as green fluorescent material to trap the electrons to build up the space charge and decrease the free electron distribution in the host Alq3 [47], subsequently reducing the current density of the device. Moreover, Cs_2CO_3 and LiF were chosen as electron injection buffer layer (EIL). And the strip electron multiplier was fabricated from the strip of Al (80 nm) and strip of CsI (110 nm) and integrated with the organic material of FEOLEDs. The entire organic material layer was prepared by using a high vacuum thermal evaporation system.

Such a structure (Fig.10) is used to examine the current-voltage (J-V) characteristics of OLEDs with a different electron injection layer to prove the electron injection capability of Cs_2CO_3. Additionally, the J-V characteristics and luminance of the FEOLED are measured by Keithly-2400, Keithly-237 and TOPCON PR-650, respectively. All the measurements are performed in a high vacuum ambient of 6×10^{-6} torr at room temperature.

4.6.2. *Results and discussion*

In this section, the current density (J)-applied voltage (V) characteristics and the optical performances of a FEOLED with Electron Multiplier are studied experimentally.

Fig.11 (a) shows the J-V characteristics of two OLEDs, one with LiF and the other with CS_2CO_3 as the electron injection layer when S_{oled} is closed and S_{ext} is open (Fig10). To enhance the OLED performance, the CS_2CO_3 is used instead of LiF as the electron injection material. The optimal thickness of CS_2CO_3 has been characterized and found to be 1 nm. Under the same driving voltage of 10 V, the OLED with the electron injection layer of CS_2CO_3 (1.0 nm) can get the current density of 93 mA/cm^2 which was higher than that of OLED with the electron injection layer of LiF (0.7 nm). The better performance of OLEDs with Cs_2CO_3 can be attributed to Cs having a low-work function of 2.14 eV relative to Li (2.9 eV) [50]. The electron injection layer of Cs_2CO_3 in OLEDs seems to have induced strong n-doping effects in Alq3 and ultimately increases the electron concentrations in the electrons-transport layer of Alq3. Moreover, OLED with an increased thickness of Cs_2CO_3 to 2 nm have shift the J-V curve to a lower current density. Notably, the OLEDs performance depends on the thickness of the Cs_2CO_3 layer.

Fig.11 (b) shows the luminance (L)-applied voltage (V) characteristics of the same two OLEDs with LiF and CS_2CO_3 as the electron injection layer when S_{oled} is closed and S_{ext} is open (Fig10). According to Figure (b), the OLED with Cs_2CO_3 (1 nm) as the electron injection layer can achieve a high luminance of 10,820 cd/m^2 and a high EL efficiency of 12 cd/A at 10 V. In contrast, OLED with a LiF layer can achieve a luminance of 5,821 cd/m^2 and an EL efficiency of 10.2 cd/A at 10 V. This demonstrates very clearly that the OLEDs, in which Cs_2CO_3 is used as an electron injection layer, show an excellent performance. It indicates that electrons are effectively injected from the cathode to the organic layer due to the lower electron injection barrier, which improves the charge carrier balance and subsequently increases the device efficiency.

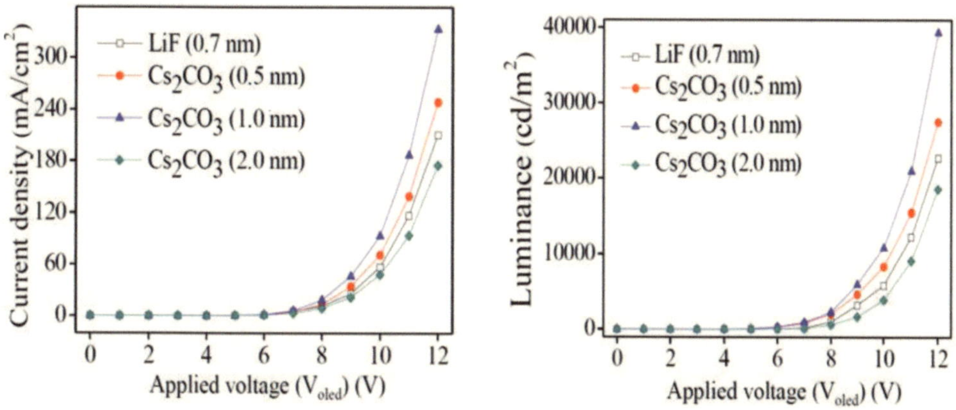

(a) (b)

Figure 11. (a) Current density (J)-Applied voltage (V) at **SOLED** closes and **Sext** is open and (b) Luminance (L)-Applied voltage (V) at **Soled**=closes and **Sext** is open(Fig.10)

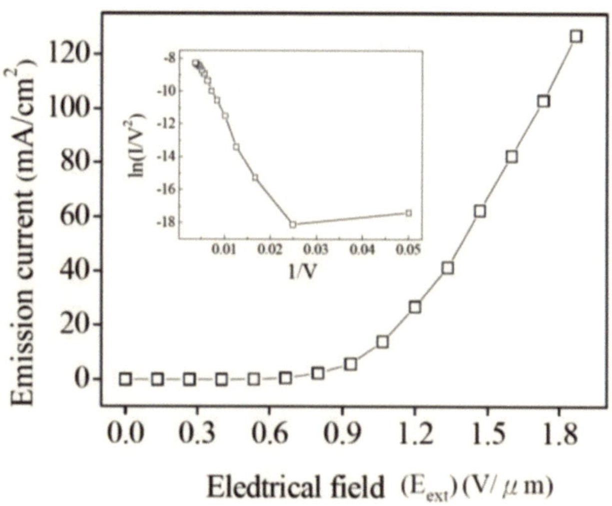

Figure 12. The field emission current versus electric field (J-E) characteristics of a CNTs template when S_{oled} is open and Sext is closed, (Eext=Vext/d) (Fig. 10)

Fig. 12 shows the field emission current versus electric field (J–E) characteristics of a CNTs template. The CNTs template is made as an external electron source for FEOLED [51]. A field emission current density of approximately 127mA/cm² is produced at an electric field of 1.86 V/μm. The enhanced current density can be attributed to the satisfactory adhesion between CNTs and the ITO glass substrate. The current density increases with the electric field. Based on the above results, the amount of electrons injected into the Al electrode can be determined by adjusting the electrical field (Fig.10, V_{ext}), which is applied to the CNTs template. Detailed operations of FEOLEDs can be described as follows. Initially, S_{oled} is turned on to drive the OLED shown in Fig.10. The OLED emits a luminance of 10,820 cd/m^2 as the driving voltage reaches 10 V and, simultaneously, S_{ext} is switched on to attract the electrons emitted from the CNT emitters.

Fig.13 shows the luminance-current density characteristics of OLEDs and FEOLEDs. The curve in section A displays the characteristics of a conventional OLED (S_{Oled}=close, S_{ext}=open, as shown in Fig.10), where sections B and C show the FEOLED (S_{oled}=close, S_{ext}=close, as shown in Fig.10). At a driving voltage of 10 V on an OLED, the luminance is enhanced from 10,820 cd/m^2 to 27,393 cd/m^2 while S_{ext} is turned on. Obviously, applying an electrical field (E_{ext}) to the CNTs template can enhance the generation of the field emission electrons into the OLED. Additionally, the current density of OLED is increased by the supplementary electrons into the multilayer of the organic light emitting layer with the external electron source (S_{oled}=close, S_{ext}=close, as shown in Fig.10). Moreover, the current density of the OLED (V_{oled}=10 V) with the external electron source increases from 93 mA/cm^2 (E_{ext}=0.8 V/μm) to 184.5 mA/cm^2 (E_{ext}=1.7 V/μm), and the luminance also increases from 10,820 cd/m^2 to 27,393 cd/m^2 simultaneously, as shown in Fig.13 (hole block line)

According to the above characteristics of FEOLED in comparison with the OLED under the same operating current density (120mA/cm^2), the FEOLED exhibits a higher luminous efficiency of 18.6 cd/A than the luminous efficiency of 11.42 cd/A for OLED, as shown in Fig.14. The FEOLED results can be attributed to the external electron injection into the multilayer organic layer of OLED, thus balancing the hole and electron. Furthermore, increasing the quantity of electrons by using an external electron source significantly increases the current density of OLED and makes the luminance efficiency higher than that of conventional OLED.

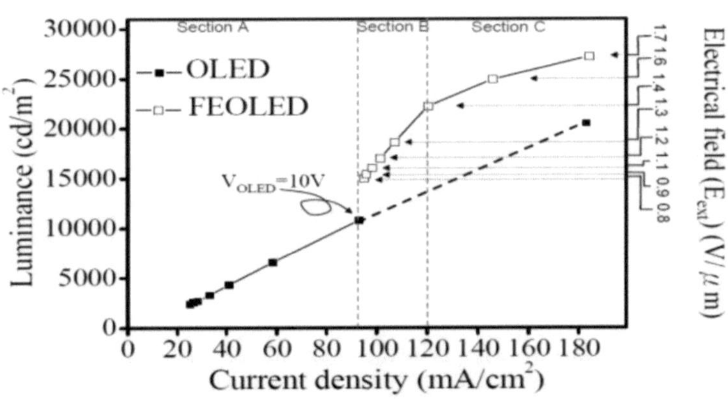

Figure 13. The Luminance (L)-Current density (J)-Electric field (E) of the OLED and FEOLED devices at both **Soled** and **Sext** close (Eext=Vext/d) (Fig. 10)

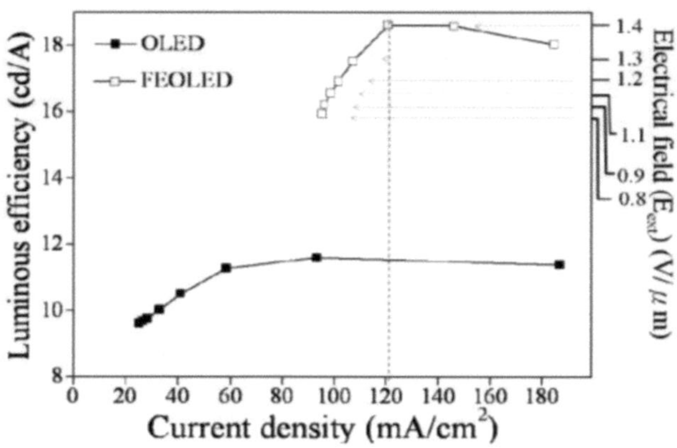

Figure 14. The Luminous efficiency (η)- Current density (J)-Electrical field (E) of the OLED and FEOLED devices at both Soled and Sext close (Eext=Vext/d)

In a FEOLED, the electrical field under vacuum condition, accelerates the electrons emitted from the CNTs cathode to affect the secondary electron material of CsI; they then pass through the Al and transport are transported through the organic EL light emitting layer. Field emission electrons with a sufficiently large electron energy supplement into OLED to increase the current density. Notably, increasing the number of the electrons that reach the organic EL light emitting layer also increases the luminous efficiency of the OLED. Therefore, the ways in which the OLED and FEOLED differ can be easily observed under the same current density. The luminance of FEOLED exceeds that of conventional OLED, as shown in Fig.13. Our results further demonstrate that the curve of the FEOLED becomes gradually saturated, especially for section C. Notably, injecting external electrons into the OLED continuously does not allow the luminance of the FEOLED to increase linearly with the current density since the quantity of electrons is larger than in the hole in section C. The carrier has become imbalanced again, subsequently decreasing the luminance. Furthermore, the electronic behavior shown in the FEOLED, it can be further demonstrates the amount of electrons is less than holes.

As describe above, we can see that the characteristics of the OLED and the FEOLED are listed by Table 1, respectively.

Table A The characteristic of the OLED.

	OLED	
	V_1 (Volt.)	V_2 (V/μm)
Driving voltage	10	0
Luminance (cd/m²)	10820	
Efficiency (cd/A)	12	

(1)

Table B The characteristic of the FEOLED.

	FEOLED	
	V_1 (Volt.)	V_2 (V/μm)
Driving voltage	10	1.4 V/μm
Luminance (cd/m²)	27393	
Efficiency (cd/A)	18.6	

(2)

Table 1. (1) The characteristics of the OLED and (2) The characteristics of the FEOLED

5. Conclusion and future direction

This work presents a novel FEOLED had further increase the luminous efficiency of OLEDs. The characteristics of an OLED constructed in the FEOLED device are optimized by inserting a thin Cs_2CO_3 electron injection layer between the Alq_3 and Al electrode. Experimental results indicate that the external field emission electrons can enhance the luminance in FEOLED efficiently owing to balanced recombination of electrons and holes. Additionally, FEOLED achieves a higher luminous capability than that of OLED under the same current density. Mechanism detection of the FEOLED further reveals that the amounts of holes are more than that of electrons in the emission layer of an OLED device. Furthermore, the secondary electron material CsI deposited onto the Al electrode in a FEOLED can provide multiple electrons as well as prevent the organic layer from electrons bombardment. The proposed device's construction is extremely important for characterizing the emission mechanism of the FEOLED.

Another objective of this chapter is to provide background knowledge to readers from the different fields to stimulate new ideas. For example, the flexible photovoltaic OLED (PVOLED) and a tandem of organic solar cell (OSC) and white organic light emitting diode (WOLED), although not addressed here, are now emerging. In PVOLEDs, the power recycling efficiency of 10.133 % is achieved under the OLED of PVOLED operated at 9V and at a brightness of 2110 cd/m^2,when the conversion efficiency of OSC is 2.3%.[52]. In a tandem of OSC and WOLED, which can be fabricated to generate electricity as well as lighting for domestic and commercial uses [53].

Author details

Meiso Yokoyama
Department of Electronic Engineering, I-Shou University, Kaohsiung City, Taiwan

6. References

[1] Meiso Yokoyama, U.S. Pat. 7,456,562 B2, (2008)

[2] Meiso Yokoyama, Japan Patent 3879101, (2006)

[3] Meiso Yokoyama, Chi-Shing Li, and Shui-Hsiang Su,"Novel Field Emission Organic Light Emitting Diodes" IEICE TRANS. ELECTRON.,VOL. E94-C,NO.3, (2011)

[4] Chi-Shing Li, Meiso Yokoyama, and Shui-Hsiang Su, "Efficiency Enhancement of Field-Emission Organic Light Emitting Diodes Using a Dynode Structure" Electrochemical and Solid-State Letters, vol. 11, pp. J1-J3,(2008)

[5] Meiso Yokoyama et al. The 6tsh internal Symposium on Organic Molecular Electronics (ISOME 2010),(2010)

[6] G. T. Chen, S. H. Su, and M. Yokoyama, "Field-Emission Organic Light-Emitting Device Using Oxide-Coated Cathode as Electron Source" Electrochem. Solid State Lett., vol. 10, no. 3, pp. J41-J44,(2006)

[7] G. N. Fursey, "Field emission in vacuum micro-electronics" Applied Surface Science, vol. 215,pp. 113-134, (2003)

[8] L. Ozawa, and M. Itoh, "Cathode Ray Tube Phosphors" Chemical Reviews, Vol. 103,pp. 3837-3855,(2003)

[9] T. Oyama, H. Ohsaki, Y. Tachibana, et al "A new layer system of anti-reflective coating for cathode ray tubes" Thin Solid Films, vol. 351,pp. 235-240, (1999)

[10] Y. Gao, X. Zhang, W. Lei, M. Liu, et al "Ion bombardment in a normal-gate FED" Applied Surface Science, vol. 243, pp. 19-23, (2005)

[11] K. R. Shoulders, "Microelectronics using electron beam activated machining techniques" Adv. Comput., vol.2, pp.135. (1961)

[12] C. A. Spindt, "A Thin Film Field Emission Cathode" J Appl Phys, vol.39, p. 3504, (1968)

[13] C. J. Summers "Phosphors for field emission displays" Tech Digest of the 10th Int Vac Microelectron Conf, Kyongju, Korea, p. 244, (1997)

[14] G. S. Choi, K. H. Son, and D.J. Kim, "Fabrication of high performance carbon nanotube field emitters" Microelectronic Engineering, vol.66, pp. 206-212, (2003)

[15] A. Ostankov, D. Paneque, E. Lorenz, et al, "A study of the new hemispherical 6-dynodes PMT from electron tubes" Nuclear Instruments and Methods in Physics Research A, vol. 442, pp. 117-123, (2000)

[16] A. J. Dekker,(1958) "Secondary Electron Emission" Solid State Physics, vol. 6, pp.251-311,.

[17] O. Groning, O. M. Kuttel, Ch. Emmenegger et al "Field Emission properties of carbon nanotubes" J. Vac. Sci. Technol. B, vol. 18, pp.665-678, (2000)

[18] M. Nikl, "Wide Band Gap Scintillation Materials: Progress in the Technology and Material Understanding" phys. stat. sol., vol. 179, pp. 595-620, (2000)
[19] J. S. Yoo and J. D. Lee, "The effects of particle size and surface recombination rate on the brightness of low-voltage phosphor" J. Appl. Phys., vol. 81, pp. 2810-2813, (1997)
[20] M. Pope, H. P. Kallmann, and P. Magnante , "Electroluminescence in Organic Crystals" J.Chem. Phys, vol. 38, pp. 2042-2043, (1963)
[21] J. H. Burroughes, D. D. C. Bradley, A. R. Brown, et al, "Light-emitting diodes based on conjugated polymers" Nature vol. 347, pp. 539-541, (1990)
[22] W. Tang, and S. A. VanSlyke, "Organic electroluminescent diodes" Appl. Phsy. Lette., vol. 51, pp. 913-915 (1987)
[23] P. Dannetun, M. Lögdlund, C. Fredriksson, et al "Reactions of low work function metals Na, Al, and Ca on α, ω diphenyl tetra decaheptaene. Implications for metal/polymer interfaces" J. Chem. Phys., vol. 100, pp. 6765-6771.(1994)
[24] R. G. Kepler, P. M. Beeson, S. J. Jacobs et al. "Electron and hole mobility in tris(8-hydroxyquinolinolato-N1,O8) aluminum" Appl. Phys. Lett., Vol. 66, pp.3618-3620, (1995)
[25] P. E. Burrows, G. Gu, V. Bulovi6 et al "Achieving full-color organic light-emitting devices for lightweight, flat-panel displays" IEEE Transactions on Electron Dev. Vol. 44, pp. 1188-1203, (1997)
[26] C. L. Lin, H. W. Lin, and C. C. Wu"Examining microcavity organic light-emittingdevices having two metal mirrors" Appl. Phys. Lett., vol. 87, pp.021101 (1-3), (2005)
[27] Y. J. Lee, S. H. K, J. Huh, et al "A high-extraction-efficiency nano patterned organic light-emitting diode" Appl. Phys. Lett., vol. 82, pp. 3779-3781,(2003)
[28] H. Kanno, Y. Sun, and S. R. Forrest,"High-efficiency top-emissive white-light-emitting organic electro-phosphorescent devices" Appl. Phys. Lett., vol. 86, pp. 263502 1-3. (2005)
[29] Y. J. Tung, M. M.-H. Lu, M. S. et al "High-Efficiency White Phosphorescent OLEDs for Lighting" Proc. of SPIE, vol. 5214, pp.114-123,(2004)
[30] Y. J. Lu, C. H. Chang, C. L. Lin, et al "Achieving three-peak white organic light-emitting devices using wavelength-selective mirror electrodes" Appl. Phys. Lett., vol. 92, pp. 123303 (1-3),(2008)
[31] C. S. Li, S. H. Su, H. Y. Chi, and M. Yokoyama "Application of highly ordered carbon nanotubes templates to field-emission organic light-emitting diodes" Journal of Crystal Growth, vol. 311, pp. 615-618, (2009)
[32] C. S. Li, S. H. Su, T. M. Lin , H. Y. Chi, and M. Yokoyama, "Luminous Efficiency Enhancement of Organic Light-Emitting Diodes by an External Electron Source", IEEE International Nano Electronics Conference (INEC).(2010)
[33] G. T. Chen, S. H. Su, and M. Yokoyama, "Field-Emission Organic Light-Emitting Device Using Oxide-Coated Cathode as Electron Source" Electrochemical and Solid-State Letters, vol. 10, pp. J41-J44, (2007)

[34] M. Pfeiffer, K. Leo, X. Zhou, et al "Doped organic semiconductors: Physics and application in light emitting diodes" Organic Electronics, vol. 4, pp. 89-103, (2003)

[35] W. A. de Heer , A. Châtelain, D. Ugarte "A Carbon Nanotube Field-Emission Electron Source" Science, Vol. 270, pp. 1179-1180, (1995)

[36] S. Iijima "Helical microtubules of graphitic carbon", Nature, vol. 354, pp. 56-58.(1991)

[37] T. 1V. Vorburger, D. Penn, and E. W. Plummer, "Field emission work functions" Surface Science, vol. 48, pp. 417-431,(1975)

[38] S. R. C. Vivekchand, L. M. Cele, F. L. Deepak, et al "Carbon nanotubes by nebulized spray pyrolysis" Chemical Physics Letters, vol. 386, pp.313-318.(2004)

[39] W. B. Choi, D. S. Chung, J. H. Kang, et al "Fully sealed, high-brightness carbon-nanotube field-emission display" Appl. Phys. Lett., vol. 75, pp. 3129-3131,(1999)

[40] L. Zhu, J. Xu, Y. Xiu, et al "Growth and electrical characterization of high-aspect-ratio carbon nanotube arrays" Carbon, vol. 44, pp.253-258, (2006)

[41] Z. Wu, Z. Chen, Xu Du, et al "Transparent, Conductive Carbon Nanotube Films" Science, vol. 305, pp. 1273-1276, (2004)

[42] R. H. Schmidt, I. A. Kinloch, A. N. Burgess, and A. H. Windle "The Effect of Aggregation on the Electrical Conductivity of Spin-Coated Polymer/CarbonNanotube Composite Films" Langmuir, vol. 23, pp. 5707-5712 (2004)

[43] H. J. Jeong, H. K. Choi, G. Y. Kim, et al "Fabrication of efficient field emitters with thin multi-walled carbon nanotubes using spray method" Carbon, vol. 44, pp. 2689-2693. (2006)

[44] L.S.Hung, C.W.Tang, M.G.Mason, P.Raychaudhuri, and J. Madathil, "Application of an ultrathin LiF/Al bilayer in organic surface-emitting diodes", J.Appl.Phys., vol.78,no.4,pp.544-546, (2000).

[45] Q.Liu, L.Duan,Y.Li,J.Qiao, Z. Yu, D. Zhang, L. Wang, G, Dong, and Y. Qiu, "Study on the electron injection mechanism of thermally decomposable Cs2CO3", Jpn. J. Appl. Phys., vol.48, pp.102302-4, October (2009).

[46] X. L. Zhu, J. X. Sun, H. J. Peng, et al "Vanadium pentoxide modified polycrystalline silicon anode for active-matrix organic light-emitting diodes" Appl Phys Lett., vol. 87, pp. 153508 (3 pages),(2005)

[47] C. H. Chen, and C. W. Tang, "Efficient green organic light-emitting diodes with stericly hindered coumarin dopants" Appl. Phys. Lett., Vol. 79, No. 22, pp.3711-3713. (2001)

[48] A. Buzulutskov, A. Breskin, and R. Chechik, "Field enhancement of the photoelectric and secondary electron emission from Csl" J. Appl. Phys. vol. 77 no.5, pp. 2138-2145,(1995)

[49] H. J. Jeong, H. K. Choi, G. Y. Kim, et al "Fabrication of efficient field emitters with thin multi-walled carbon nanotubes using spray method" Carbon, vol. 44, pp. 2689-2693,(2006)

[50] Q. Liu, L. Duan, Y. Li, J. et al "Study on the Electron Injection Mechanism of Thermally Decomposable Cs2CO3" Jpn. J. Appl. Phys., vol. 48, pp. 102302 (4 pages), (2009)

[51] W. A. de Heer, A. Chatelain, D. Ugarte "A Carbon Nanotube Field-Emission Electron Source" Science, vol. 270, pp. 1179-1180, (1995)

[52] Meiso Yokoyama*, Wu Chung-Ming and Shui-Hsiang Su" Enhancing the efficiency of white organic light-emitting diode using energy recyclable photovoltaic cells" Jpn. J. Appl. Phys. Vol. 51, pp.032102, (2012)

[53] J. Singh, " Developing a tandem of organic solar cell and light emitting diode" Phys. Status Solidi C 8, No. 1, 189–192 (2011)

Chapter 3

Polarized Light-Emission from Photonic Organic Light-Emitting Devices

Byoungchoo Park

Additional information is available at the end of the chapter

http://dx.doi.org/10.5772/54233

1. Introduction

Since the early pioneering work on efficient Organic Light-Emitting Devices (OLEDs) that was based on both small molecules and polymers, OLEDs have attracted a great deal of research interest due to their promising applications in full-color flat-panel displays and solid-state lighting [1-5]. Intensive research has been conducted into the development of OLEDs for realizing strong and efficient electroluminescent (EL) emission. To date, almost all previous work carried out on organic EL emission has involved unpolarized EL emission. Nevertheless, a number of researchers have reported the results of experiments in which linearly polarized EL emissions have been observed [6-17]. This particular avenue of research has been considered to be important because polarized EL emission from OLEDs is of potential use in a range of applications, not just those limited to high-contrast OLED displays, but also in efficient backlight sources in liquid crystal (LC) displays, optical data storage, optical communication, and stereoscopic 3D imaging systems [17]. In order to design and manufacture these novel light-emitting devices, a high degree of polarization ratio (*PR*) of emitting light is required, which has to be at least 30 ~ 40:1, between the brightness of two linearly polarized EL emissions that are parallel and perpendicular to the polarizing axis. Most cases of linearly polarized EL emission have been achieved through the use of uniaxially oriented materials, such as LC polymers or oligomers, incorporated within emissive layers. Methods that are commonly used for the uniaxial alignment of such layers include the Langmuir-Blodgett technique [6], rubbing/shearing of the film surface [7, 8], mechanical stretching of the film [9, 10], orientation on pre-aligned substrates [11, 12], precursor conversion on aligned substrates [13], epitaxial vapor deposition [14], and the friction-transfer process approach [15, 16]. Although there have been a number of such efforts to achieve linearly polarized EL emission, the polarization ratio and the device performance (in terms of brightness and efficiency) reported are still insufficient for most applications.

Here we introduce an approach different from the conventional methods using uniaxially oriented materials. As an alternative, for the purpose of improving device performance, we suggest a technique to control the polarization of light emitted from OLEDs that are achieved using an anisotropic photonic crystal (PC) film. It has been predicted that in anisotropic PCs, the photonic band structure splits with respect to the state of polarization of the interacting light, in contrast to the degenerated band structure of conventional isotropic PCs, in which a certain energy range of photons is forbidden, giving rise to a photonic band gap (PBG) [18-20]. Of these applications, the study of light emission at the PBG edge is particularly attractive, as a result of the fact that the group velocity of photons approaches zero and the density of mode changes dramatically at the PBG edge [21-24]. The combination of PCs with OLEDs has also been reported to achieve high out-coupling emission efficiency, as achieved in the micro-cavity OLEDs or multi-mode micro-cavity OLEDs [25-27]. Moreover, by employing the anisotropic photonic structure, one may also obtain the polarized emission of EL light.

In this chapter, we describe in brief a technique to control the polarization of EL light emitted from photonic OLEDs that make use of a *Giant Birefringent Optical* (GBO) [28] multilayer reflective polarizer [29-31] as the anisotropic PC film. When a large degree of birefringence is introduced into the in-plane refractive index between adjacent material layers of a multilayer photonic system, GBO effects begin to occur [28]. Pairs of groupings of adjacent layers (unit cells) can produce constructive interference effects when their thicknesses are scaled properly to the wavelength of interest. These interference effects in multilayered structures result in the development of alternating wavelength regions of high reflectivity (reflection bands) adjacent to wavelength regions of high transmission (pass bands) [28]. A significant optical feature of these multilayer interference stacks is the difference in the refractive index in the thickness direction (z axis) relative to the in-plane directions (x and y directions) of the film. By appropriate adjustment of the refractive indices of the adjacent layers, it is possible to construct a GBO multilayer reflecting polarizer using an interference stack that is composed of multiple layers of transparent polymeric materials [28]. The reflection band of the GBO polarizer exhibits a unique optical property, where the reflectivity of interference polarizers either remains constant or increases with increasing the angle of incidence. Furthermore, a graded unit cell thickness profile is normally used to create a wider reflective band that accommodates wavelengths from the blue through to the green and red color regions [28]. Such a multilayer polymer polarizer may routinely be used for optical applications that require high reflectivity and wavelength selectivity. As an example of this application, GBO multilayer polarizers have been used to create reflective polarizers that make LC displays brighter and easier to view. By using this property of the GBO polarizer, one might obtain highly linearly polarized EL light emission over a wide range of optical wavelengths. These anisotropic photonic effects of GBO cause the reflecting band structure to be polarized, and thus make it possible to show that such a combined OLED device can achieve polarized light-emission with high brightness and efficiency, resulting in a high *PR* value even for wideband EL emission from white light-emitting OLEDs (WOLEDs).

2. Polarized photonic OLEDs with GBO films

Three kinds of polarized photonic OLEDs are presented here to demonstrate the use of the GBO film in the highly polarized OLEDs, exhibiting high brightness and efficiency.

2.1. OLEDs on the GBO polarizer substrates

In this section, we describe the polarization of EL light emitted from OLEDs that use a flexible GBO multilayer reflecting polymer polarizer substrate, instead of the conventional isotropic glass substrate. By using such a substrate, we demonstrate the potential for highly polarized light emission from OLEDs. Luminous EL emissions are produced from the polarized photonic OLEDs, and the direction of polarization for the emitted EL light corresponds to the polarizing axis (transmission axis or passing axis) of the GBO reflecting polarizer. The estimated polarization ratio between the brightness of two linearly polarized EL emissions parallel and perpendicular to the polarizing axis can be achieved as high as 25 for the OLEDs on GBO substrates.

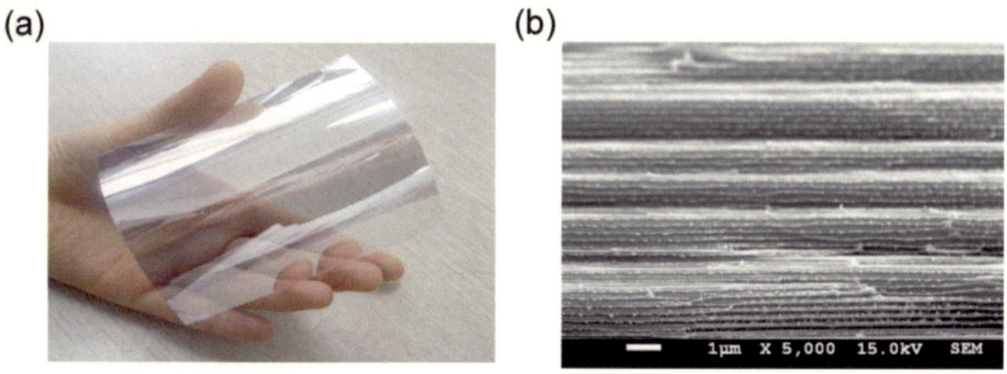

Figure 1. (a) Photograph showing the flexible transparent GBO reflecting polymer polarizer film and (b) SEM image of the cross-section of the studied GBO film.

2.1.1. *Device fabrication and materials used*

Sample OLEDs were prepared by placing an EL layer between an anode and a cathode on a flexible GBO reflecting polarizer film in the following sequence: GBO reflecting polarizer film substrate / thin semi-transparent Au anode / hole-injecting buffer layer / EL layer / electron-injecting layer / Al cathode. For the GBO reflecting polarizer film, a commercial multilayer reflecting polymer polarizer film (3M) has been used. The film is approximately 90 µm thick, and the wavelength of the reflection band is found to be in an approximate range of 400 ~ 800 nm. This film is normally used in an LC display backlight unit as a reflecting polarizer film. After routine cleaning of the GBO reflecting polarizer film using ultraviolet-ozone treatment, a flexible semi-transparent thin Au layer was deposited (90 nm, 40 ohm/square) by sputtering onto the GBO reflecting polarizer to form the anode. This Au anode is used in preference to the typical rigid indium-tin-oxide (ITO) anode in order to preserve the flexibility of the GBO polarizer substrate. The optical transmittance of the Au electrode is about 60 % in the visible wavelength region. A solution of PEDOT:PSS (poly(3,4-ethylenedioxythiophene): poly(4-styrenesulphonate), Clevios PVP. Al 4083, H. C. Starck Inc.) is spin-coated onto the Au anode in order to produce the hole-injecting buffer layer. Subsequently, to form an EL layer, a blended solution is also spin-coated onto the PEDOT:PSS layer. This blended solution consists of a host polymer of poly(vinylcarbazole) (PVK), an electron-transporting 2-(4-biphenylyl)-5-(4-tert-butylphenyl)-1,3,4 oxadiazole (Butyl-PBD), a hole-transporting N,N'-diphenyl-N,N'-bis(3-methylphenyl)-1, 1'biphenyl-4,4'-diamine (TPD), and a phosphorescent guest dye of Tris(2-phenylpyridine) iridium (III) ($Ir(ppy)_3$), whose emission peak wavelength is ~510 nm with a full width at half maximum (FWHM) of ~85 nm [32]. A mixed solvent of 1,2-dichloroethane and chloroform (mixing weight ratio 3:1) is used for the solution. The thicknesses of the PEDOT:PSS and EL layers are adjusted to be about 40 nm and 80 nm, respectively. In order to form the electron-injecting layer, a ~1 nm thick Cs_2CO_3 interfacial layer is formed on the EL layer using thermal deposition (0.02 nm/s) at a base pressure of less than 2×10^{-6} Torr with a shadow-mask that had 3×3 mm^2 square apertures. Finally, a pure Al (~50 nm thick) cathode layer is deposited on the interfacial layer using thermal deposition under the same vacuum conditions. For comparison, we have also fabricated a reference device using a glass substrate in place of the GBO polarizer substrate. Apart from using different substrate materials, the reference devices are fabricated in exactly the same way as the sample OLED on the GBO polarizer substrate. Once the fabrication of OLEDs thus completed, the optical transmittance and reflectance spectra are measured using a Cary 1E (Varian) UV-vis spectrometer and a multichannel spectrometer (HR 4000CG-UV-NIR, Ocean Optics Inc., 0.25 nm resolution). A combination of a polarizer and an analyzer is also used to investigate the polarization of the light emitted from the sample device. A Chroma Meter CS-200 (Konica Minolta Sensing, INC.) and a source meter (Keithley 2400) have been used for measuring the EL characteristics.

2.1.2. Results and discussion

Figure 1(a) shows a photograph of the flexible GBO reflecting polarizer substrate used in this study. As shown in Fig. 1(a), the GBO substrate is easy to bend and quite transparent, in contrast to conventional linear dichroic polarizer film made from light-absorptive materials. Figure 1(b) shows a scanning electron microscopy (SEM) image of the cross-sectional structure of the GBO polarizer film. The SEM image shows clearly that the uniform layers of two alternating layered elements [a/b] are formed in multiple stacks with different refractive indices, (n_{ax}, n_{ay}, n_{az}) and (n_{bx}, n_{by}, n_{bz}). The optical anisotropy of the GBO polarizer may be seen by inspecting the polarized microphotograph of the GBO film between crossed polarizers at four angles of sample rotation of the GBO film substrate, as shown in Figure 2(a). This figure shows that the GBO film has a clear optical birefringence. We can define the orientation of the two optical axes, x and y, for the GBO film from the darkest views of the polarized microphotographs. The polarized transmittance spectra from the GBO polarizer film have then been observed for the two incident lights polarized linearly along the x and y axes, as shown in Figure 2(b). From this figure, it is clear that the nature of the reflection bands depends strongly on the polarization of the incident light, and the polarized transmission spectra are thus quite different from each other. When measured in the *y* direction, the transmission spectrum shows a strong and broad reflection band, while in the *x* direction, there is no reflection band in the wide visible wavelength range (350 ~ 800 nm) that incorporates red, green, and blue light. This significant difference between the reflection bands clearly indicates that in a GBO reflecting polarizer film, the refractive indices of alternating layers are matched along both the *x*- and *z*- axes and mismatched along the *y*-axis. It is thus evident that the birefringence causes the reflecting band structure to be polarized and that the *x* and *y* axes represent the ordinary (*o*) and extraordinary (*e*) axes, respectively. Note that the *o* axis is consistent with the polarizing axis (or passing axis) and the *e* axis represents the blocking axis of the GBO reflecting polarizer. The average extinction ratio of the GBO reflecting polarizer used was estimated to be about 16:1 in the wavelength region between 470 and 700 nm.

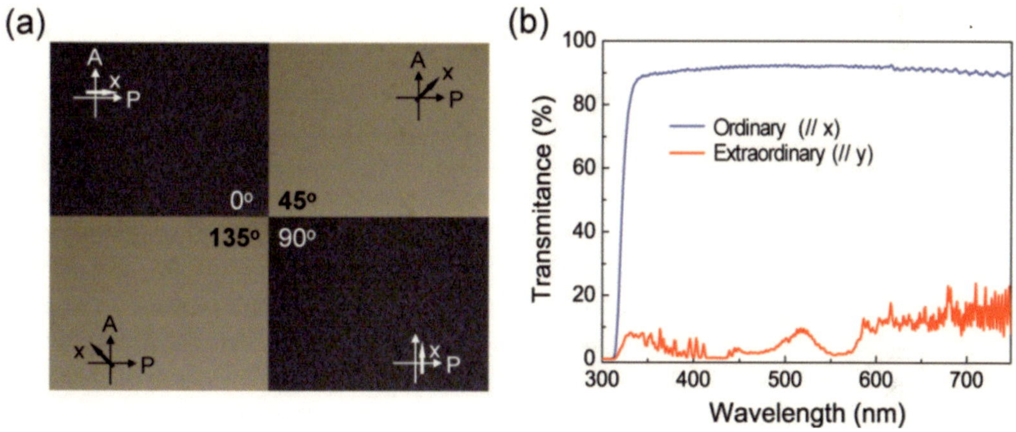

Figure 2. (a) Polarized microphotographs under crossed polarizers at four angles of sample rotation of the flexible GBO reflecting polymer polarizer film. (b) Polarized transmittance spectra for incident light polarized linearly along the *x* (ordinary) and *y* (extraordinary) axes.

On the design outlined above, we have prepared samples of OLEDs on the GBO reflecting polarizer substrate. In order to study the EL characteristics of the sample OLEDs, we have observed the current density-luminance-voltage (*J-L-V*) characteristics, as shown in Figure 3(a). It is clear from this figure that both the charge-injection and turn-on voltages are below 4.0 V, with sharp increases in the *J-V* and *L-V* curves. The EL brightness reaches ~4,500 cd/m^2 at 14.5 V. This performance of the sample OLED with respect to luminescence is nearly the same as that of the reference device using a conventional linear dichroic polarizer film, which shows ca. 5,000 cd/m^2 at 14.5 V. In contrast, as shown in Figure 3(b), the peak efficiencies (6.1 cd/A and 2.0 lm/W) of the sample OLED are much higher than those of the reference device (2.3 cd/A and 0.6 lm/W). The relatively high efficiencies of the sample device may be caused by the improved transition probability of exciton (singlet and triplet) relaxation with respect to the polarization along the transmission axis due to the reduced transition probability of exciton relaxation with respect to the polarization perpendicular to the transmission axis [20, 33].

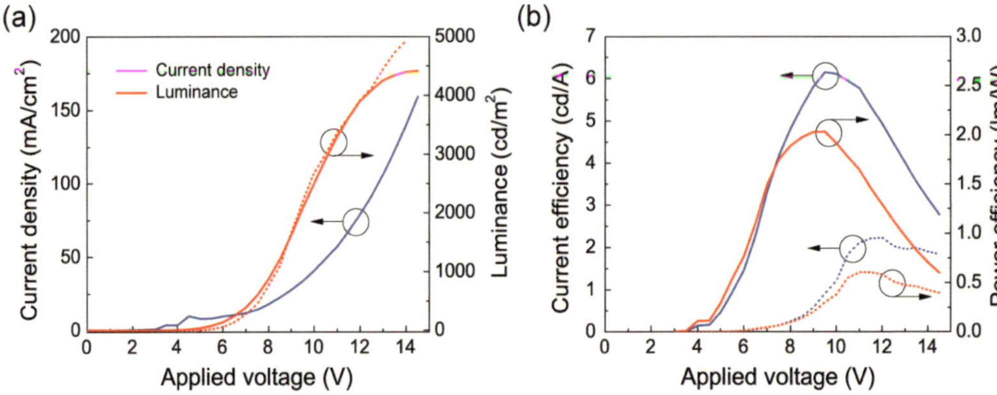

Figure 3. (a) Current density-voltage and luminance-voltage characteristics and (b) current efficiency-voltage and power efficiency-voltage characteristics of the sample OLED on the flexible GBO reflecting polarizer. The dotted curves show the characteristics of the reference device.

In order to interpret the observed EL characteristics of our sample device, we have also measured its polarization characteristics, as shown in Figure 4. Figure 4(a) shows the polarized EL emission spectra for the polarizations along the *o* ($EL_{||}$) and *e* ($EL\perp$) axes at normal incidence (0°). The curves represented by the dotted lines show the total spectra (*o* + *e*). It may be seen that the broad emission spectra are quite similar to that of the reference device, which coincides with the EL emission spectra of conventional OLED devices that have been reported elsewhere [32]. This figure also shows that polarized EL emission spectra strongly depend on the polarization state ($EL_{||}$ and $EL\perp$), and that the sample OLED exhibits highly polarized EL emission over the entire range of emission from 470 nm to 650 nm. The EL polarization ratio (*PR*) of the integrated intensities of the parallel ($EL_{||}$) and perpendicularly ($EL\perp$) polarized EL emission is approximately 25. This ratio is significantly higher than that of the reference device which shows a *PR* of 1 (unpolarized light emission). Here, the *PR* is deduced using the ratio of the intensities, which were measured with polarization parallel and perpendicular to the passing axis of the GBO film, respectively, *i.e.* $PR = EL_{||} / EL\perp$. These results show that this technique for assembling polarized OLEDs, which utilizes a GBO reflecting polarizer, is at least as good as the previous approach, which uses the alignment of uniaxially oriented polymers or oligomers.

Figure 4(b) shows the relative polarized *L-V* characteristics of the same OLED for the polarizations along the *o* and *e* axes. This figure also gives quantitative results for polarized light emissions that were observed along the *o* (||) and the *e* (⊥) axes. The highly polarized *L-V* characteristics give a high averaged *PR* value of 25 over the whole brightness range. (See Figure 5)

Next, as shown in Figure 6 are photographs of the operating polarized OLED sample (3 × 3 mm^2, 10 V) with the polarization along the *o* ($EL_{||}$, left) and *e* ($EL\perp$, right) axes of the flexible GBO reflecting polarizer substrate. It may be seen from the figure that under a rotatable linear dichroic polarizer (left), the OLED is relatively more luminous and highly polarized along the ordinary axis of the GBO polarizer substrate. From these results, we may conclude that a flexible polarized OLED with a high polarization ratio can be fabricated successfully using the GBO reflecting polarizer substrate.

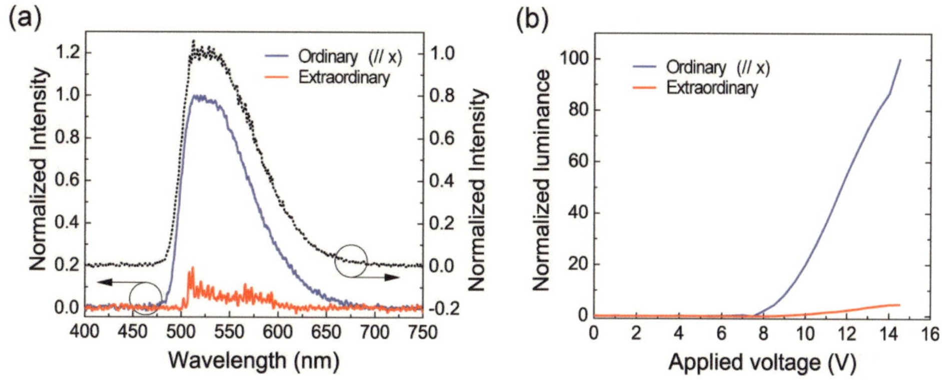

Figure 4. (a) Polarized EL emission spectra along the *o* ($EL_{||}$, blue solid curves) and *e* ($EL\perp$, red solid curves) axes for the fabricated polarized OLED. The dotted curves show the total emission spectra (*o* + *e*). (b) The relative *L-V* characteristics for polarization along the *o* ($EL_{||}$) and *e* ($EL\perp$) axes of EL emission.

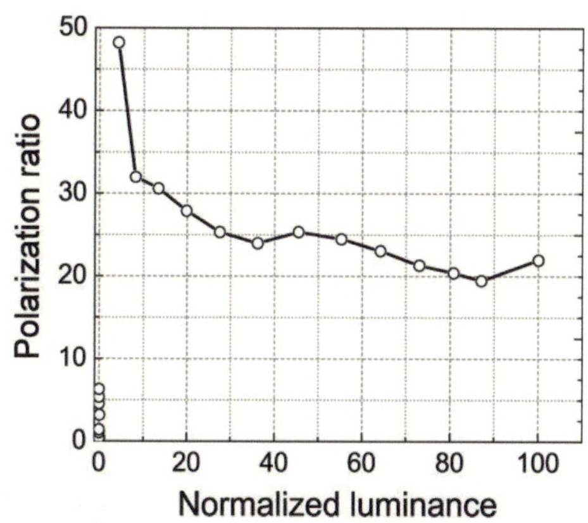

Figure 5. The polarization ratio characteristics obtained using the *L-V* characteristics shown in Fig. 4(b).

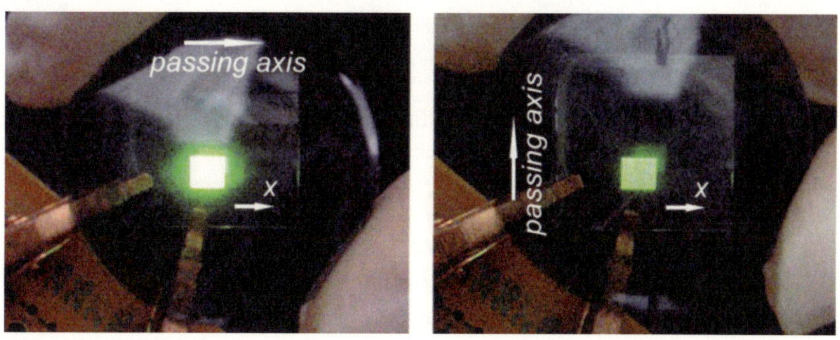

Figure 6. Photographs showing the operating polarized OLED sample (3 × 3 mm², 10 V) for the polarizations along the *o* ($EL_{||}$, left) and *e* ($EL\perp$, right) axes of the flexible GBO reflecting polarizer substrate under a rotating linear dichroic polarizer film. The *passing axis* represents the polarizing axis (or transmission axis) of the linear dichroic polarizer.

2.1.3. Summary

In this section, we have presented the results of a flexible, polarized, and luminous OLED using a flexible GBO substrate. It is shown that EL brightnesses over 4,500 cd/m^2 can be produced using the sample OLED, with high peak efficiencies in excess of 6 cd/A and 2 lm/W. The polarization of the emitted EL lights from the sample OLED corresponds to the passing axis of the GBO polarizer substrate used. Furthermore, it is also shown that a high polarization ratio of up to 25 can possibly be achieved over the whole emission brightness range. These results show that use of GBO reflector enables the development of flexible OLEDs with highly polarized luminescence emissions.

2.2. OLEDs with a quarter waveplate film and a GBO polarizer film

We present here an alternative approach to achieving highly linearly polarized EL emission by resorting again on GBO films. We present a simple polarized OLED that can be driven by a 'photon recycling' concept, which is similar to that developed by Belayev et al [34]. We apply a quarter-wave retardation plate (QWP) film and a GBO reflective polarizer to a non-uniaxial OLED. The QWP film used in our study is a sheet of a birefringent (double refracting) material, which creates a quarter-wavelength (λ/4) phase shift and can change the polarization of the light from linear to circular and *vice versa*. Our combination of the QWP film with a GBO reflective polarizer has enabled us to achieve a high degree of linear polarization with high brightness and efficiency.

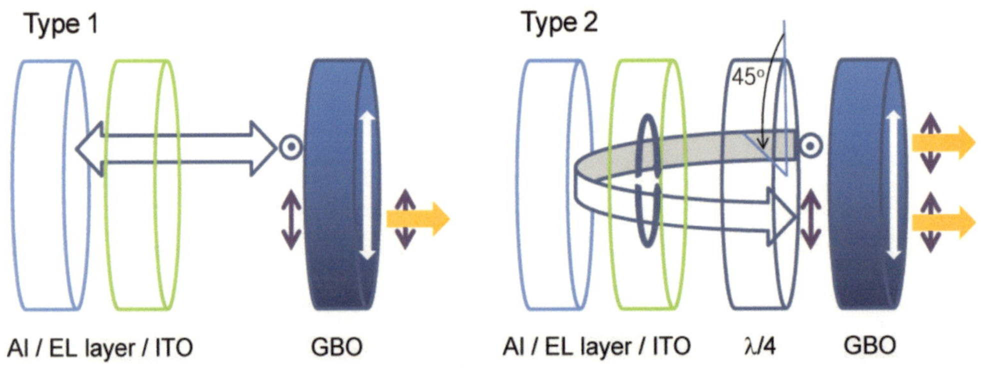

Figure 7. Schematic structures of polarized EL emitting OLEDs. Type 1: simple structure of a polarized OLED with a GBO reflective polarizer, Type 2: Combined structure of a polarized OLED with a QWP (λ/4 plate) film and a GBO reflective polarizer.

A schematic configuration of the device structure, designed to achieve highly linearly polarized EL emission is shown as Type 2 in Figure 7. For comparison, we have also shown the Type 1 device in Fig. 7, which is presented above in section 2.1. In this Type 2 device a QWP film and a GBO reflective polarizer are assembled on an OLED device, at an angle of 45° between the fast optic axis of the QWP film and the passing axis (↕) of the GBO polarizer, as shown in Fig. 7. Then the unpolarized EL light generated from the OLED gets linearly polarization state by QWP and GBO polarizer, as follows; The EL emission that is polarized along the direction parallel to the passing axis (↕) of the GBO polarizer is transmitted through GBO, whereas the other EL polarized perpendicular (⊙) to the passing axis of the GBO polarizer is reflected back selectively as a result of the photonic band of the GBO polarizer. This reflected light changes its polarization to circular (*i.e.*, right-handed circularly polarized light) after its transmission through the QWP film. The sense of the rotation of this right-handed circularly polarized EL light is then changed by reflecting it from the surface of the metal cathode, *i.e.*, it now becomes left-handed circularly polarized light. Finally, by retransmitting it through the QWP film, the polarization of the light is again changed from left-handed circularly polarized to linearly polarized (↕). Now as the direction of polarization becomes parallel to the passing axis of the GBO it is transmitted through the GBO reflective polarizer. By this method, all generated EL light can be transmitted through the GBO reflective polarizer, has linear polarization (↕) along the passing axis of the GBO polarizer.

2.2.1. Device fabrication and materials used

The polarized OLEDs are prepared by placing an organic EL layer between an anode and a cathode on a glass substrate, together with a QWP film and a GBO reflective polarizer, in the following sequence: GBO reflective polarizer / QWP film / glass substrate / transparent ITO (80 nm, 30 Ω/square) anode / hole-injecting buffer layer / EL layer / electron-injecting layer / Al cathode (Type 2). A commercial QWP film (Edmund Sci.) approximately 110 μm thick and with a central operating wavelength of about 500 nm has been used. After a routine cleaning of the ITO substrate using wet (acetone and isopropyl alcohol) and dry (UV-ozone) processes, a solution of PEDOT:PSS is spin-coated onto the ITO anode in order to produce the hole-injecting buffer layer. Subsequently, in order to form an EL layer, a blended solution is also spin-coated onto the PEDOT:PSS layer. This blended solution consisted of a host PVK polymer, an electron-transporting butyl-PBD, a hole-transporting TPD and a phosphorescent guest dye of $Ir(ppy)_3$. A mixed solvent of 1,2-dichloroethane and chloroform (mixing weight ratio 3:1) was used for the solution. The thicknesses of the PEDOT:PSS and EL layers were adjusted to about 40 nm and 80 nm, respectively. In order to form the electron-injecting layer, a ~2 nm thick Cs_2CO_3 interfacial layer was formed on the EL layer using thermal deposition (0.02 nm/s) at a base pressure of less than 2×10^{-5} Torr. Finally, a pure Al (~50 nm thick) cathode layer was formed on the interfacial layer using thermal deposition by means of a shadow-mask that had square (3 mm × 3 mm) apertures under the same vacuum conditions. After the Al cathode had been formed, the QWP and the GBO films were attached sequentially to the ITO glass substrate using index-matching oil. In order to assess the effectiveness of our device, we also fabricated unpolarized conventional reference devices, using exactly the same method as for the polarized OLEDs but without the GBO and QWP films (1st reference device). For further comparison, 2nd reference device was also fabricated using only the GBO film Type 1, Fig. 7). It may be noted that in both type 1 and type 2 devices, the organic layer structure and organic materials used are identical, and thus, electrical characteristics such as the current density-voltage (*J-V*) curve are identical.

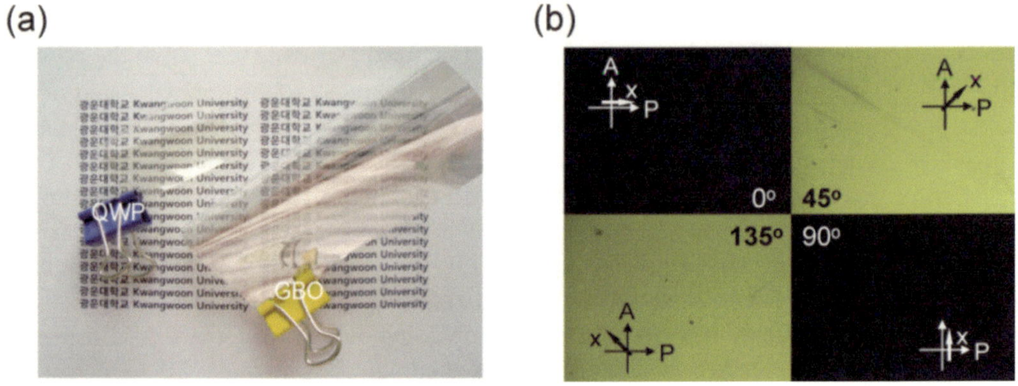

Figure 8. (a) Photographs of the transparent QWP film (left) and GBO reflective polarizer (right). (b) Polarized microphotographs at four angles of sample rotation of the QWP film.

2.2.2. Results and discussion

Figure 8(a) shows a photograph of the QWP film and the GBO reflective polarizer used in this study. As it can be seen the QWP film and GBO reflective polarizer are quite transparent. In Fig. 8(b), the optical anisotropy of the QWP film is shown in the polarized microphotograph obtained between crossed polarizers for four angles of rotation. Figure 8(b) shows that the QWP film has a clear optical birefringence. The two darker views of the polarized microphotographs enable us to obtain the orientation of the two optical axes for the QWP film.

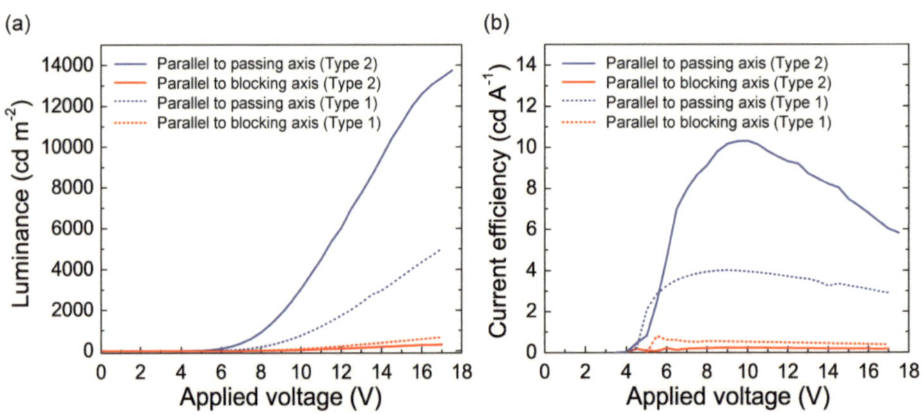

Figure 9. (a) Polarized *L-V* characteristics. ($EL_{||}$ blue and $EL_{\perp}$ red) (b) Current efficiency-voltage characteristics of the polarized OLEDs (solid curves). The dotted curves show the characteristics of the 2nd reference device.

The performance of the polarized OLEDs thus fabricated with the QWP film and the GBO reflective polarizer are presented here. Figure 9(a) shows the polarized *L-V* characteristics of the fabricated OLEDs for the polarizations along the passing ($EL_{||}$, blue curves) and the blocking ($EL_{\perp}$, red curves) axes. The figure indicates that the turn-on voltages are below 4.0

V (1 cd/m^2 in Fig. 9(a)) with sharp increases in the *L-V* curve for polarization parallel to the passing axis. The polarized EL brightness ($EL_{||}$) reaches ~13,400 cd/m^2 at 17.0 V. This performance with respect to luminescence approaches the luminescence of ca. 18,500 cd/m^2 at 17.0 V for the unpolarized 1st reference device, in which the QWP film and the GBO polarizer are omitted. The polarized *L-V* curves shown here also give quantitative results for the polarized light emissions observed along both the passing and blocking axes. As shown in Fig. 9(a), the highly polarized *L-V* characteristics for the polarized OLEDs give a high average *PR* of at least 40 over the whole voltage range (4.0 ~ 17 V). This ratio is significantly higher than that of the 1st and 2nd reference devices, which show *PR* of 1 (*i.e.*, unpolarized light) and 7.53, respectively. Fig. 9(a) also shows that the EL emission polarized along the passing axis reaches only ca. 5,000 cd/m^2 at 17.0 V for the 2nd reference OLED which only has the GBO polarizer. This performance of the 2nd reference OLED with respect to polarized luminescence along the passing axis of the GBO is only about the half. This relatively low brightness of the 2nd reference device is due to the absence of the 'photon recycling' effect mentioned above. It may also be seen that the EL brightness polarized along the blocking axis for the polarized OLEDs is further reduced compared with that of the 2nd reference OLEDs, as shown in Fig. 9(a). This is due to the reduced light intensity polarized along the blocking axis in the polarized device, following the change in the polarization to a direction parallel to the passing axis. Similarly, as shown in Figure 9(b), the peak efficiencies (10.3 cd/A and 3.63 lm/W) of the EL emission polarized along the passing axis for the polarized OLED are nearly double of that of the 2nd reference device (4.0 cd/A and 1.71 lm/W), while the efficiency of the EL emission polarized along the blocking axis for the polarized device is further reduced compared with that of the 2nd reference OLED.

We also measured the polarization characteristics and Fig. 10(a) shows the polarized EL emission spectra for polarizations along the passing ($EL_{||}$, blue solid curves) and blocking ($EL_{\perp}$, red solid curves) axes at normal incidence, for an applied voltage of 10 V. It may be seen that the broad emission spectra are almost the same as those of the reference devices and conventional OLED devices reported elsewhere [32]. This figure also shows that the polarized EL emission spectrum depends very much on the polarization state, and that the polarized OLED shows highly polarized EL emission over the whole emission spectrum range. For the polarized device, *PR* of the integrated intensities of the $EL_{||}$ and $EL_{\perp}$ lights is always greater than 40. It may therefore be concluded that our polarized OLEDs a QWP film and GBO reflective polarizer incorporated perform extremely well. Fig. 10(b) shows the *PR-L* characteristics of our polarized OLEDs. As shown in Fig. 10(b), in comparison with *PR* = 7.5 of the 2nd reference device our polarized device has a *PR* of over 40 in the whole brightness range.

The operation of the 2nd reference and polarized OLEDs (3 mm × 3 mm, 10 V) for polarizations along the passing and blocking axes of the GBO reflective polarizer is shown in Fig. 11. It may be seen from the figure that under a rotation of linear dichroic polarizer, right OLED is more luminous (left fig.) and more highly polarized along the passing axis of the GBO polarizer in comparison to the left 2nd reference device (right fig.). All these results demonstrate a successful fabrication of a highly polarized OLED with a high *PR* (> 40), using a QWP film and a GBO reflective polarizer.

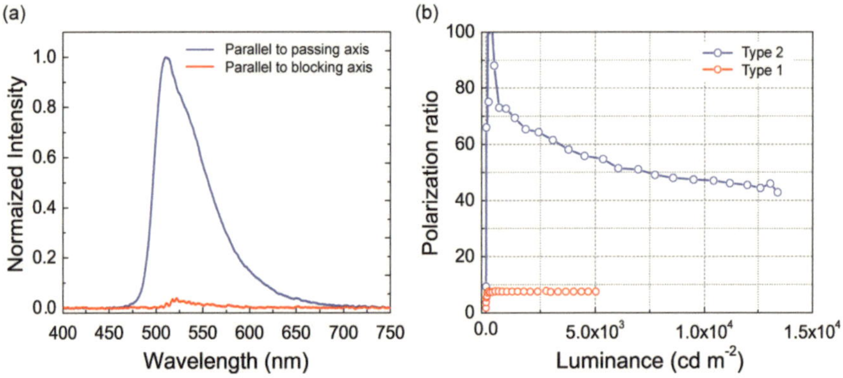

Figure 10. (a) Polarized EL emission spectra along the passing and blocking axes. (b) Polarization ratios of the polarized OLED (blue curve) against luminance. The points in red show the characteristics of the 2nd reference device.

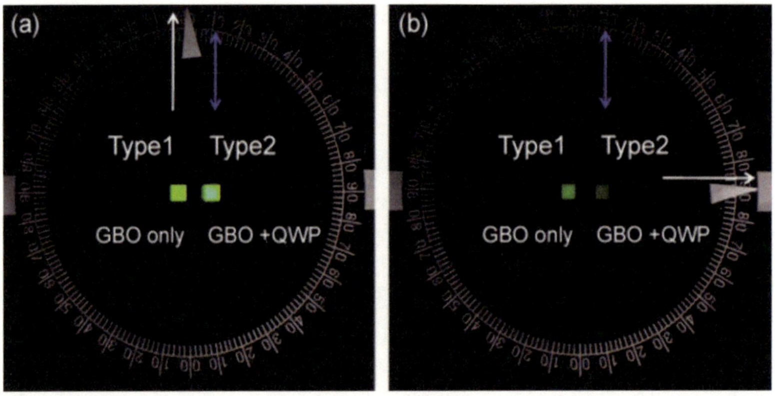

Figure 11. Photographs of the operating 2nd reference (left) and polarized (right) OLEDs for polarizations along the passing ($EL_{||}$) (a) and the blocking ($EL\perp$) (b) axes of the GBO reflective polarizer under a rotating linear dichroic polarizer film. The white arrow represents the transmission axis of the linear dichroic polarizer and the blue arrow represents the transmission axis of the GBO polarizer.

2.2.3 Summary

In summary, we have described the fabrication and operation of a polarized and luminous OLED using the combination of a QWP retardation film and a GBO reflective polarizer. A peak polarized EL brightness of over ca. 13,000 cd/m^2 is produced from the polarized OLED, with high peak efficiencies in excess of 10 cd/A and 3.5 lm/W. The polarization direction of the EL light emitted from the polarized OLED corresponds to the passing axis of the GBO polarizer used. Furthermore, it has also been shown that a high polarization ratio greater than 40 is possible over the whole emission brightness range. These results show that using the QWP film and GBO reflective polarizer we can develop bright OLEDs with highly polarized luminescence emissions.

2.3. Polarized white OLEDs with achromatic QWP films on GBO substrates

Here we describe the third technique that can be used to achieve high linearly polarized white EL emission based on the 'photon recycling' concept [34] for a wide visible wavelength range including red, green, and blue light. We apply a GBO reflective polarizer to a WOLED with a broadband (achromatic) QWP film whose phase retardation is maintained at $\pi/2$ for a wide range of wavelengths, in contrast to the narrow band QWP used in section 2.2. The applied achromatic QWP film also creates a phase shift of a quarter of a wavelength ($\lambda/4$), and can change the polarization of the broad EL emission from linear to circular, and *vice versa*.

The configuration of the device is shown in Figure 12(a), which is nearly identical to Type 2 in presented in section 2.2 as shown in Figure 7. Here an achromatic QWP film and a GBO reflective polarizer are attached to a WOLED with an angle of 45° between the fast optic axis of the QWP film and the passing axis (↕) of the GBO polarizer. From the unpolarized EL light generated from the WOLED EL ($EL_{||}$) polarized along the direction parallel to the passing axis (↕) of the GBO polarizer is transmitted through the GBO polarizer. The EL ($EL\perp$) polarized perpendicular (⊙) to the passing axis of the GBO polarizer is reflected selectively by the wide photonic band of the GBO polarizer. The polarization of this reflected light is changed to right-handed circular (*R*) after its transmission through the achromatic QWP film. The sense of rotation of this circularly polarized EL light is then reversed to left-handed circular (*L*) by reflecting it from the surface of the metal cathode. Then by retransmission of this light through the achromatic QWP film changes its polarization again from circularly to linearly polarized (↕), which can be transmitted through the GBO reflective polarizer. This method allows nearly all the generated white EL light to be transmitted through the GBO reflective polarizer with a direction of linear polarization (↕) parallel to the passing axis of the GBO polarizer.

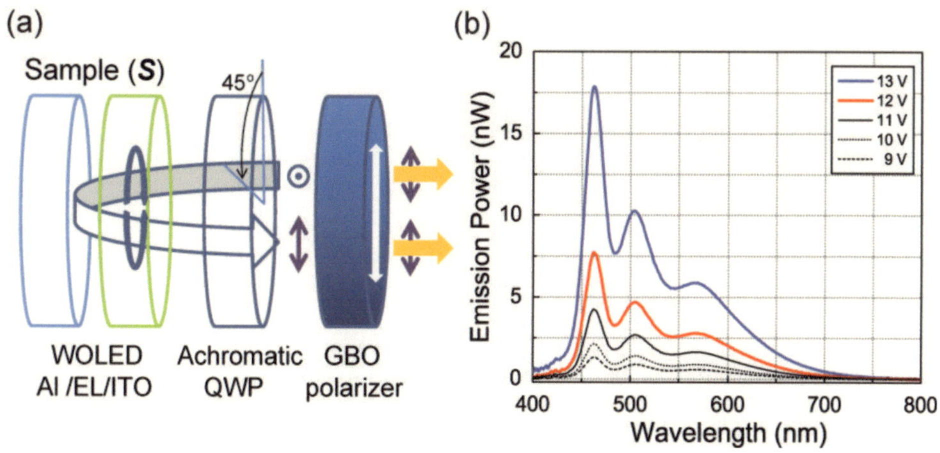

Figure 12. (a) Polarized WOLED (*S*) combined with an achromatic QWP ($\lambda/4$ plate) film and a GBO reflective polarizer: Type 2 and (b) Unpolarized EL spectra of the WOLED used for the polarized WOLEDs.

2.3.1. *Device fabrication and materials used*

The polarized WOLEDs were prepared by fabricating organic layers between an anode and a cathode on a glass substrate, together with a commercially available achromatic QWP film and a GBO reflective polarizer. The QWP film was approximately 110 μm thick, and the range of its operating wavelengths were approximately 420 ~ 650 nm. After routine cleaning of the ITO (150 nm, 10 Ω/square) substrate using both wet (acetone and isopropyl alcohol) and dry (O_2 plasma) processes, the organic layers were deposited on the ITO anode to form the structure of the tandem hybrid WOLED: ITO anode / short reduction layer (5 nm) / hole injection layer 1 (10 nm) / hole injection layer 2 (25 nm) / fluorescent blue-light emitting material layer (10 nm) / 8-hydroxy-quinolinato lithium (Liq)-doped electron injection layer (20 nm) / Li doped electron injection layer (20 nm) / hole injection layer 3 (10 nm) / hole transporting layer (55 nm) / phosphorescent green- and red-light emitting material layer (25 nm) / hole blocking layer (10 nm) / Liq doped electron injection layer (30 nm) / Al cathode. This is similar to the structure reported in reference [35]. The organic layers and Al cathode (150 nm) were deposited consecutively by thermal evaporation in a chamber with a base pressure of less than 1×10^{-6} Torr by means of a shadow-mask with square (1 mm × 1 mm) apertures. When the cathode was ready, the achromatic QWP and the GBO films were combined sequentially to the fabricated WOLEDs (device *S*). For comparison, we also fabricated reference devices, using exactly the same method as for the WOLEDs, but with only the GBO film (reference device *R*). The structure of the organic layer and the organic materials used were identical for each of the devices described herein. Figure 12(b) shows the white-light EL spectra (unpolarized) observed for the fabricated WOLED, in which three balanced emission peaks may be seen at 463 (blue), 503 (green), and 563 (red) nm. The spectral shape of the EL spectrum emitted from the device did not change significantly with applied voltage, and the color coordinates varied by less than 10% for the applied voltages between 7 ~ 14 V.

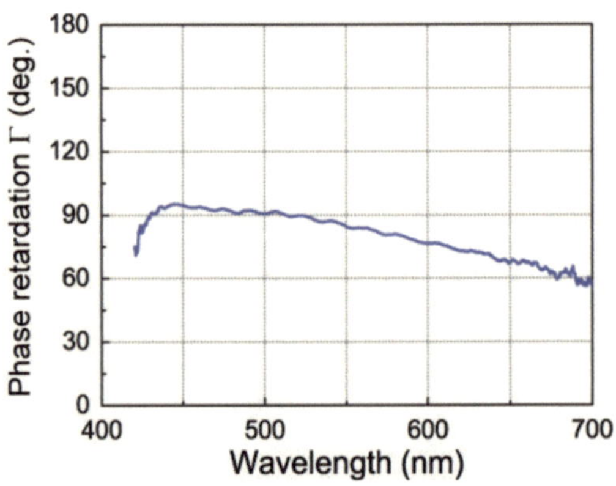

Figure 13. Phase retardation (*Γ*) of the broadband achromatic QWP film used in this study.

2.3.2. *Results and discussion*

The phase retardation (*Γ*) of the achromatic QWP film used in this study is measured by observing the transmission *T* ($T = 1/2 \sin^2(2\phi) \sin^2(\Gamma/2)$) through the QWP film placed between crossed polarizers. Here, ϕ represents the angle between a fast axis of the achromatic QWP film and a transmitting axis of the polarizer. The measured results are shown in Figure 13. This figure shows clearly that the phase retardation of the achromatic QWP film is about $\pi/2$. Although the retardation decreases slightly as the wavelength increases, the QWP film has a nearly uniform phase retardation of a quarter of a wavelength ($\lambda/4$) in a wide visible range of wavelengths (420 ~ 650 nm) that includes blue, green, and red light.

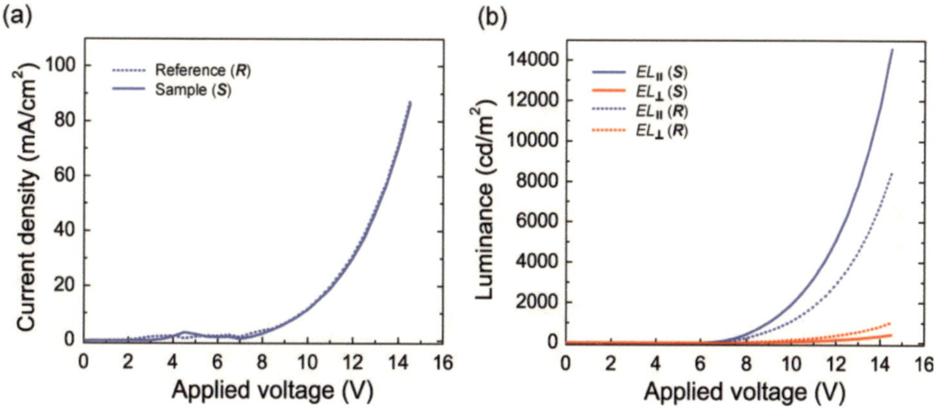

Figure 14. (a) *J-V* and (b) polarized *L-V* characteristics of Reference (*R*, dotted curves) and Sample (*S*, solid curves) WOLEDs.

Figure 14(a) shows the *J-V* curves of the fabricated WOLED devices *S* and *R*. For all the WOLEDs described herein, the organic layers used are the same, and the electrical characteristics (such as *J-V* curves) are therefore found to be identical for each device as shown in Fig. 14 (a). Figure 14(b) shows the *L-V* characteristics of the WOLEDs for $EL_{||}$ (blue curves) and $EL_{\perp}$ (red curves). Figure 14(b) shows clearly that the devices operate at relatively low turn-on voltages (~ 6 V) and have bright EL emissions, which indicate the efficient emission of white EL from the WOLEDs. It is noteworthy that even without full operational optimization of the polarized WOLED, its performance shows its potential attractiveness. In particular, the WOLED with both the GBO and achromatic QWP films (device *S*) exhibits excellent performance, in which operating voltages of about 7.0 and 8.7 V are required to obtain brightnesses ($EL_{||}$) of 100 cd/m^2 and 1,000 cd/m^2, respectively, with a peak luminescence of ca. 14,600 cd/m^2 at 14.5 V. It may be seen that the peak luminance ($EL_{||}$) of the device *S* under test is much higher than that of a previously reported polarized WOLED (ca. 850 cd/m^2 in Ref. 8) that used a uniaxially oriented polymeric material as an EL layer. Figure 14(b) also shows that the $EL_{||}$ reaches only ca. 8,400 cd/m^2 at 14.5 V for device *R*, whose performance with respect to $EL_{||}$ is only about half as good as that of device *S*.

In Fig. 15 (a), we have shown the current efficiencies of S and R WOLEDs. For the $EL_{||}$ of device *S*, a current efficiency (η_C) of 16.5 cd/A is obtained at 100 cd/m^2 (7.0 V), reaching η_C = 18.3 cd/A at 1,000 cd/m^2 (8.7 V) and η_C = 16.5 cd/A at 14,600 cd/m^2 (14.5 V). We have also determined the power efficiency η_P for the $EL_{||}$ of device *S*, which increases and reaches a maximum of 7.4 lm/W before slowly decreasing, with increasing bias voltage as shown in

Figure 15(b). These results indicate that the peak efficiencies (18.3 cd/A and 7.41 lm/W) for the $EL_{||}$ of device *S* are nearly double those of device *R* (9.63 cd/A and 3.71 lm/W). These relatively high brightness and efficiency values of the $EL_{||}$ of device *S* are achieved by the 'photon recycling' effect. It is noted that the brightness and efficiency of $EL\perp$ for device *S* are further reduced compared with those of device *R*, as shown in Figures 14 and 15. This is due to the reduced $EL\perp$ in device *S* that occurred after the change in polarization to the direction parallel to the passing axis.

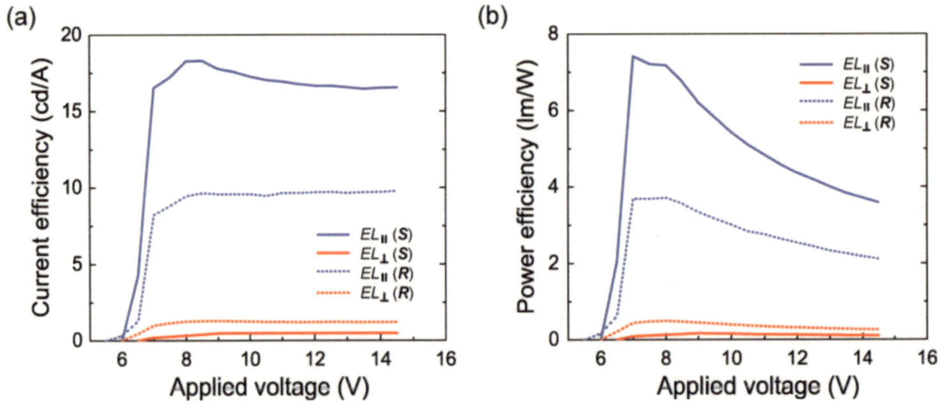

Figure 15. (a) Current efficiency-voltage and (d) power efficiency-voltage characteristics of Reference (*R*, dotted curves) and Sample (*S*, solid curves) WOLEDs.

Next, we have estimated the relationship between polarization ratio and luminance *PR-L* for the polarized WOLED *S*, thereby presenting quantitative results for the polarized emissions. Figure 16(a) shows that the highly polarized characteristics of the polarized WOLED *S* give a high average value of *PR* ($EL_{||}$: $EL\perp$) of at least ~35:1 over the whole range of brightness. It should be noted that this value of *PR* is significantly higher than that of device *R* (8.21:1). In order to understand the characteristics of the polarized EL, we have measured the polarized emission spectra for $EL_{||}$ (blue curves) and $EL\perp$ (red curves) for the device *S* under an applied voltage of 10 V (Figure 16(b)). It may be noted that the spectral shape of the $EL_{||}$ for device *S* is very similar to that for device *R*. The observed color rendering index (CRI) of $EL_{||}$ for device *S* is about 80.0, and the CIE XYZ color space is (0.285, 0.363, 0.353), with a correlated color temperature (CCT) of about 7,600 K. These characteristics are also similar to those of $EL_{||}$ from device *R*, which has a CRI of about 74.0, CIE XYZ color space of (0.275, 0.342, 0.383), and CCT of about 8,500 K. At the same time Figure 16(b) also shows that the polarized EL emission spectrum depends very much on the polarization state and that device *S* produces highly polarized EL emission over the whole spectrum. For the device *S*, the highest value of *PR* calculated from the integrated intensities of the parallel and perpendicularly polarized EL spectra is approximately 35:1, which is significantly higher than that of the white-light emitting devices that use uniaxially oriented materials [17]. These results prove that our polarized WOLED (*S*), which incorporates an achromatic QWP film with a GBO reflective polarizer, outperforms all other similar devices.

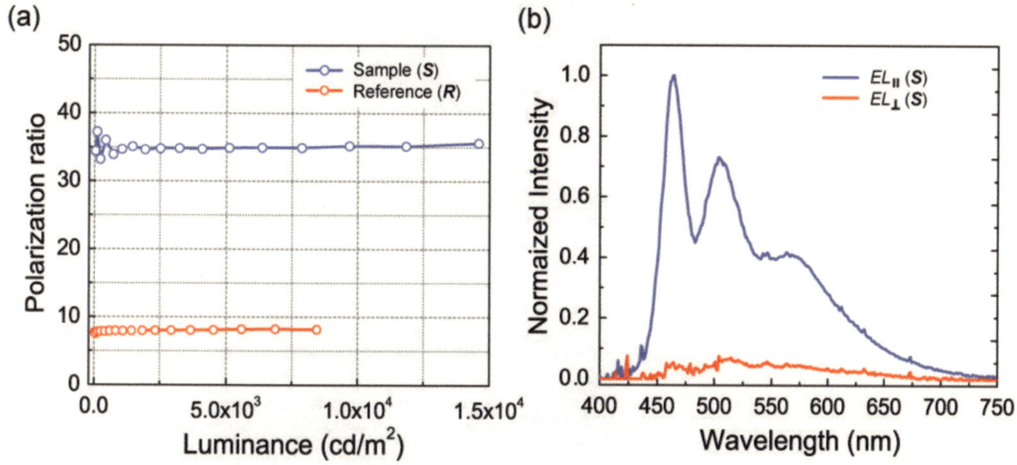

Figure 16. (a) Polarization ratios of the polarized WOLEDs against luminance. (b) Polarized EL spectra of $EL_{||}$ (blue curve) and $EL_{\perp}$ (red curve) of the sample WOLED *S* at 10 V.

Finally, we have shown in Fig. 17 the photographs of the performance of WOLEDs (1 mm × 1 mm) operating under the same bias voltage of 8 V for polarizations along the passing (upper) and blocking (lower) axes of the GBO reflective polarizer. Fig. 17 shows clearly that under a rotating linear dichroic polarizer, the EL emission from device *S* is fairly brighter and more highly polarized along the passing axis of the GBO polarizer, compared to that of device *R*.

2.3.3. *Summary*

In summary, we have described the fabrication and investigation of the properties of a polarized WOLED using a combination of an achromatic QWP and a GBO reflective polarizer. By applying the achromatic QWP and the GBO polarizer to the WOLED, polarized EL brightnesses in excess of ca. 14,600 cd/m^2 can be obtained from the polarized WOLED, together with high peak efficiencies of more than 18 cd/A (7.4 lm/W), which are almost double of those obtained from the polarized WOLED with only the GBO polarizer. We have also found that a high polarization ratio of ca. 35:1 is possible over the whole range of brightness of the emissions. Although the *PR* value of the polarized WOLED is slightly lower than that of polarized narrow band (green) OLED in section 2.2, it may be noted that only the polarized WOLED can provide a polarized light source for a wide range of wavelengths from the blue through to the green and red color regions.

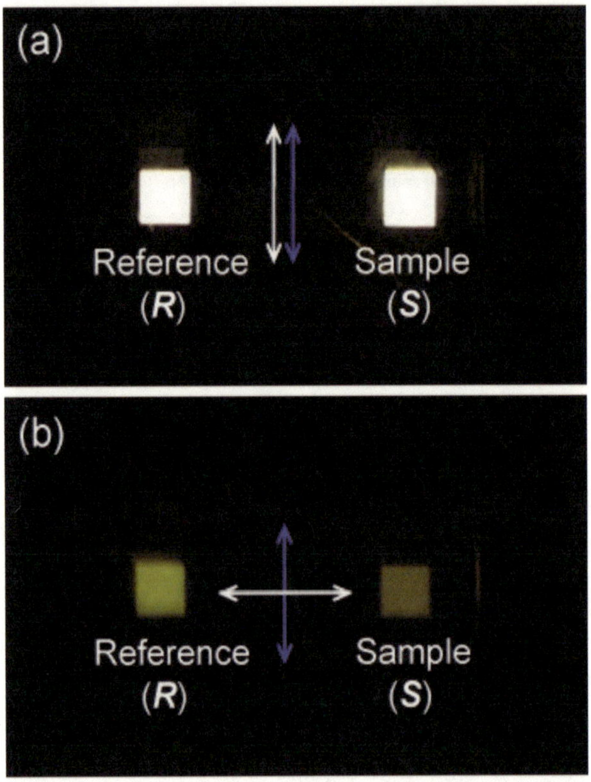

Figure 17. Photographs of the brightness obtained from the reference (*R*, left) and sample (*S*, right) WOLEDs (at 8 V) for $EL_{\parallel}$ (a) and $EL_{\perp}$ (b) under a rotating linear dichroic polarizer film. The white and blue arrows represent the transmission axes of the linear dichroic polarizer and the GBO polarizer, respectively. The active areas of the polarized WOLEDs were 1 mm × 1 mm. (It may be noted that the device *R* appears to be brighter than device *S* in Fig (b) because device *S* is more highly polarized along the passing axis of the GBO polarizer, compared to that of device *R*.)

3. Conclusions

We have fabricated flexible, polarized, and luminous OLEDs using a flexible GBO reflecting polarizer substrate. We have also described the fabrication and investigation of a polarized and luminous OLED and WOLED using a combination of a QWP retardation film and a GBO reflective polarizer. Polarized EL brightnesses of over 10,000 cd/m^2 can be produced from the polarized OLED, with high peak efficiencies in excess of 10 cd/A, which are almost double those obtained from the polarized WOLED with only the GBO polarizer. The polarization direction of the EL light emitted from the polarized OLED corresponds to the passing axis of the GBO polarizer used. Furthermore, we have also shown that a high polarization ratio of greater than 35~40 is possible over the whole emission brightness range. These results show that using the (achromatic) QWP film and the GBO reflective polarizer one can develop bright (W)OLEDs with highly polarized luminescence emissions. It is also noted that the polarization ratio of the polarized WOLED can be further improved by introducing a high quality achromatic QWP film for a wide range of wavelengths including red, green, and blue light. By combining the devices presented here with the luminous EL layers reported elsewhere [35], it may be possible to develop highly efficient polarized OLEDs with a wide range of optical applications. For example, the device structure used in this study can be applied to the design of special light-emitting devices, such as polarized backlights for LC displays. Such devices can also be used for the development of a new class of polarized OLEDs such as polarized surface emitting devices for 3-D displays and/or the polarized light sources of optical waveguide devices.

Author details

Byoungchoo Park
Department of Electrophysics, Kwangwoon University, Seoul, Korea

Acknowledgement

This work was supported by the National Research Foundation of Korea(NRF) grant funded by the Ministry of Education, Science and Technology, Republic of Korea (20120003831 and 2012015654). This research was also supported by the Converging Research Center Program through the Ministry of Education, Science and Technology (2012K001303) and the leading industry of NEW-IT and equipments of the Chungcheong Leading Industry Office of the Korean Ministry of Knowledge Economy (A002200104).

4. References

[1] C. W. Tang, S. A. Van Slyke, Organic electroluminescent diodes. Applied Physics Letters 51, 913 (1987).

[2] R. H. Friend, R. W. Gymer, A. B. Holmes, J. H. Burroughes, R. N. Marks, C. Taliani, D. D. C. Bradly, D. A. Dos Santos, J. L. Bredas, M. Logdlund, W. R. Salaneck, Electroluminescence in conjugated polymers. Nature (London) 397, 121 (1999).

[3] M. A. Baldo, S. Lamansky, P. E. Burrows, M. E. Thompson, S. R. Forrest, Very high-efficiency green organic light-emitting devices based on electrophosphorescence. Applied Physics Letters 75, 4 (1999).

[4] M. Ikai, S. Tokito, Y. Sakamoto, T. Suzuki, Y. Taga, Highly efficient phosphorescence from organic light-emitting devices with an exciton-block layer. Applied Physics Letters 79, 156 (2001).

[5] C. Adachi, M. E. Thompson, S. R. Forrest, Architectures for efficient electrophosphorescent organic light-emitting devices. IEEE Journal of Selected Topics in Quantum Electronics 8, 372 (2002).

[6] V. Cimrova, M. Remmers, D. Neher, G. Wegner, Polarized light emission from LEDs prepared by the Langmuir-Blodgett technique. Advanced Materials 8, 146 (1996).

[7] M. Jandke, P. Strohriegl, J. Gmeiner, W. Brutting, M. Schwoerer, Polarized electroluminescence from rubbing-aligned poly(p-phenylenevinylene). Advanced Materials 11, 1518 (1999).

[8] D. X. Zhu, H. Y. Zhen, H. Ye, X. Liu, Highly polarized white-light emission from a single copolymer based on fluorine. Applied Physics Letters 93, 163309 (2008).

[9] P. Dyreklev, M. Berggren, O. Inganas, M. R. Andersson, O. Wennerstrom, T. Hjertberg, Polarized electroluminescence from an oriented substituted polythiophene in a light emitting diode. Advanced Materials 7, 43 (1995).

[10] C. C. Wu, P. Y. Tsay, H. Y. Cheng, S. J. Bai, Polarized luminescence and absorption of highly oriented, fully conjugated, heterocyclic aromatic rigid-rod polymer poly-p-phenylenebenzobisoxazole. Journal of Applied Physics 95, 417 (2004).

[11] M. Grell, D. D. C. Bradley, Polarized luminescence from oriented molecular materials. Advanced Materials 11, 895 (1999).

[12] K. Sakamoto, K. Miki, M. Misaki, K. Sakaguchi, M. Chikamatsu, R. Azumi, Very thin photoalignment films for liquid crystalline conjugated polymers: Application to polarized light-emitting diodes. Applied Physics Letters 91, 183509 (2007).

[13] K. Pichler, R. H. Friend, P. L. Burn, A. B. Holmes, Chain alignment in poly(p-phenylene vinylene) on oriented substrates. Synthetic Metals 55, 454 (1993).

[14] M. Era, T. Tsutsui, S. Saito, Polarized electroluminescence from oriented p-sexiphenyl vacuum-deposited film. Applied Physics Letters 67, 2436 (1995).

[15] M. Misaki, Y. Ueda, S. Nagamatsu, M. Chikamatsu, Y. Yoshida, N. Tanigaki, K. Yase, Highly polarized polymer light-emitting diodes utilizing friction-transferred poly(9, 9-dioctylfluorene) thin films. Applied Physics Letters 87, 243503 (2005).

[16] M. Misaki, M. Chikamatsu, Y. Yoshida, R. Azumi, N. Tanigaki, K. Yase, S. Nagamatsu, Y. Ueda, Highly efficient polarized polymer light-emitting diodes utilizing oriented films of β-phase poly(9, 9-dioctylfluorene). Applied Physics Letters 93, 023304 (2008).

[17] A. Liedtke, M. O' Neill, A. Wertmoller, S. P. Kitney, S. M. Kelly, White-Light OLEDs Using Liquid Crystal Polymer Networks. Chemistry of Materials 20, 3579 (2008).

[18] I. H. H. Zabel, D. Stroud, Photonic band structures of optically anisotropic periodic arrays. Physical Review B 48, 5004(1993).

[19] Z. Y. Li, J. Wang, B. Y. Gu, Creation of partial band gaps in anisotropic photonic-band-gap structures. Physical Review B 58, 3721 (1998).

[20] G. Alagappan, X. W. Sun, P. Shum, M. B. Yu, M. T. Doan, One-dimensional anisotropic photonic crystal with a tunable bandgap. Journal of the Optical Society of America B 23, 159 (2006).

[21] J. P. Dowling, M. Scalora, M. J. Bloemer, C. M. Bowden, The photonic band edge laser: A new approach to gain enhancement. Journal of Applied Physics 75, 1896 (1994).

[22] T. Matsui, R. Ozaki, K. Funamoto, M. Ozaki, K. Yoshino, Flexible mirrorless laser based on a free-standing film of photopolymerized cholesteric liquid crystal. Applied Physics Letters 81, 3741 (2002).

[23] J. Hwang, M. H. Song, B. Park, S. Nishimura, T. Toyooka, J. W. Wu, Y. Takanishi, K. Ishikawa, H. Takezoe, Electro-tunable optical diode based on photonic bandgap liquid-crystal heterojunctions. Nature Materials 4, 383 (2005).

[24] M. H. Song, B. Park, S. Nishimura, T. Toyooka, I. J. Chung, Y. Takanishi, K. Ishikawa, H. Takezoe, Electrotunable Non-reciprocal Laser Emission from a Liquid-Crystal Photonic Device. Advanced Functional Materials 16, 1793 (2006).

[25] H. Zhang, H. You, W. Wang, J. Shi, S. Guo, M. Liu, D. Ma, Organic white-light-emitting devices based on a multimode resonant microcavity. Semiconductor Science and Technology 21, 1094 (2006).

[26] A. Dodabalapur, L. J. Rothberg, T. M. Miller, Color variation with electroluminescent organic semiconductors in multimode resonant cavities. Applied Physics Letters 65, 2308 (1994).

[27] J. Lim, S. S. Oh, D. Y. Kim, S. H. Cho, I. T. Kim, S. H. Han, H. Takezoe, E. H. Choi, G. S. Cho, Y. H. Seo, S. O. Kang, B. Park, Enhanced out-coupling factor of microcavity organic light-emitting devices with irregular microlens array. Optics Express 14, 6564 (2006).

[28] M. F. Weber, C. A. Stover, L. R. Glbert, T. J. Nevitt, A. J. Ouderkirk, Giant birefringent optics in multilayer polymer mirror. Science 287, 2451 (2000).

[29] B. Park, C. H. Park, M. Kim, M.-Y. Han, Polarized organic light-emitting device on a flexible giant birefringent optical reflecting polarizer substrate. Optics Express 17, 10136 (2009).

[30] B. Park, Y. H. Huh, H. G. Jeon, Polarized electroluminescence from organic light-emitting devices using photon recycling. Optics Express 18, 19824 (2010).

[31] B. Park, Y. H. Huh, S. H. Lee, Y. B. Kim, Linearly-polarized White-light-emitting OLEDs Using Photon Recycling. Journal of the Korean Physical Society 59, 341 (2011).

[32] B. Park, M. Y. Han, S. S. Oh, Solution processable ionic p-i-n phosphorescent organic light-emitting diodes. Applied Physics Letters 93, 093302 (2008).

[33] J. M. Bendickson, J. P. Dowling, M. Scalora, Analytic expressions for the electromagnetic mode density in finite, one-dimensional, photonic band-gap structures. Physical Review E 53, 4107 (1996).

[34] S. V. Belayev, M. Schadt, M. I. Barnik, J. Funfschilling, N. V. Malimoneko, K. Schmitt, Large Aperture Polarized Light Source and Novel Liquid Crystal Display Operating Modes. Japanese Journal of Applied Physics 29, L634 (1990).

[35] Y.-S. Tyan, Y. Q. Rao, X. F. Ren, R. Kesel, T. R. Cushman, W. J. Begley, N. Bhandari, Tandem Hybrid White OLED Devices with Improved Light Extraction. SID Symposium Digest of Technical Papers 40, 895 (2009).

Chapter 4

Effect of Photonic Structures in Organic Light-Emitting Diodes – Light Extraction and Polarization Characteristics

Soon Moon Jeong and Hideo Takezoe

Additional information is available at the end of the chapter

http://dx.doi.org/10.5772/54669

1. Introduction

The development in organic light emitting diodes (OLEDs) has been one of the fastest growing research areas because of their potential applications in lighting and flat panel displays. Some commercialization of OLED devices such as lightings and displays has already been made. In particular, OLED displays are awaiting true commercialization toward large market. However, there are still several problems to be solved. Particularly two areas require ongoing improvements, 1) light extraction and 2) polarization. In this chapter the research activities in these two areas are summarized.

2. Light extraction

As far as the material for organic light-emitting diodes (OLEDs) is concerned, semiconductor-based organic light emitters are the obvious choice because semiconducting organic light-emitting materials have reached a high level with internal quantum efficiencies of ~100% [1]. Unfortunately, however, most of this light is trapped inside OLEDs, and only 20% can be outcoupled because of the total internal reflection [2-7]. In this section, various light extraction technologies are reviewed to suppress guided light loss. In particular, the enhanced light extraction efficiency by means of photonic structures onto OLEDs is discussed in depth.

2.1. Limited light extraction and improvement strategy

OLEDs suffer from poor external efficiency that arises from Snell's law; i.e., light generated in a high-refractive-index layer tends to remain trapped in the layer due to total internal reflection [2,3]. In fact, whatever the internal quantum efficiency might be, the light extraction efficiency of OLEDs with flat multi-layered structures including no additional surface modifications is typically only about 20% of the internal quantum efficiency [4,5]. In such OLEDs, the extracted emission cone to air is very small and only a small fraction of the light generated in the material can be outcoupled from the device but the rest is trapped within by the total internal reflection. The emitted light has to travel from the emissive layer ($n_{organic}$ = 1.6~1.7) through the ITO contact (n_{ITO} = 1.8~2.0) and the glass substrate ($n_{substrate}$ = ~1.5) and finally into air (n_{air} = 1). By using ray optics, simply the amount of extracted light out of incident light to the air η can be estimated by the following relation [8].

$$\eta = \frac{1}{2n^2} \tag{1}$$

where n is the effective refractive index of the emissive layer with respect to the that of outcoupled space (air) By taking account of Fresnel reflections at the glass/air interface [9] the calculated extraction of light to the air is only about ~17%.

Resumption of at least part of the remaining trapped light (~80%) has been one of the most important issues in fabricating OLEDs for practical applications over the past years. The intense research efforts have been focused on, e.g., substrate surface roughening [10], microlenses [11,12], monolayer of silica spheres as a scattering medium [13], insertion of low-refractive-index materials [5], distributed Bragg reflectors (DBRs) [14-20], and one-dimensional (1-D) or two-dimensional (2-D) photonic structures [21-28]. The research developments in these areas are described below:

Light extraction by scattering: One of the low-cost methods for enhanced light extraction efficiency is roughening the substrate surface as shown in Fig. 1(a). The roughness causes random light scattering at the interface between substrate and air, and thus a guided light can be extracted from OLEDs. Schnitzer *et al.* have demonstrated a 30% increase of external efficiency in GaAs light-emitting diodes (LEDs) using nanotextured surface [29]. In case of OLEDs, however, this surface roughening techniques have not been used widely because semiconducting organic materials have relatively low refractive index of 1.7 compared with inorganic materials (n=3.5). Although a randomly rusted surface can extract guided modes, it also interrupts the external air modes (within critical angle) by non-transparent surface. Therefore, inorganic LEDs with larger amount of guided modes have better extraction ratio than organic LEDs.

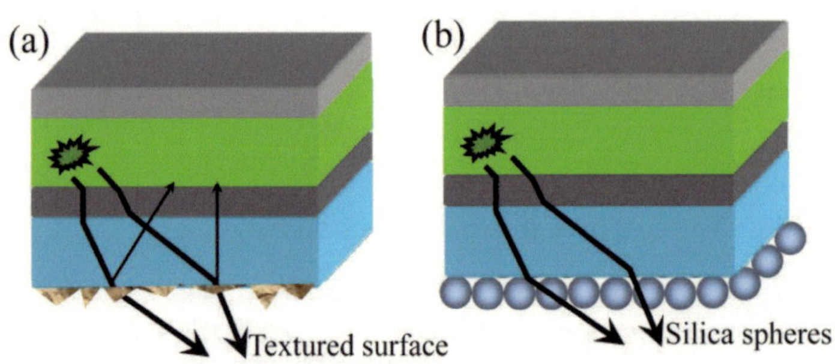

Figure 1. Schematic structure of OLEDs with modified substrate surface and light extraction.

To overcome this problem, an ordered monolayer of silica microspheres with a diameter 550 nm as a scattering medium has been used by Yamasaki *et al.* as shown in Fig. 1(b) [13]. They have used hexagonally closed packed silica microspheres to enhance the light extraction by a 2-D diffraction lattice. However, in this structure a spectral change is inevitable by the periodic configuration of microspheres, and the device does not show Lambertian emission with the increase of output polar angle from 0° to 90°.

Microlens array: Figure 2 shows a microlens that has been used as an array for light extraction from the substrate modes [11,12]. The range of diameter of the semi-spherical lens is typically from a few μm to hundreds of μm. Without such lenses, light emitted into substrate with large incidence angles is totally reflected internally by the flat interface between the substrate and air. However, by introducing semi-spherical lenses, the considerable degree of light is transmitted to the air with less total internal reflection because the incidence angle with respect to the surface normal of the lens is now lower than the critical angle. Recently this technology has mostly been used in OLEDs fabricated for lighting because low-cost fabrication of light extraction films is possible via imprint processes. However, the enhanced light extraction efficiency achieved so far is only 1.5-1.7 times of that of reference devices without lenses. To make the matter worse, the color variation with increase of angle is inevitable, preventing the Lambertian emission which is very essential for lighting. It is shown by Lim *et al.* that randomly fabricated microlenses can solve this problem to a certain extent and improve the angle dependence [30].

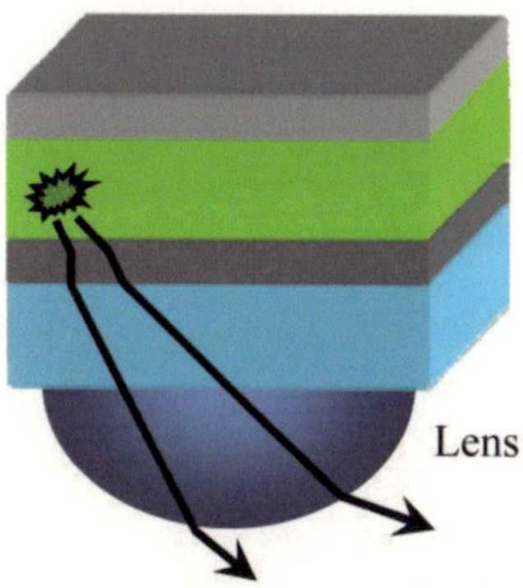

Figure 2. Schematic structure of OLEDs with a microlens and light extraction.

Insertion of low-refractive-index materials: Tsutsui *et al.* [5] have used ultrathin organic emissive layers as very poor waveguides with only a very few allowed modes. This allows a considerable amount of light to leak into the substrate, and eq. (1) is no longer valid. In addition, if the index of refraction of the substrate can also be lowered, the light output can be improved significantly as shown in Fig. 3. Tsutsui *et al.* have proposed the use of aerogels with a refractive index close to that of air ($n_{aerogel}$~1) and demonstrated that the out-coupling efficiency gets doubled.

Distributed Bragg reflectors (Microcavity): Another promising light extraction technique is the use of microcavity structures [18-20,31,32]. In microcavity devices, the internal emission can be effectively extracted via interference effects. In addition, microcavity provides us with spectral narrowing and spatial redistribution of the emission. Microcavity using inorganic distributed Bragg reflectors (DBRs) consisting of alternating inorganic layers with different refractive indices has been extensively studied over the past several decades. The advantage of DBRs is that they may have very high reflectivity and very low loss. These reflectors have a selective reflectivity in a specific wavelength range, which can form constructive interference, and effectively suppress other modes and induce a high reflectance over a certain range of wavelengths depending on the difference in refractive indices of constituting layers. This leads to spectral narrowing and intensity enhancement of spontaneous emission in microcavity OLEDs.

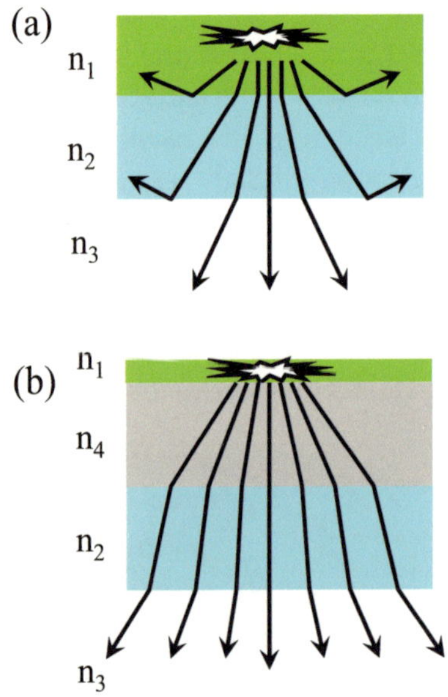

Figure 3. Schematic diagram of light extraction from OLED with (a) conventional structure and (b) structure with a thin active layer and a low-refractive index layer.[5] Copyright 2001, Wiley-VCH.

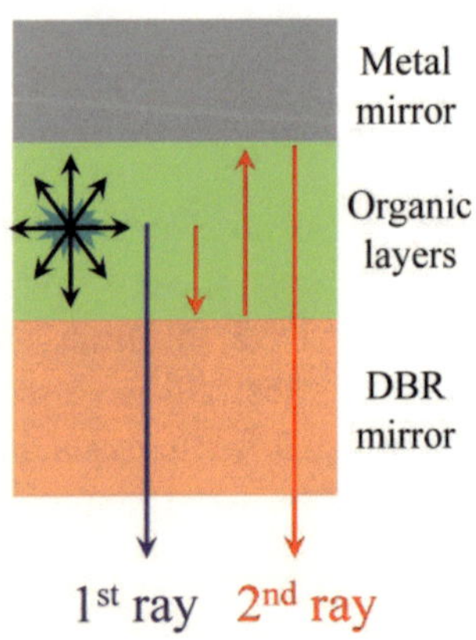

Figure 4. Schematic structure of microcavity OLEDs.

The schematic structure of a microcavity is shown in Fig. 4. The total optical path difference between direct emission and emission after single-round reflection L (see Fig. 4) is given by:

$$L = \frac{\lambda}{2}\left(\frac{n_{eff}}{\Delta n}\right) + \sum_i n_i L_i + \left|\frac{\phi_m}{4\pi}\lambda\right| \quad (2)$$

where n_{eff} is the effective refractive index, n_i is refractive index of ith layer, φ_m denotes phase shift. The first, second, and third terms stand for effective penetration depth, summation of the optical thickness in each layer, and phase shift, respectively. Usually, the total thickness of the organic materials in OLED structure is about 100 nm and the ITO thickness is determined by considering good electrical conductance and high transparence. Therefore, the optical length of the cavity can be modified by varying the second term in eq. (2). If resonant condition for constructive interference is $2L=m\lambda$, the light with λ is selectively enhanced and the color purity is also enhanced.

Diffractive resonators: The application of Bragg grating to OLEDs has been reported by Matterson *et al.* [21] and Lupton *et al.* [22]. They have demonstrated an increase in the light extraction by Bragg-scattering of waveguided light using a corrugated photoresist layer. However, in this device structure the light must transmit through the absorptive gold and photoresist layers, which limit its absolute efficiency [24]. Later, Ziebarth *et al.* [23] have demonstrated a more conventional ITO-based electroluminescent (EL) device using a stamped Bragg grating into poly(3,4-ethylenedioxythiophene)-poly(styrenesulfonate) (PEDOT:PSS) layer. Even though the soft-lithography is beneficial in fabricating large-scale devices at low cost, waveguide absorption is strong in ITO and patterned PEDOT layers [24]. This results in reduction of grating effect particularly in the shorter wavelength region. Fujita *et al.* have also shown the improved electroluminescence from a corrugated ITO device using this concept [27,28]. They have used vacuum evaporated EL materials and square-shaped pattern substrates. Using vacuum evaporation, the organic materials deposited on patterned substrates retain their pattern shape due to the low adatom mobility of deposited organic molecules. This corrugated shape through all device structure enhances the light extraction efficiency by not only waveguided light diffraction but also surface plasmon. However, a square type pattern is not suitable for fabricating stable EL devices because of possible electrical short problems.

The enhanced light extraction from OLEDs with diffractive resonators can be explained as follows. First, waveguided light propagation along the in-plane direction of the device is emitted to the surface direction by Bragg diffraction in the grating device. Figure 5 illustrates the mechanism of how the diffracted light can be extracted by Bragg diffraction in a 1-D grating sample for simple consideration. If the light incident on a material with a refractive index of n_2 from that with n_1, the diffraction condition is given by

$$m\lambda = d_c\left(n_1 \sin\theta_1 - n_2 \sin\theta_2\right)\left(\theta_1 > 0, \theta_2 < 0\right) \quad (3)$$

where λ is the wavelength, n_i (i=1 and 2) is the effective refractive index, d_c is the grating period, and m is the diffraction order. If we consider a waveguided light to have a high incident angle (θ_1~90°) and diffracted light has the lowest diffraction angle (θ_2~0°), then eq. (3) can be simplified as:

$$m\lambda = n_{eff} d_c \tag{4}$$

where n_{eff} is the effective refractive index. For m=1, first-order diffracted light can be extracted to surface normal direction resulting in an increased light extraction.

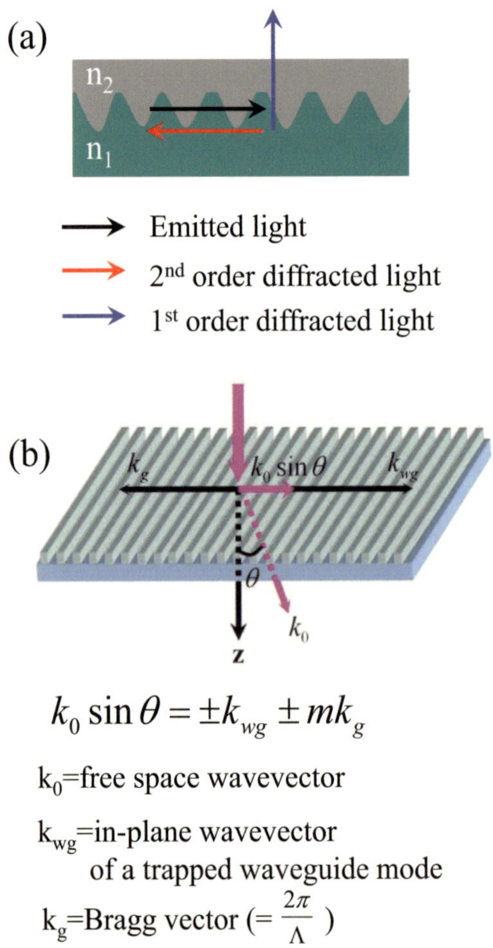

Figure 5. Typical structure of a diffractive resonator along one axis in the plane of the waveguide.[33] Copyright 2008, American Institute of Physics.

This diffraction relation can also be applied to the surface-emitting distributed feedback laser. Here we consider a situation where the waveguided and diffracted lights have high incident angles, i.e., θ_1 = 90° and θ_2 = - 90°. By substituting these angles into eq. (3), one obtains:

$$m\lambda = 2n_{eff}d_c \tag{5}$$

For m=2, second-order diffracted light makes the counter propagating mode, which results in an optical feedback for lasing. Lasing does not occur along the guided direction because of the low quality of side surfaces but it is outcoupled to the surface normal direction by first-order Bragg diffraction.

Hence, the physical meaning of this relation is that both first- and second-order Bragg diffractions of waveguided light occur simultaneously in different directions. That means waveguided light satisfying the Bragg condition is perfectly extracted toward the surface normal direction by the first-order Bragg diffraction until the second-order Bragg diffracted light decays along in-plane direction.

2.2. Device fabrication with nano-patterned structures

Nano-patterned structures are prepared to fabricate corrugated OLEDs. For periodic and quasi-periodic nano-patterned substrates, 1-D and 2-D grating and buckling structures, respectively, are prepared as described below:

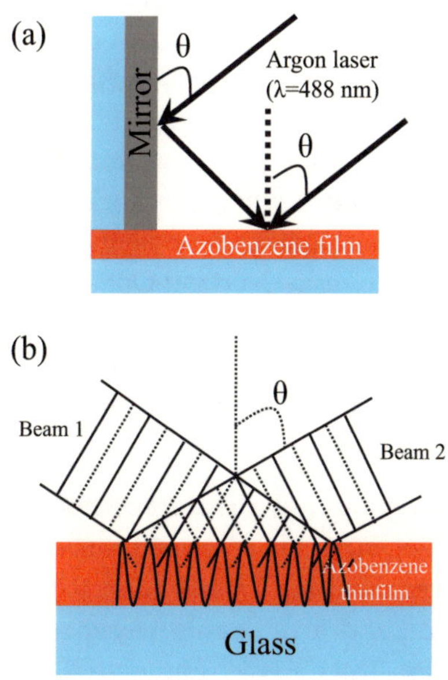

Figure 6. Schematic illustration of surface relief grating fabrication process.

To form a 1-D surface relief grating (SRG) structure, an azobenzene polymer thin film is irradiated using two Ar^+ laser (488 nm) beams. This is achieved due to the mass transport from the region of constructive interference to that of destructive interference region in azobenzene polymer, leading to a volume decrease in the highly irradiated (constructive intereference) region with the increase in the irradiation time as shown in Fig. 6. This process is quite different from the other more conventional microscopic processes such as laser ablation and chemical etching. The major advantage of this photo-fabrication approach is the possible precise control of grating depth by adjusting the light exposure energy and polarization states of writing beam.

The two recording lights are circularly polarized. Based on the following equation, the periodicity of the SRG can be controlled by changing the crossing angle between two recording laser beams.

$$\Lambda = \frac{\lambda_{laser}}{2\sin\theta} \tag{6}$$

where λ_{laser} is the wavelength of the laser and θ is an incidence angle. The formed SRG pattern is transferred onto a UV curable epoxy or a CYTOP (perfluoropolymer, Asahi Glass Co. Ltd.,), so that a patterned substrate with an SRG is obtained.

For a buckling fabrication, standard commercial poly(dimethylsiloxane) (PDMS) materials (Wacker ELASTOSIL RT 601) are mixed with a curing agent in a weight ratio of 9:1 and then spin-coated on a pre-cleaned glass substrate. The coated PDMS is cured at 100 °C for 1 h and then an aluminum layer is deposited on it. A structure thus prepared is heated to 100 °C with an external radiation source by thermal evaporation at a pressure below 1×10^{-3} Pa. It is then cooled to the ambient temperature by keeping it in a chamber for more than 30 min and venting to atmosphere. The difference in the thermal expansion coefficients between the aluminum film and PDMS generates a buckled structure on the PDMS film. The PDMS replica is formed by pouring PDMS over the buckled PDMS master and by curing it at 100 °C for 1 h. The PDMS replica can be easily peeled off from the PDMS master. For the second deposition of a 10-nm-thick aluminum layer, the buckled PDMS replica is used as the substrate. After deposition, the buckling pattern of the replica is transferred to a glass substrate after UV curing in curable resin (Norland Optical Adhesive 81) for 10 min. For the third deposition of a 10-nm-thick aluminum layer, the buckled PDMS replica thus fabricated is used as the substrate, and the above process is repeated. After the deposition, the buckled resin master is again produced, from which, finally, the buckled PDMS replica mould for making devices is fabricated.

The method of replicating nano-patterned structures (azobenzene film and buckling) onto substrates is shown in Fig. 7 [34]. First, the patterned azobenzene polymer film or buckling pattern is converted to a master mould of PDMS. After the heat treatment at 50 °C for 1 h the silicon rubber becomes firm and it can easily be separated from azobenzene polymer substrate as a free-standing film. The CYTOP solution (CTL-109A, Asahi Glass Co. Ltd.) is drop-cast on a glass substrate and is heated under the patterned silicon rubber mould pressed on it at 60 °C for 1 h. For the UV curable epoxy (Norland Optical Adhesive 81), UV light is irradiated to harden the epoxy layer under silicon rubber mould. The patterned structure is thus replicated onto a substrate.

We fabricated OLEDs on patterned nano-structured substrates. The device structures introduced in this chapter are classified in two parts, organic and polymeric devices. The organic layers in the organic device are coated by thermal evaporation (2-D grating & buckling devices), whereas spin coating (1-D grating device) is used in the polymeric device.

However, the basic concept is almost the same in both cases except for deposition method and materials used.

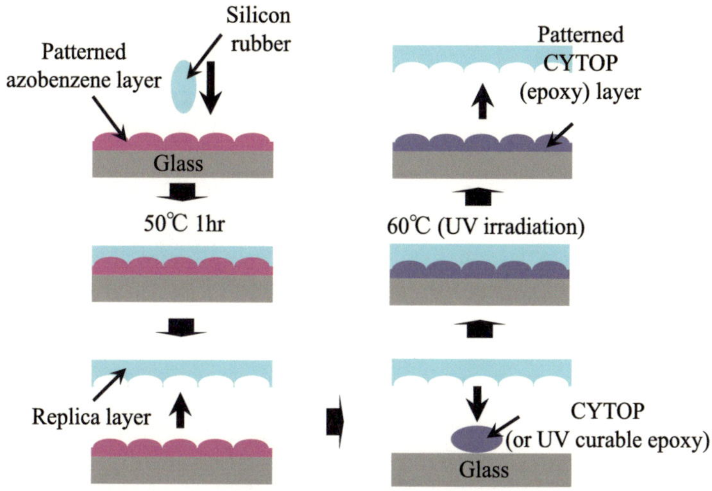

Figure 7. Schematic diagram of nano-imprint process.[34] Copyright 2008, American Institute of Physics.

We prepared ITO or Au stripes as an anode using sputter deposition or thermal evaporation. In case of organic EL devices, a two or three-layered structure consisting of (CuPc: used in 2-D grating device))/TPD/Alq3 is deposited successively by vacuum evaporation. For making the polymeric EL device, PEDOT doped with poly-(styrenesulfonate) PSS is spin coated as a hole transport layer (HTL) and heated at 60 °C for 20 min (T_g of CYTOP : 108 °C) in an oven. Then MEH-CN-PPV is spin coated as an emissive layer (EML) on the HTL successively.

After coating organic or polymeric layers using a metal mask, lithium fluoride (LiF) as an electron injection layer (EIL) and Al as a cathode are deposited. The detailed experimental conditions including thickness of each layer were explained elsewhere [33,35,36]. The pixel with a size of 3 mm×3 mm was used for electroluminescence measurements.

2.3. Effects of periodic DFB grating structures in OLEDs

In this section, we examine light extraction characteristics from OLED devices with 1-D or 2-D DFB grating substrates. The waveguided light is extracted to normal direction by an imprinted low-refractive index layer (1-D DFB grating). Also, electrical characteristics in OLEDs with 2-D hexagonally nano-imprinted periodic structures are investigated to confirm the enhanced light extraction from this device (2-D DFB grating). We review previously reported results in view of light extraction characteristics and electrical characteristics from periodically corrugated OLEDs.

Optical characterization of corrugated OLEDs with periodic structures (1-D grating): Figure 8(a) shows a schematic illustration of the Bragg diffraction process of waveguided light in periodic structures. When waveguided light is incident on the grating structure, the light is reflected by a photonic band gap and simultaneously diffracted in the direction perpendicular to the photonic crystal surface because the Bragg condition is satisfied in this direction. The angle θ of the emission direction with respect to the surface normal is governed by the conservation of momentum in the plane of the waveguide [21-24,37,38] given by

$$k_0 \sin\theta = \pm k_{wg} \pm k_g = \pm \frac{2\pi n_{eff}}{\lambda} \pm m\frac{2\pi}{\Lambda} \tag{7}$$

where λ is the wavelength, k_{wg} and k_g are respectively the wavenumbers of the waveguided light and the grating with a period Λ, n_{eff} is the effective refractive index of the waveguide mode, k_0 is the free-space wavenumber of the diffracted light, and m is the diffraction order. If Λ and n_{eff} are known, eq. (7) gives the emission angle of extracted light as a function of wavelength for the first-order Bragg diffraction of waveguided light.

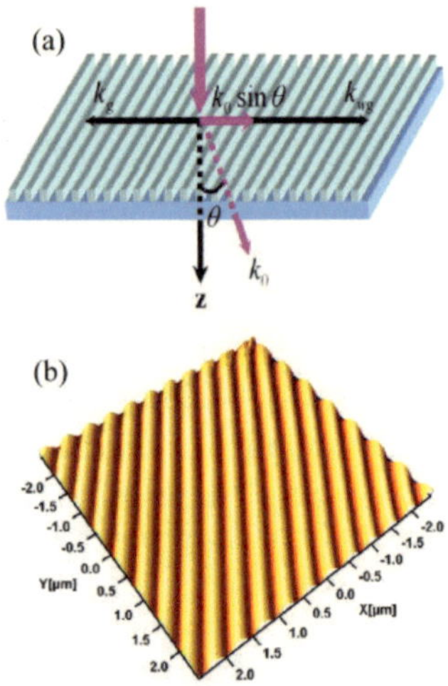

Figure 8. (a) Schematic illustration of Bragg diffraction of waveguided light. (b) AFM image of patterned azobenzene polymer. [33] Copyright 2008, American Institute of Physics.

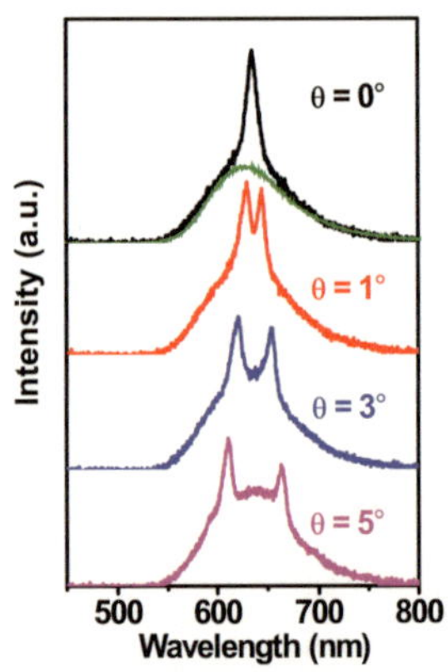

Figure 9. Angular dependence of electroluminescence spectra from OLED devices with flat (green curve at θ =0°) and patterned CYTOP layer. [33] Copyright 2008, American Institute of Physics.

Figure 9 shows the angle dependence of EL spectra for the two EL devices with and without the grating structure. For measuring the EL spectra, a detector with a diameter of 5 mm is located 10 cm apart from the device surface. A sharp peak (632 nm) has been observed in the normal direction (θ =0°), and a peak splitting has been found to occur by increasing the detection angle because the grating diffracted waveguided light travels in the opposite direction. The wavelengths of the separated two peaks have been measured as a function of the detection angle as shown in Fig. 10(a). The measured wavelength positions agree well with the lines given by eq. (8) with θ_3, which is given by

$$\theta_3 = \arcsin\left(\frac{n_2}{n_3}\sin\theta_2\right) = \arcsin\left(\frac{n_2}{n_3}\sin\left(\arcsin\left(\frac{n_1}{n_2}\sin\theta_1\right)\right)\right) \quad (8)$$

considering all refractions in PPFVB/glass/air, as shown in Fig. 10(b). This means that extraction angle of Bragg diffracted light is closely related to refractive indices of stacked materials.

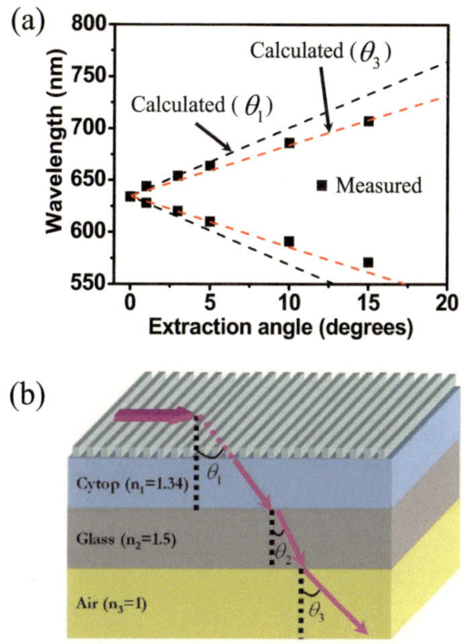

Figure 10. (a) Measured extraction angles of Bragg diffracted light. θ_1 and θ_3 denote calculated extraction angles in CYTOP and air, respectively. (b) Schematic illustration for refraction of Bragg diffracted light in each stacked material. [33] Copyright 2008, American Institute of Physics.

To compare the effects of refractive indices of imprinted materials, we have calculated light extraction angles from materials with various refractive indices, as shown in Fig. 11. The wavelength of vertically emitted light is assumed to be 632 nm. According to eq. (8), as the refractive index becomes high, the light extraction angle becomes wider. For n=1.80, the extraction angle of light with wavelength of 800 nm is 65°, whereas the total light extraction angle is only 23° for n=1.00. In the case of CYTOP (n=1.34), the total light extraction angle is

35°. Thus nano-imprinted CYTOP layer can extract waveguided light with high directionality. Such characteristics provide an advantage for small- or medium-size OLEDs, which are mainly viewed from the forward direction [39].

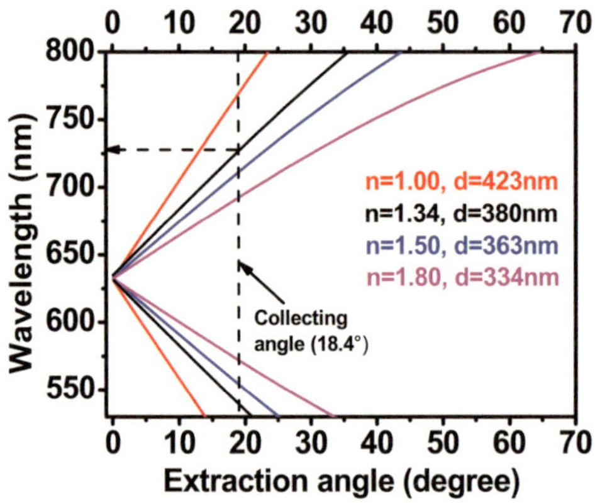

Figure 11. Calculated extraction angles in imprinted materials with various refractive indices as a function of wavelength.[33] Copyright 2008, American Institute of Physics.

The effect of a grating on normally-directed EL has been observed by collecting EL spectra between ±18.4°, as shown in Fig. 12. The enhancement of EL spectra in the device with patterned CYTOP layer has been observed over the wavelength range from 540 nm to 728 nm. However, it should be noted that the highest EL intensity is observed only around 650 nm, whereas vertically directed emission peak position is 632 nm. This results from different transmittance of ITO at various wavelengths as shown in Fig. 12. Because the wavelength of the highest EL intensity is closely related to both natural fluorescence and waveguide absorption in ITO layer, the light can be extracted more efficiently due to the high transmittance and fluorescence at 650 nm. Hence, the grating effect is higher in longer wavelength region which has higher transmittance. If ITO with high transparency is possible to be deposited at room temperature, the grating effect of CYTOP with high transmittance will be increased.

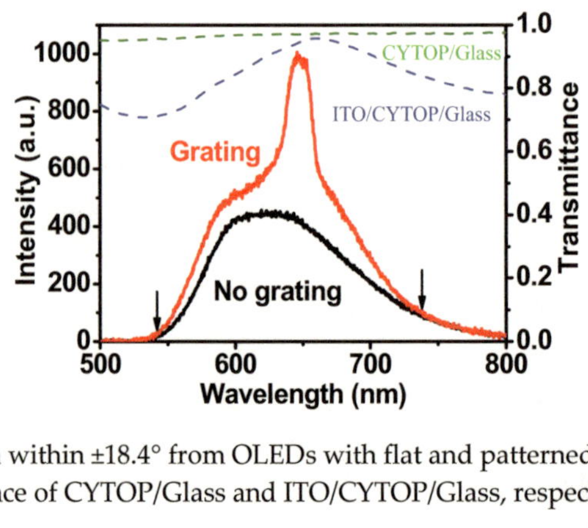

Figure 12. Overall EL spectra within ±18.4° from OLEDs with flat and patterned CYTOP layers. Two dotted lines show transmittance of CYTOP/Glass and ITO/CYTOP/Glass, respectively.[33] Copyright 2008, American Institute of Physics.

Electrical characterization of corrugated OLEDs with periodic structures (2-D grating): The current–voltage (I-V) characteristics of an EL device with a 2-D grating (2-D grating device) has been measured and compared with those of an EL device without grating (non-grating device). The 2-D grating device shows a higher current level compared to the non-grating device, as shown in Fig. 13(a). Both EL devices show a power-law dependence of $I \sim V^{6\text{-}7}$ over a large current and voltage range. Because of large trap concentration and low mobility in organic semiconductors, the carrier transport in OLEDs is trap-charge-limited current (TCLC) [40], which is known to show power law dependence.

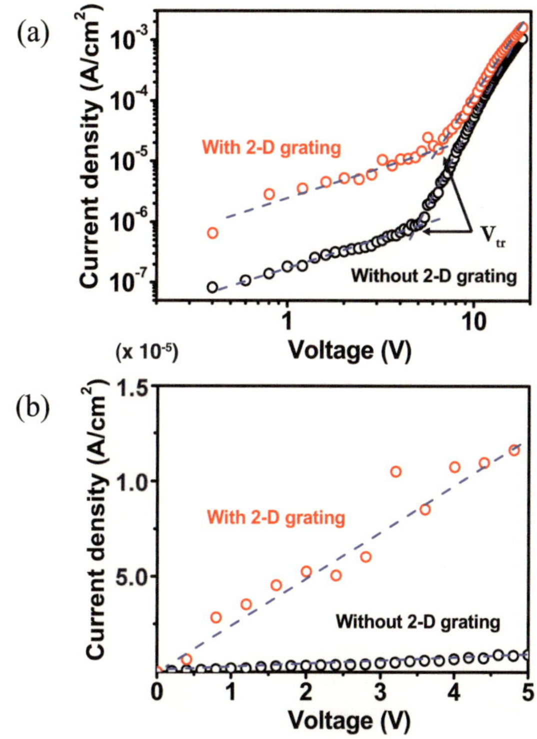

Figure 13. (a) Current-voltage plot measured from a 2-D grating and non-grating devices. (b) Magnified current-voltage plots in low voltage region.[35] Copyright 2008, The Japan Society of Applied Physics.

One may intuitively think that higher current effect in 2-D grating devices is simply due to the increase of interface contact area by corrugation between electrode and organic semiconductors. However, this cannot explain the increase of transition voltage (V_{tr}) at which the conduction model changes from ohmic to TCLC, as indicated by two arrows in Fig. 13(a). If the increase in the interface contact area is a major effect, V_{tr} in the grating device must be shifted to a voltage lower than that of the non-grating device because higher current must satisfy TCLC conduction more quickly.

At low voltages, low-mobility ohmic conduction via thermally generated free charge is observed. In this case, the current density J is described by

$$J = q\mu_n n_0 V / d_t, \quad (9)$$

where q is the electronic charge, μ_n is electron mobility, n_0 is a thermally generated background free charge density, V is the applied voltage, and d_t is the organic layer thickness. In order to find what induces the low voltage ohmic current, we have examined the *I-V* plot in the low voltage range. According to Fig. 13(b), the 2-D grating device shows a higher ohmic current than that of the non-grating device. This means that the 2-D grating device has a lower total resistance R_{total} which is a sum of junction resistance (R_J), bulk resistance (R_B) of organic layers and electrode resistance (R_{EL}) and is given by:

$$R_{total} = R_J + R_B + R_{EL} \quad (10)$$

Here the ohmic resistance induced by Al and Au (R_{EL}) and the junction resistance (R_J) induced by interfacial barrier between electrode and organic layer are the same in both samples. Hence R_B in the 2-D grating device must be smaller than that in the non-grating device. This result may be understood from the concept of 'partial reduction thickness of organic layers' proposed by Fujita *et al.* [27,28]. They have observed improved electroluminescence from a corrugated ITO where the reduction of thickness of organic layers is effectively induced by each edge of Al and ITO square-shape patterned electrodes shown as black areas in Fig. 14(a). At the edge of each patterned electrode, a higher electric field develops (See Fig. 14 (a)) and this results in reduction of operating voltage. Thus, the increased low voltage current in the 2-D grating device may be explained due to the lower bulk resistance (R_B) and hence the lower total serial resistance.

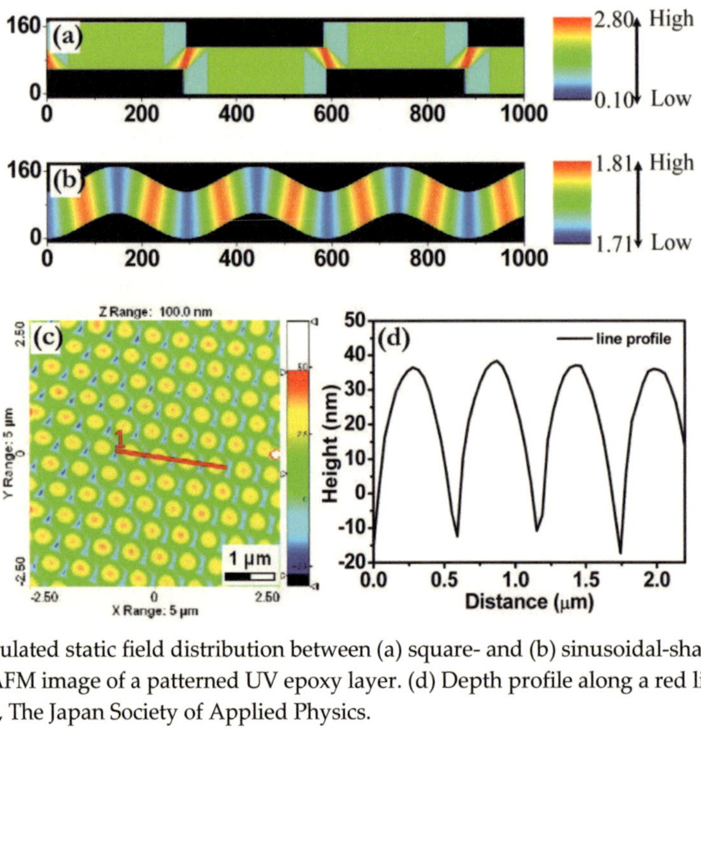

Figure 14. Calculated static field distribution between (a) square- and (b) sinusoidal-shape patterned electrodes. (c) AFM image of a patterned UV epoxy layer. (d) Depth profile along a red line in (c).[35] Copyright 2008, The Japan Society of Applied Physics.

Even though the depth of patterned shape is only 50 nm, the 2-D grating device does not show any breakdown during applying voltage. Generally thin EL devices can easily suffer from breakdowns because the internal field distribution is very sensitive to interface roughness and dust particles. It is therefore very important that the patterned electrode shape must be optimized for the stability of EL devices. Otherwise, the patterned electrode may result in worse device condition without realizing any high light extraction efficiency. Because of this reason, one should use those patterned electrode structures which give minimal 'partial reduction thickness of organic layers'. This means that if the field distribution between cathode and anode is uniform, the possibility of breakdown may be reduced even when the depth of patterned shape is high. For studying the effect of the shape of patterned electrodes, we have calculated the static field distribution in EL devices for patterned electrodes of square and sinusoidal shapes. Although the light is diffracted by 500-nm-pitched lines, which has the same width as the interfered periodicity of the two Ar^+ laser beams (Fig. 6(b)), the electric field distribution is related to the distance between closest protrusions. Hence the distance used for the calculation is 580 nm. (See Fig. 14(c)). As shown in Fig. 14(a), a high electric field gets localized at the edges of square-shaped cathode and anode electrodes. However, if the patterned electrodes are of sinusoidal shapes the field distribution becomes almost uniform. Figure 14(b) shows static field distribution in sinusoidal-shaped electrode. Note that we use different ranges of relative field intensity in Fig. 14(a) and (b) to clearly visualize the field distribution as color variations. Although the field is concentrated in the intermediate regions between the top and bottom of the patterned electrodes (see Fig. 14(b)), the field distribution becomes much more uniform compared with the case of Fig. 14(a). Figure 14(d) represents the depth profile of a patterned azobenzene film obtained along a red line in the AFM image shown in Fig. 14(c). The shape at the upper region is approximated as sinusoidal. This shape results in no breakdown of 2-D grating devices even though leakage current is high.

Next, we describe the relationship among the reduction of thicknesses, current efficiency, and diffraction effects. Figure 15(a) displays the external current efficiency versus current density. In the 2-D grating device, a higher efficiency is obtained in a high current density region. However, below a current density of 3×10^{-5}A/cm^2, the efficiency of the non-grating device is found to be slightly higher than that of the 2-D grating device, as shown in Fig. 15(b). How can we explain this? As mentioned above, the major difference between a 2-D grating device and a non-grating device is in R_B or effective thickness of the bulk layer; i.e., R_B is lower and the layer is thinner in the 2-D grating device than that in the non-grating device. Hence, we should discuss the dependence of current efficiency on the emitter thickness [41]. For this purpose, the recombination probability (P_{rec}), which is directly proportional to the EL yield, is considered. P_{rec} is defined by the ratio of the recombination time τ_{rec} and the transit time τ_t of the charge carriers as:

$$P_{rec} = \tau_t / \left(\tau_t + \tau_{rec}\right) = 1 / \left(1 + \tau_{rec} / \tau_t\right) \tag{11}$$

This gives $P_{rec} = 1$ when $\tau_{rec}/\tau_t=0$ and P_{rec} decreases with increasing τ_{rec}/τ_t. The thickness dependence in τ_t comes only from:

$$\tau_t = d_t / \mu F, \tag{12}$$

where d_t is the emitter layer thickness, μ the carrier mobility, and F the applied electric field operating on the sample. According to eqs. (11) and (12), at a given electric field, increasing emitting layer thickness will increase τ_t and hence P_{rec}. Employing this theory, 2-D grating device must show lower current efficiency due to the short transit time by reduction of thickness.

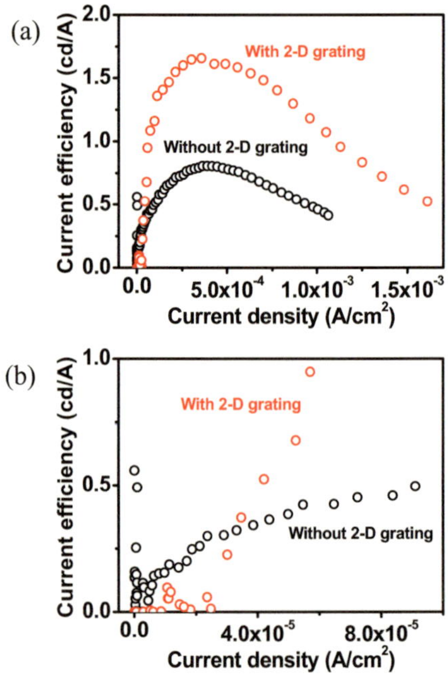

Figure 15. (a) Extracted current efficiency against current density measured from a 2-D grating and non-grating devices. (b) Magnified extracted efficiency vs current density in low current density region. [35] Copyright 2008, The Japan Society of Applied Physics.

It should be noted, however, that a higher current efficiency in the 2-D grating device increases even though the non recombined current is higher. The enhanced current efficiency in the 2-D grating device can be explained as follows. First, the waveguided light propagating along the in-plane direction of the device is emitted to the surface direction by Bragg diffraction in the 2-D grating device. Figure 16(a) shows how the diffracted light can be extracted by Bragg diffraction in a 1-D grating sample for simple consideration. If the light incident on a material with the refractive index of n_2 from that of n_1, the diffraction condition is given by

$$m\lambda = d_c(n_1\sin\theta_1 - n_2\sin\theta_2) \tag{13}$$

where d_c is a periodic distance and m is a diffraction order. Because the refractive indices are n_1=1.7 and n_2=1.5 in EL layer and epoxy/glass, respectively, the 1st-order diffracted light (m = 1) can be emitted, as shown in Fig. 16(b) and (c). In flat devices without grating, only EL

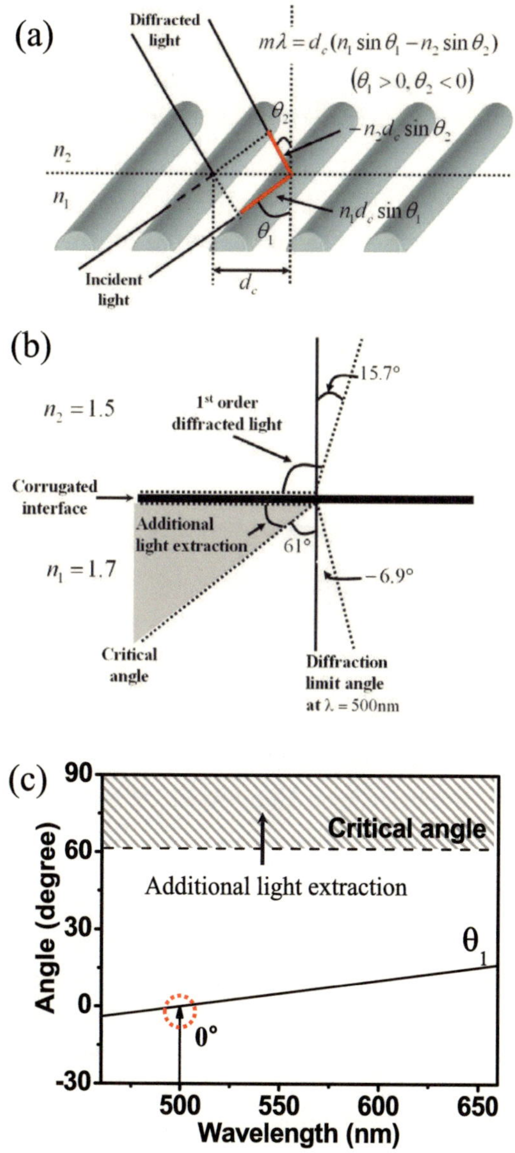

Figure 16. (a) Schematic illustration of diffraction of the guided light. (b) Schematic illustration of the angle range for total reflection and diffraction at 500 nm wavelength. (c) The diffraction limit of incidence angle θ_1 for the first diffraction as a function of wavelength. [35] Copyright 2008, The Japan Society of Applied Physics.

light to an angle below the critical angle (61°) can be emitted due to the total internal reflection. However, in grating devices, the incident light within the angle range between θ_1 and 90° is diffracted and then emitted at an angle between θ_2 and -90°. For example, in the case of a light with 500 nm wavelength, θ_1 and θ_2 are 0° and 7.6°, respectively. This means that the incident light within the angle range between 0° and 90° can be emitted at an angle in the range between 7.6° and -90°. It should be noted that an incident light within 61°~90° cannot be emitted in flat devices without a grating. In other words, the light diffracted from the incidence angle in the range between 61° and 90° contributes to additional light extraction in 2-D grating devices obtained, resulting in an increase in the of light output.

Another aspect of enhancing the light extraction in a 2-D grating device is by recovering the quenched light coupled with surface plasmon mode. This effect can be observed in an Alq_3-based system because the excitons have no preferred orientation in an Alq_3 layer, whereas conjugated polymer systems show a lower effect because the dipole moments lie in the plane of the film due to spin casting. Hobson *et al.* [42] have found that a further recovery of the trapped light can be obtained by the surface plasmon with the help of a periodic grating formed on substrate particularly in Alq_3-based EL devices. This effect can also explain the increased current efficiency in Alq_3-based EL devices because the corrugation remains intact on the Al electrode layer.

2.4. Effects of quasi-periodic buckling structure in OLEDs

Buckling patterns are produced spontaneously by thermal evaporation of Al films on poly(dimethylsiloxane) (PDMS) substrates preheated to 100 °C using an external heat source. Al layers with a thickness of 10 nm are deposited on thermally expanded PDMS. After cooling to ambient temperature, the buckling process spontaneously occurs, releasing the compressive stress induced by the difference between the thermal expansion coefficients of PDMS and Al films [43-45]. Figure 17(a), (b) and (c) shows atomic force microscopy (AFM) images of buckles formed by a 10-nm-thick Al layer applied once, twice and three times, respectively. The vague symmetric ring in the fast Fourier transform (FFT) pattern shows that the buckling structure has a characteristic wavelength with a wide distribution and without preferred orientation of the periodic structure. The characteristic wavelength can be obtained by the power spectrum of FFT as a function of wavenumber $k=2\pi/\lambda$. Figure 17(d) presents the power spectra of various buckles plotted against the wavelength instead of the wavenumber for direct comparison with outcoupled spectra of OLEDs. The buckling structure of the 10-nm-thick Al layer shows a peak periodicity at wavelengths of ~400 nm (Fig. 17d), resulting in a ~1.4% increase in the surface area ratio of the buckled to flat PDMS with a buckle depth as low as 25–30 nm as shown in Fig. 17(a).

In general, the depth of buckling structure D depends on the buckling periodicity λ, which is proportional to the thickness d of thin films and the imposed compressive strain (stress) Δ as $D \sim \lambda\ \Delta^{1/2}$ [46,47]. The buckles need to have a large depth for efficient diffraction and require a shorter buckling periodicity than that shown in Fig. 17(a) to be effective for an emission peak at a wavelength of ~525 nm. However, there is a trade-off between these factors, because D is proportional to λ. We have therefore adopted an alternative method assuming that the larger the compressive stress, the deeper are the buckles at a constant wavelength [48]. We have introduced additional compressive stresses by further deposition of a 10-nm-thick Al layer, once or twice more, on a buckled PDMS replica fabricated from a buckled PDMS mould after the first deposition of an Al layer (Figs. 17(b) and (c)).

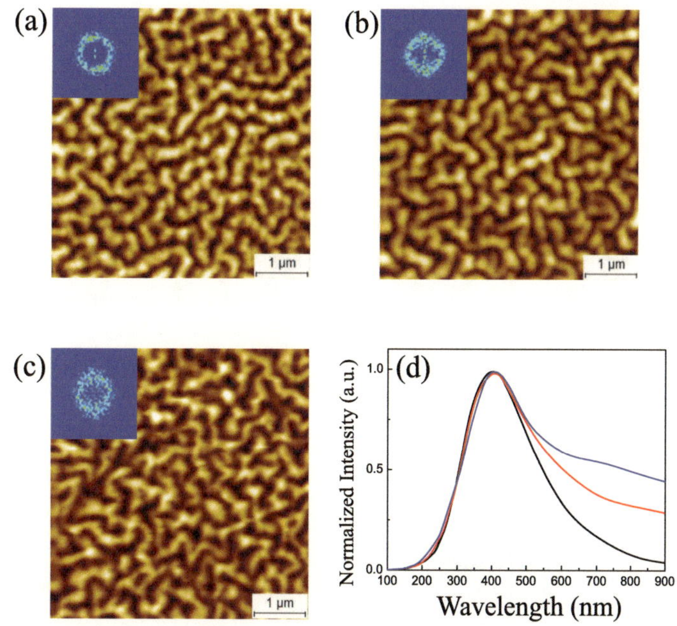

Figure 17. AFM analysis of buckling pattern. (a) Buckled structure formed by a 10-nm-thick Al layer. (b),(c) Buckled structures formed by deposition of a 10-nm-thick Al layer twice and three times, respectively. Resin layers imprinted with a buckled PDMS replica were used for measurement. Inset: FFT patterns of each image. (d) Power spectra from FFTs as a function of wavelength for buckled patterns obtained with deposition of a 10-nm-thick Al layer once (black), twice (red) and three times (blue).[36] Copyright 2010, Nature Publishing Group.

The observation that the FFT ring patterns are of similar size indicates that the characteristic wavelength does not change after redeposition. Moreover, the FFT ring patterns after multiple deposition processes display more diffuse patterns, indicating a broader distribution. The power spectra in Fig. 17(d) represent the unchanged peak wavelengths at ~410 nm and the broader distributions in the long wavelength side for the multiple depositions. In addition, the surface area ratio after deposition twice and three times significantly increases from ~1.4% to ~9.0% and 11.3% corresponding to depths of 40–70 nm and 50–70 nm, respectively.

The devices with buckling show higher current density (J) and luminance (L) than those without buckling and a device with triple buckling shows higher J and L than that with only double buckling (Fig. 18(a)). It has been reported that the larger J in the corrugated device mainly results from a stronger electric field because of the partially reduced organic layer thickness in the intermediate region between the peak and valley of the sinusoidal patterned gratings [27,35]. Measurements have also been made on devices without buckling but with the organic layer thickness decreased by 20% and 40%. As mentioned in 2.3, current density (J) for these devices is shown by dotted and dashed curves in Fig. 18. The current density in the device with triple buckling lies between that in the reference devices and in devices with thinner organic layers. This suggests that the thickness of the organic layers on buckling is partially reduced by ~20–40%. In the devices with double, triple and without buckling, the

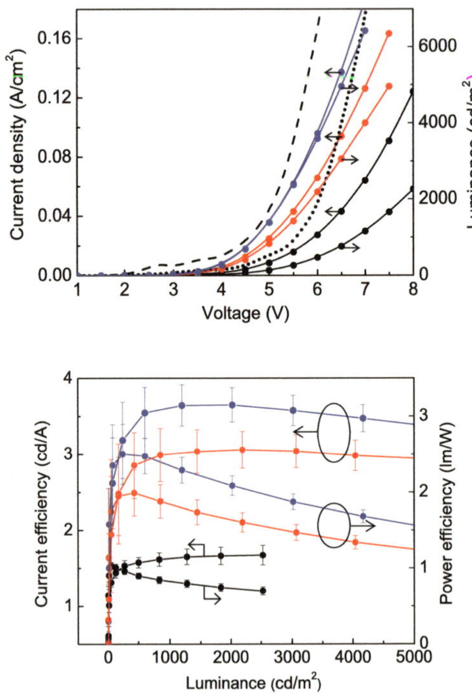

Figure 18. Device performance. (a) Current density–luminance–voltage characteristics of typical OLEDs without buckling (black) and with double (red) and triple (blue) buckling. The dotted and dashed lines represent the current density of devices without buckling but with the organic layer thickness decreased by 20% and 40%, respectively. (b) Current efficiency (cd/A) and power efficiency (lm/W) as a function of luminance (cd/m²) for OLEDs without buckling (black) and with double (red) and triple (blue) buckling. [36] Copyright 2010, Nature Publishing Group.

current efficiencies are found to be 3.05 cd/A (double buckling), 3.65 cd/A (triple buckling) and 1.67 cd/A (without buckling), and the power efficiencies 1.64, 2.1 and 0.73 lm/W, respectively, at a luminance of 2,000 cd/m². These efficiency increases correspond to enhancements of ~83% with double buckling, and 120% with triple buckling in the current efficiency and 120% with the double buckling and 190% with triple buckling in the power efficiency (Fig. 18(b)). We attribute the greater enhancement of efficiencies in the devices with triple buckling than those in double buckling to an increase in the optical confinement factor due to the greater buckling depth [28]. The observed enhancement in the power efficiency higher than in the current efficiency may be attributed to the reduction in operating voltage due to the partial decrease in the organic layer thickness in the corrugated structure (see Fig. 18(a)). One may expect that the decreased thickness of the N,N'-bis(3-methylphenyl)-N,N'-diphenylbenzidine (TPD) and Alq_3 layers may lead to a better charge balance with a better internal quantum efficiency because of the stronger electric field dependence of electron mobility in the Alq_3 layer than that of hole mobility in the TPD layer. However, the devices without buckling but with decreased thickness of the organic layer show no improvement in the current efficiency. The device with a decrease in thickness of 40% shows a significantly decreased current efficiency of 0.86 cd/A at 2,000 cd/m². This is consistent with the reported studies in which, as the Alq_3 layer thickness decreases below 30 nm, the carrier recombination probability decreases and the exciton-quenching effects at the Al cathode increase, thereby decreasing the internal quantum efficiency of the devices [25,40,49,50]. Therefore, the great enhancement of current and power efficiency in the devices with buckling is obviously caused not by a change of internal quantum efficiency, but by an increase in the outcoupling efficiency, that is, enhanced extraction of the waveguide light.

To investigate the outcoupling of the TE_0 and TM_0 modes, we have measured the electroluminescence spectra of these devices. Contrary to the enhancement emerging as new sharp peaks in conventional corrugated OLEDs [21-24,27,28,33], our buckled devices exhibit enhancement over the entire electroluminescence spectrum (Fig. 19(a)). We have evaluated

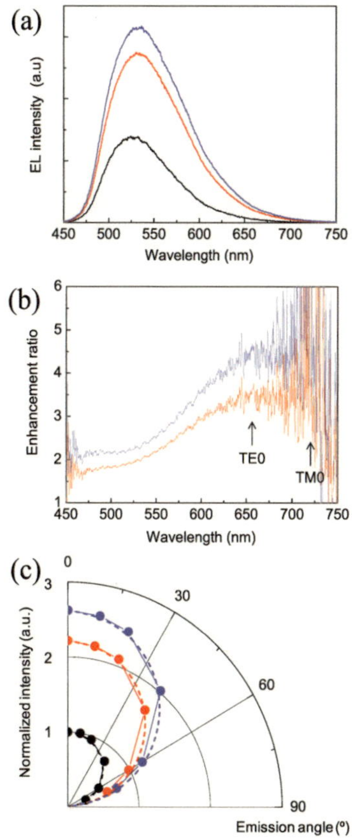

Figure 19. Electroluminescence spectral characteristics. (a) Electroluminescence spectra of devices without buckling (black) and double (red) and triple (blue) buckling, measured from the surface normal at a current density of 5 mA/cm^2. (b) Enhancement ratio of intensity by buckling as a function of emission wavelength, obtained by dividing the spectrum of the device with double (red) and triple (blue) buckling by that without buckling. The wavelengths of the TE_0 and TM_0 modes are indicated by arrows at 655 and 720 nm, respectively. (c) Angular dependence of light intensity for devices without buckling (black) and with double (red) and triple (blue) buckling. All data were normalized with the intensity of the devices without buckling in the normal direction. Each dashed line represents a guide to the ideal Lambertian emission pattern. All devices with and without buckling show the Lambertian emission pattern with a maximum intensity in the normal direction.[36] Copyright 2010, Nature Publishing Group.

the wavelength dependence of the enhanced emission by considering the intensity ratio of the two spectra in the devices with and without buckling (Fig. 19(b)). The calculated peak wavelengths of the TE_0 and TM_0 modes for the first-order diffraction are consistent with the broad peak intensities in Fig. 19(b), although the enhancement due to the TM_0 mode is not distinct because of the weak emission intensity above 700 nm. The relatively flat enhancement by a factor of ~2.2 around λ_0=525 nm in the devices with triple buckling is partially due to the relatively weak first- and second-order diffraction TE_0 and TM_0 modes, whereas the remarkable enhancement (factor of 4.0) around 655 nm is mainly due to the strong first-order diffraction in TE_0 and TM_0 modes (see Fig. 19(b)). These results indicate that a further enhancement of more than a factor of at least 2.2 can be expected if the peak wavelength of the buckles is optimized for the TE_0 and TM_0 modes to be diffracted at around 525 nm in the normal direction. Moreover, the broad distribution of periodicity in the buckling structure suggests that the entire emission wavelength range over blue, green and red in white OLEDs can be simultaneously outcoupled by only one grating structure. The angular dependence of the light intensity for the devices is shown in Fig. 19(c). It is interesting to note that all devices with and without buckling show a Lambertian emission pattern with a maximum intensity in the normal direction. According to the Bragg equation, the first-order diffraction angles of the TE_0 and TM_0 modes around the main emission wavelength of 525 nm by the grating period of 410 nm are expected to be between 20° and 40°. However, because k_G has random orientation and broad periodicity due to the buckling, it is distributed over all azimuthal directions in contrast to one- or two-directional k_G in conventional corrugated OLEDs [21-24,27,28,33]. Thus, the outcoupled emission concentrates into the normal direction, resulting in the Lambertian emission pattern.

3. Polarized electroluminescence

The polarization control of light is important for optical information processing, display and storage devices. Although linearly polarized light has already been applied to various optical devices, there are only a few reports on circularly polarized devices. However, the potential applications of circularly polarized light have been suggested for optical data storages and flat panel displays. Recently, the research for active devices that can emit polarized light has gained attention [51-58]. Peeter *et al.* [56] have first demonstrated circularly polarized (CP) EL from a polymer LED using a chiral π-conjugated poly(*p*-phenylenevinylene) (PPV) derivative as an active layer, although the degree of circular polarization was very low. Later, Oda *et al.* [57] have succeeded in obtaining a high CP-EL using main-chain polymer liquid crystals (LCs) and chiral-substituted polyfluorenes (PF) as an active layer. However, the degree of circular polarization was still insufficient for applications in optical devices. More recently, Grell *et al.* [58] have proposed a new idea for CP-EL without using chiral active materials and succeeded in achieving high degree of circular polarization. They used a simple CP-EL device that can be driven by nonchiral polymer LED using "photon recycling" concept developed by Belayev *et al.* [59]. Belayev *et al.* and Grell *et al.* used a chiral nematic liquid crystal (cholesteric liquid crystal; CLC) cell attached to the glass side of polymer LED and obtained a high degree of circular polarization at the center of the stop band. However, the degree of circular polarization outside of the stop band rapidly decreased, because the emissive material had wider emission band than the stop band width formed.

For evaluating the degree of circular polarization at a certain wavelength λ, a g-factor is used which is defined as:

$$g(\lambda) = 2\frac{I_L(\lambda) - I_R(\lambda)}{I_L(\lambda) + I_R(\lambda)} = 2\frac{r(\lambda) - 1}{r(\lambda) + 1} \tag{14}$$

where $I_{L/R}$ is the intensity of left/right-handed CP (L-CP, R-CP) light, and r is the left/right-handed intensity ratio, $I_L(\lambda)/I_R(\lambda)$. It is evident that $|g(\lambda)|$ is zero for nonpolarized light ($r(\lambda)$=1) and is equal to -2 for pure, single-handed circularly polarized light ($r(\lambda)= \infty$ or 0). The $g(\lambda)$ values found were 0.001 [56], 0.25 [57], and 1.6 [58], but only in a narrow wavelength range. Woon *et al.* and Geng *et al.* respectively reported circularly polarized PL [60] and EL [61] with a constant $g(\lambda)$ value over a wide spectral range covering most of the emission band. However, the bandwidth [60] and $g(\lambda)$ value [61] were still insufficient for application to commonly used emissive materials with wide emission band.

To achieve a tunable polarization of electroluminescence, we have used combination of voltage dependent nematic liquid crystal (NLC) phase retarders and photon recycling concept [62,63]. The phase retardation arises between two optical eigenmodes during light propagation in an anisotropic medium as a phase retarder. Upon emerging from the phase retarder, the relative phase of the two eigenmodes is found to be different from that at the incidence, and thus their polarization state becomes different as well [64-66]. Now suppose we apply a voltage (V) across the cell filled with NLC, by which the liquid crystal molecules change their orientation toward the field direction, if the NLC has positive dielectric anisotropy. With increasing the voltage, the birefringence $\Delta n(E) = n_e(E) - n_0$ decreases, where n_e and n_o are refractive indices for extraordinary-(e-) and ordinary-(o-) light waves, respectively, and the retardation ($\Delta\varphi$) decreases as well. Hence, as the e- and o-waves propagate through the NLC cell, their relative phase difference changes, and the state of polarization of the wave also changes.

We have introduced another polarization characteristics, namely polarization conversion in surface plasmon (SP) coupled emission by buckling structures. In section 2.4, we have demonstrated that the quasi-periodic buckling structures with broad distribution and directional randomness can effectively enhance the light-extraction efficiency by outcoupling the waveguide modes without introducing spectral changes and directionality [36]. In this study, however, we could not differentiate the outcoupling of transverse electric (TE) mode from that of the surface-plasmon (SP) mode (transverse magnetic (TM) mode) by buckles because of the broad periodicity of the buckling structure and the similar propagation vectors of the TE and SP modes. The explanation of polarization conversion in the surface-plasmon-coupled emission presented here is based on a trial method for distinguishing TE and TM modes in light enhancement in OLEDs with buckling pattern. However in this trial approach, an interesting phenomenon of polarization conversion in SP coupling has been observed.

In this section, we have summarized and introduced our studies regarding not only circularly polarized EL and its tunability but also the polarization conversion in surface coupled emission from corrugated OLEDs with buckling structures.

3.1. Device fabrication

We have fabricated multi-layered polymer CLC (PCLC) films for using them as wide-band reflectors or single-layered films for polarization-tunable OLEDs. As an experimental method for fabricating single-layered PCLC films is a part of fabricating multi-layered PCLC films, we introduce here only the fabrication of multi-layered films and skip the fabrication of single-layer films.

The fabrication process of multi-layered polymer PCLC films is shown in Fig. 20. Mixtures of two aromatic polyester liquid crystalline polymers (Nippon Oil Corporation; currently, JX Nippon Oil & Energy Corporation) are used to make PCLCs. One of the polymers (chiral polymer) contains 25% chiral units in its chemical composition and the other contains no chiral unit. By changing the ratio of the amounts of the two polymers, the helical pitch of PCLC (photonic band gap wavelength) is controlled.

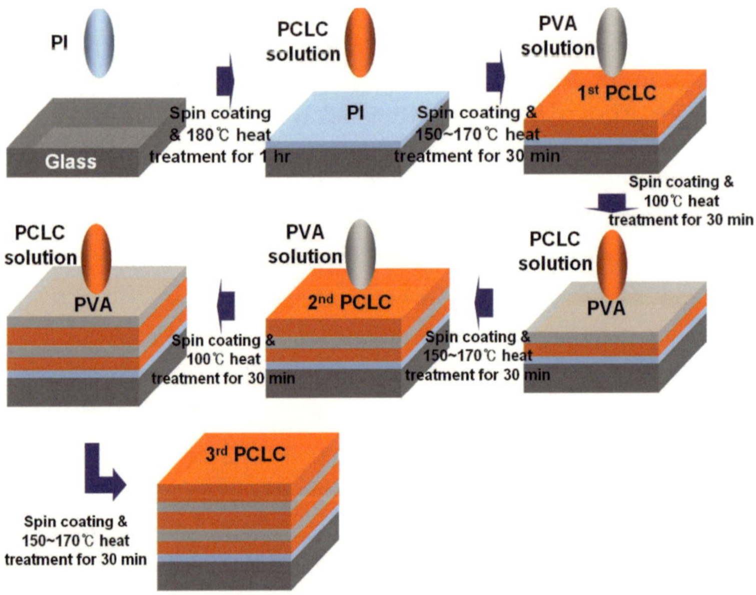

Figure 20. Fabrication process of multi-layered PCLC films.

For fabricating three-layered PCLC films for use as a wide band reflector, the PCLC (λ_p=610 nm; chiral polymer 72 wt%) is spin-cast on glass substrates with unidirectionally rubbed polyimide (PI ; AL1256, JSR). Then, aqueous solution of polyvinyl alcohol (PVA) is spin-cast and the film surface is rubbed again unidirectionally. Another PCLC (λ_p= 510 nm; chrial polymer 87 wt%) is spin-cast on the rubbed PVA surface. The same procedure is repeated for preparing the third PCLC film (λ_p= 530 nm; chrial polymer 82 wt%). Finally PCLC films thus fabricated are cured for 30 min at 160 °C.

The fabrication method of a tunable phase retarder is as follows. The single-layered PCLC films are fabricated by spin coating the solution onto ITO glass substrates coated with PI rubbed unidirectionally at room temperature. The coated PCLC films are cured for 30 min at a temperature over 160 °C in a bake oven, and then quenched to room temperature. The sample cell is made of L-PCLC and PI coated glass substrates and is sustained by spacer. The NLC (ZLI2293, Merck) is introduced into an empty cell using capillary action. The illustration of the fabrication of the final cell is shown in Fig. 21

The OLED structure used here is fabricated in the same way as described in section 2.2. The vacuum evaporated OLEDs with structure of ITO/CuPc/TPD/Alq_3/LiF/Al are described in section 3.2, an spin-coated OLEDs with structure of ITO/PEDOT:PSS/MEH-CN-PPV/LiF/Al are given in section 3.3.

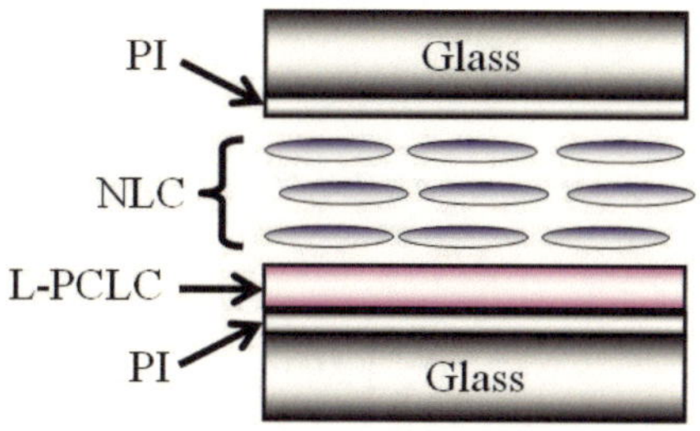

Figure 21. Schematic illustration of tunable phase retarder. [63] Copyright 2008, American Institute of Physics.

3.2. Highly circularly polarized electroluminescence

The device configuration for highly CP-EL from OLEDs is illustrated in Fig. 22. We have simply attached an L-PCLC reflector to an OLED device. After generating the unpolarized light by electrical pumping, the R-CP-EL transmits through the PCLC reflector, whereas L-CP-EL is reflected by the selective reflection of the L-PCLC. This reflected light by the PCLC is still L-CP and changes the polarization to R-CP by getting reflected at the metal surface, and gets transmitted trough the L- PCLC reflector. Thus all the transmitted light has the same sense of rotation, R-CP.

In comparison with the previous work [58] here the band width of the reflector is wider. For fabricating a wide-band CLC reflector, the use of PCLCs has two advantages compared with general low-molecular-weight CLCs. First advantage is that PCLCs used here have higher optical anisotropy (n_e-n_o=0.22), resulting in a wider photonic band gap (PBG). The second advantage is that PCLC films can be easily stacked to multi-layered films by spin-casting. Using these technical advantages we have fabricated a wide-band PCLC reflector using multi-layered PCLC films with different selective reflection bands.

The structure of a three-layered PCLC film with a wide stopband width is shown in Fig. 23(a). The fabrication method of multi-layered PCLC films is already explained in section 3.1. Figure 23(b) shows the reflectance spectra of single-layered and three-layered PCLC films and the emission spectrum of the active EL material, Alq_3 (see below). A wider selective reflection band formed due to the overlap of the selective reflection bands of the three-layered PCLC films extends to the whole emission band, although the selective reflection band of the single-layered PCLC film covers only the emission peak region.

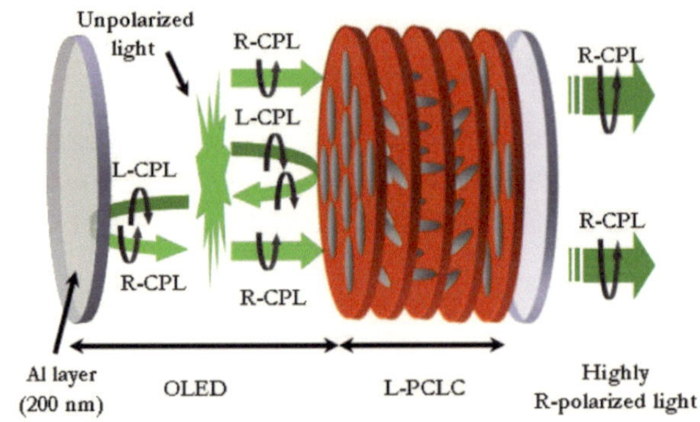

Figure 22. Schematic illustration of a 'photon recycling' device.[62] Copyright 2007, American Institute of Physics.

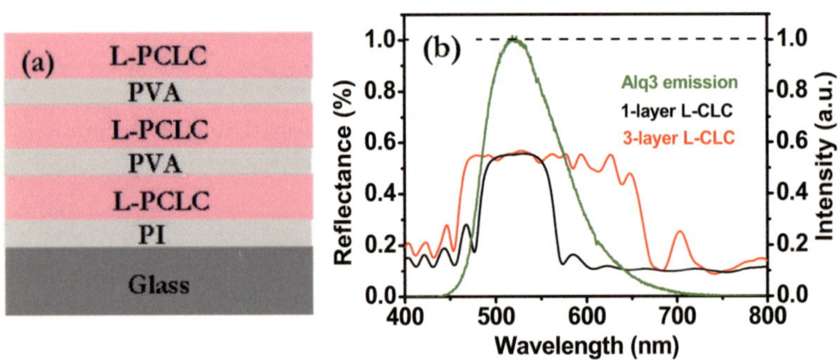

Figure 23. (a) The structure of a three-layered PCLC film. (b) Normalized electroluminescence spectrum of Alq_3 and reflectance spectra of single-layered and three-layered PCLC films.[62] Copyright 2007, American Institute of Physics.

To evaluate the degree of circular polarization quantitatively, R-polarizer and L-polarizer are inserted between the EL device and detector. We confirmed that R- and L-polarized EL intensities are almost the same in OLEDs without a PCLC reflector. In contrast, OLEDs with narrow (single-layered)-PCLC and with wide (three-layered)-PCLC reflectors emit high intensity R-circularly polarized EL within the stopband of PCLC as shown in Fig. 24(a) and 24(b). The R- and L-CP-EL spectra from OLEDs with the narrow-PCLC film are almost the same as that with the wide-PCLC film in the selective reflection region of narrow-PCLC (480nm–560nm) as a result of 'photon recycling'. Outside of the stopband of narrow-PCLC, however, both R- and L-CP components from the narrow-PCLC device do not show any prior circularly polarization characteristics due to the lack of 'photon recycling' (Fig. 24(a)), whereas the wide-PCLC device shows the highly R-circularly polarized light over the whole emission spectrum range, as shown in Fig. 24(b).

It is also noted that the degree of circular polarization is high in the wide-PCLC device over the whole emission band. Figure 25 shows the wavelength dependence of the *g*-factor [eq. (14)] for light emitted from each device. At the center of the stopband, $|g(\lambda)|$ approaches to 1.67 in both the devices with PCLC films. However, the difference is that $|g(\lambda)|$ remains same over the whole emission band in the wide-PCLC device but it suddenly decreases outside of the stopband in the narrow-PCLC device.

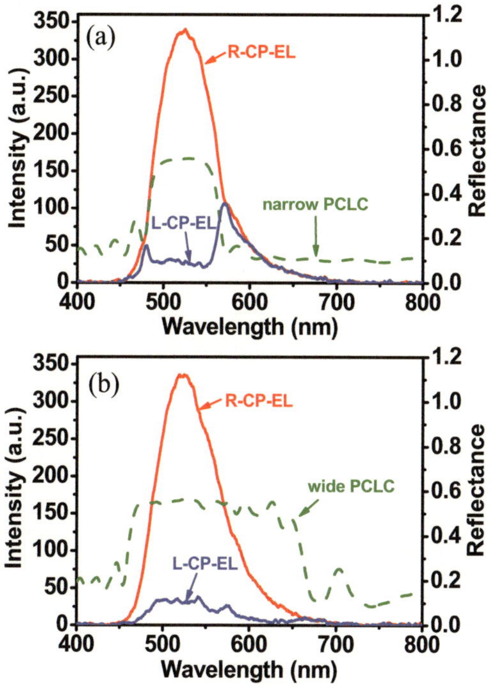

Figure 24. R- and L-CP-EL spectra from OLED devices with (a) narrow- and (b) wide-PCLC films. Reflection spectra for the narrow- and wide-PCLC films are also shown using dotted curves.[62] Copyright 2007, American Institute of Physics.

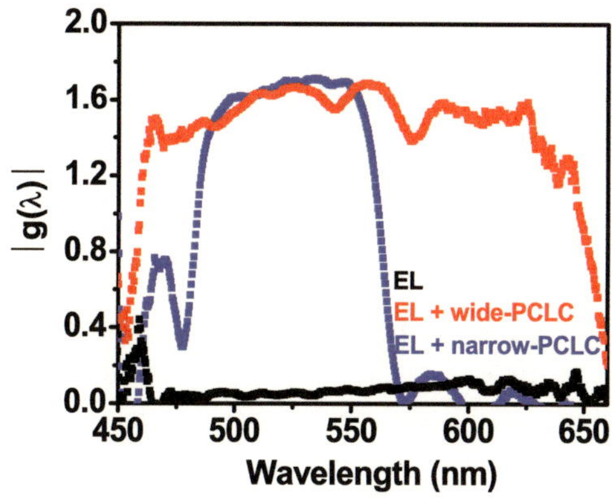

Figure 25. Calculated *g*-factor values in each device over the whole wavelength range of the emission band of Alq_3.[62] Copyright 2007, American Institute of Physics.

3.3. Polarization-tunable organic light-emitting diodes

In this section, we examine the electro-tunable polarization of electroluminescence by combination of circularly polarized OLEDs (same concept as explained in section 3.2) and tunable phase retarder. A voltage controllable liquid crystal cell is adopted as a tunable phase retarder for tunable polarization characteristics.

The device configuration for polarization-tunable OLEDs with a phase retarder is shown in Fig. 26. For the phase retardation, NLC is filled between the glass substrates with a rubbed PI layer. The phase retarder is simply attached to one of the glass sides of OLEDs. After the generation of unpolarized light from OLED, the whole EL light is extracted as R-CPL by photon recycling. This R-CP-EL can be transformed into arbitrary polarizations by changing the orientation of NLC through applying a voltage. The phase retardation $\Delta\varphi$ at a wavelength λ can be expressed by:

$$\Delta\varphi = \frac{2\pi}{\lambda} d\Delta n \tag{15}$$

where d is cell thickness and Δn is birefringence of NLC.

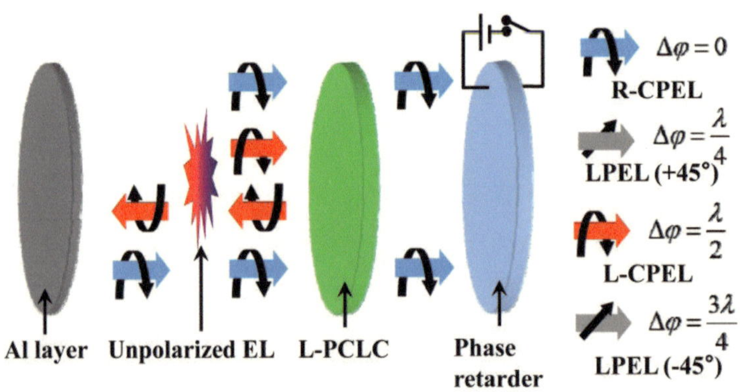

Figure 26. Schematic illustration of the principle of polarization-tunable OLED and polarized light with different polarization.[63] Copyright 2008, American Institute of Physics.

If the wavelength and cell thickness are constant, phase retardation between *e*- and *o*-waves can be varied by applying an electric field. Then the effective birefringence of NLC $\Delta n(\theta)$ is determined by the angle between the director and the substrate surface; i.e., $\Delta n(\theta=0°)= \Delta n$ and $\Delta n(\theta=90°)=0$. If $d\Delta n(\theta)$ is equal to $(2m+1)\lambda/2$ (m = 0, 1, 2…), NLC layer acts as a half-wave plate. On the other hand, if $d\Delta n(\theta)$ is equal to $(4m+1)\lambda/4$ (m = 0, 1, 2…), the NLC layer acts as a quarter wave plate. It should be noted, however, that the maximum phase retardation must be over half-wavelength ($\lambda/2$) to realize four kinds of different polarizations. Hence we have fabricated a cell with the thickness satisfying $3\lambda/2$ retardation condition in the absence of a field.

Figure 27 shows the voltage dependence of transmittance spectra through R-, L- circular and linear polarizers with the direction of +45° and -45°. The applied voltages of 0, 4.5, 6 and 7.5

V correspond to 3λ/2, 5λ/4, λ, 3λ/4 wave plates, respectively. When the R-polarizer is inserted, the spectrum shows a selective reflection band at 0 V (=3λ/2) as shown in Fig. 27(a). This is because the transmitted R-CPL changes its polarization to L-CPL through L-PCLC, and the transmittance decreases down to 0.15 within the selective reflection band. As the voltage increases up to 6 V (=λ), the spectral shape shows no selective reflection band because the phase retarder acts as a full-wave plate. On the other hand, the situation is reversed in L-polarizer as shown in Fig. 27(b).

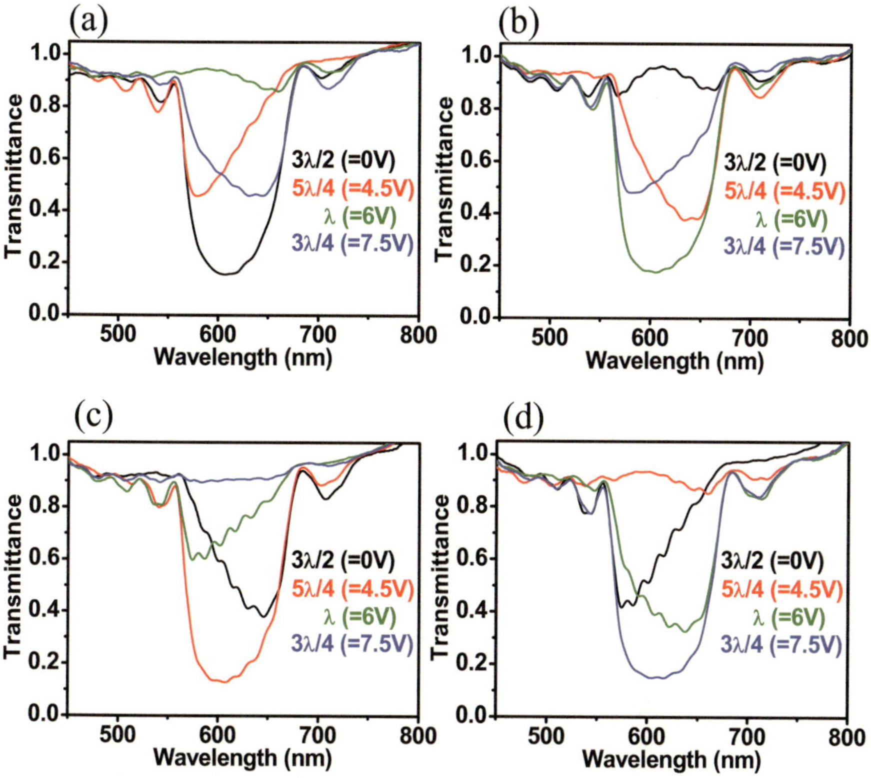

Figure 27. Polarization characteristics of voltage dependent transmittance spectra of a phase retarder. Transmittance spectra of (a) R-CPL, (b) L-CPL, (c) LPL(+45°) and (d) LPL(-45°) under fields of 0, 6, 4.5 and 7.5 V, respectively.[63] Copyright 2008, American Institute of Physics.

Conversion to linearly polarized light is also possible by 4.5 (=5λ/4) and 7.5 V (=3λ/4) applications. If the phase retardation is a quarter-wave, R-CPL changes the polarization condition to linearly-polarized light (LPL). Figure 27(c) and (d) shows LPL with electric field direction of +45° and -45°, respectively. At 4.5 V, the phase retardation is 5λ/4 resulting in a LPL (+45°) as shown in Fig. 27(c). The transmitted R-CPL changes into LPL (-45°) and shows a selective reflection band when the direction of linear polarizer is +45°. On the other hand, if the phase retardation is 3λ/4 (=7.5 V), the transmitted R-CPL changes into LPL (+45°) after transmitted through the linear polarizer. As a result, no selective reflection is observed in the transmittance spectrum. Reversed situation is also observed when the direction of linear polarizer is -45°, as shown in Fig. 27(d).

In order to apply this concept to OLEDs, we have attached a phase retarder to an EL device. This situation is different from the transmittance measurement system because here the EL device has a metallic mirror as a cathode. The output of EL light is R-CP-EL, as explained in Fig. 26. Hence different polarization states are also possible by controlling the birefringence of the NLC layer. To evaluate the degree of polarization quantitatively, R-, L-circular or linear polarizer with the direction of +45° and -45° is inserted in the emissive EL devices between the phase retarder and detector. The output of EL light transmitted from the L-PCLC is R-CP-EL within the wavelength range corresponding to the stopband. The emitted R-CP-EL can be changed into a different polarization by the phase retardation. Figure 28 shows the polarized EL spectra with different polarizations as applied voltage increases from 0 (Fig. 28(a)) to 4.5 (Fig. 28(c)), 6 (Fig. 28(b)), and 7.5 V (Fig. 28(d)). Thus EL light with different polarizations can be selectively emitted by varying the voltage. Outside of the stopband of PCLC, the intensity of opposite polarized light becomes higher because the stopband of PCLC cannot cover a wide wavelength range. It should be noted, however, if a multilayered-PCLC with different pitches is used, the polarization rate can be high over all wavelength [62,67].

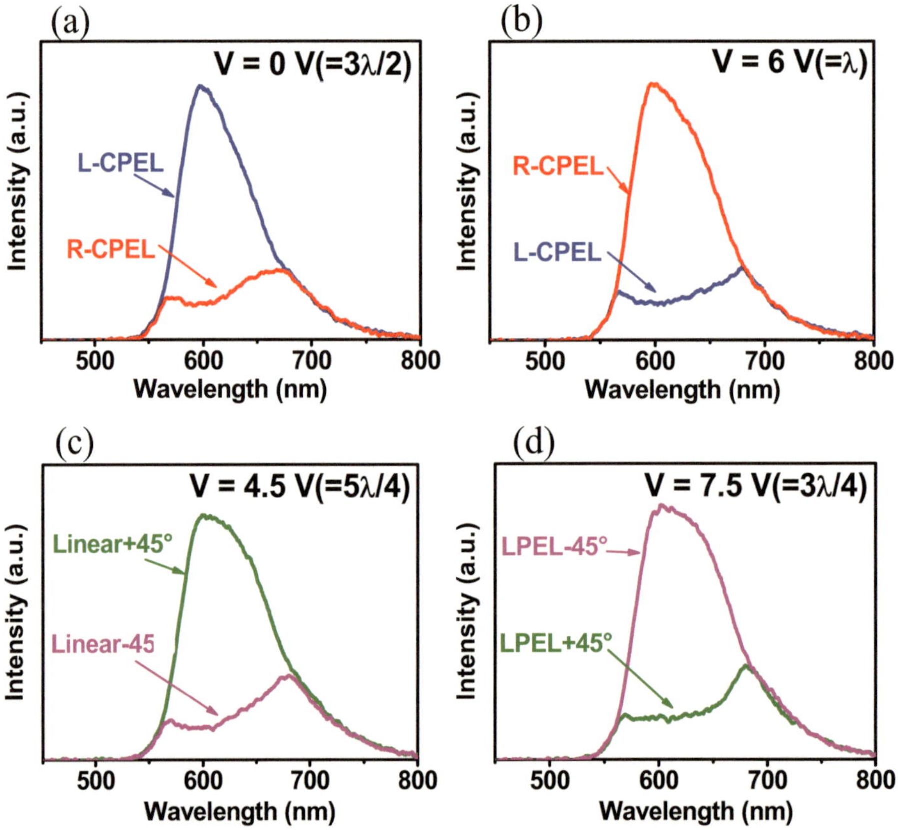

Figure 28. Measured polarized electroluminescence from OLED. Selectively emitted light of (a) L-CP-EL, (b) R-CP-EL, (c) LP-EL(+45°) and (d) LP-EL(-45°) under fields of 0, 6, 4.5, and 7.5 V, respectively.[63] Copyright 2008, American Institute of Physics.

3.4. Polarization conversion in surface-plasmon-coupled emission from corrugated OLEDs with buckling structures

The fabrication process of buckling and OLED devices is almost the same as described in section 2.2. The only difference is the use of a thinner ITO (40 nm) than previous one (120 nm) to extract TM mode preferentially by a surface plasmon coupled emission [68].

To characterize the outcoupled SP mode by buckles, we have calculated its in-plane propagation vectors and plotted the grating period for the emission angles of 0°, 20°, 40°, and 60° as a function of the wavelength of the outcoupled light in Fig. 29(a). Considering the distribution maximum of the buckling periodicity at ~410 nm, it is reasonable that the main diffraction of the SP mode for the normal direction occurs at the emission wavelength of ~690 nm. In addition the FWHM of the periodicity distribution from 300–600 nm allows outcoupling of the SP mode over the entire emission wavelengths by the first- and second-order diffractions. As the emission angle increases, the main diffraction wavelength shifts from ~690 nm for 0° to ~580 nm, ~490 nm, and ~440 nm for 20° , 40° , and 60° , respectively.

We have measured the linearly polarized electroluminescence spectra of the devices with and without buckles at the emission angles of 0°, 20°, 40°, and 60°, and then calculated the light-enhancement ratio (the intensity ratio of the two spectra in the devices with and without buckles) as a function of emission wavelength. Figure 29(b) presents the enhancement ratio of the TM-polarized light. The broad peak intensities for each emission angle are consistent with the main diffraction wavelengths calculated in Fig. 29(a), as indicated by arrows. It is very interesting to note that the TE-polarized light also gets enhanced by buckles as shown in Fig. 29(c). This enhancement is even greater than that for TM-polarized light, particularly at larger emission angles, although generally the SP mode is considered to be excited only by TM-polarized light and the diffraction gratings do not convert the polarization state of an incident light upon diffraction. However, it is also known that the polarization conversion can occur if the grating wavevector is not parallel to the plane of incidence [69-73]. So-called conical diffraction occurs at 0°–90° azimuthal angles by the grating with different wavevectors with respect to the incidence plane, where even TE-polarized light may excite the SP mode because of the existence of the electric field component parallel to the grating vector. In other words, the SP mode excited by a TM-polarized light can be outcoupled to the TE- as well as TM-modes radiation. As the azimuthal angle increases from 0° to 90°, the outcoupled TM mode decreases and the outcoupled TE mode increases by the conical diffractions [69,70]. As far as we know, this was the first report on the polarization state of the extracted SP mode, although a qualitative description on the polarization state can be found for the outcoupled SP mode from a silver cathode with a 2-D corrugated structure [74].

Because the grating vectors in a buckling structure are random over all azimuthal angles, the SP mode in the device with buckles also experiences conical diffractions at all azimuthal angles and then the polarization conversion of the outcoupled light occurs. For example, $k_0\sin\theta$, k_{SP}, and k_G for the emission wavelength at 600 nm are graphically presented in Fig. 30. Here only one grating wavevector from a 1-D grating with a periodicity of 410 nm is assumed. The radius of the solid circle (blue) corresponds to k_{SP}, the momentum space

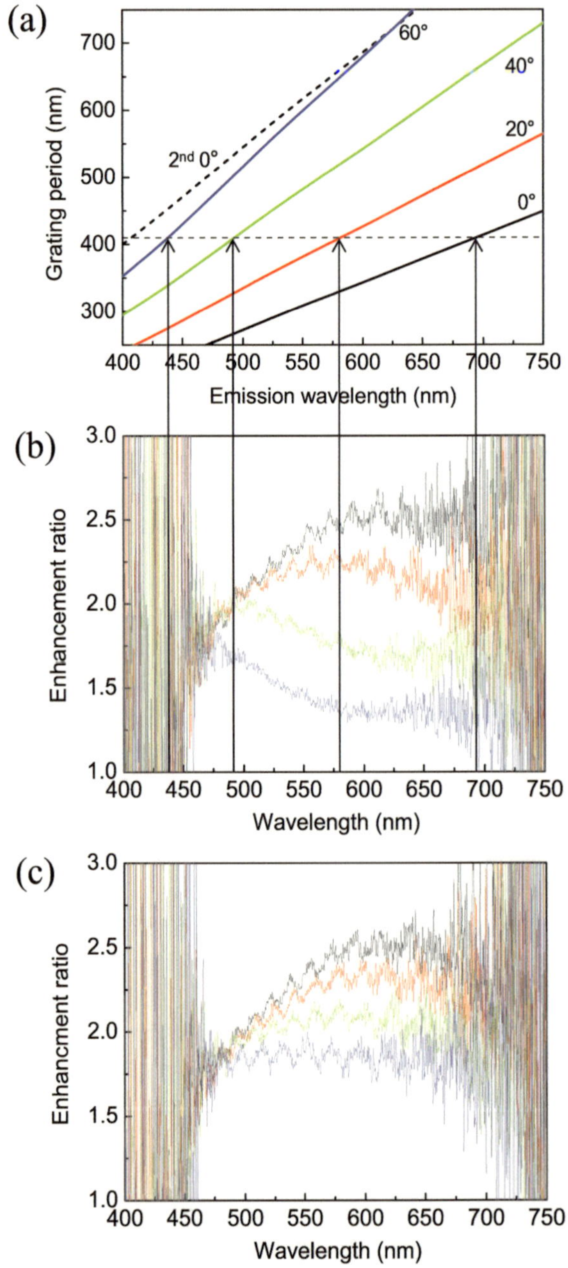

Figure 29. (a) Relation between the outcoupled emission wavelength and the grating period for the emission angles of 0° (black), 20° (red), 40° (green), and 60° (blue), satisfying the first- and second-order (only for 0°) diffractions condition. The dashed horizontal line represents the peak wavelength of 410 nm in the periodicity of the buckles used as the grating. (b) Enhancement ratios of TM-polarized light by buckles at the same angles as (a), 0° (black), 20° (red), 40° (green), and 60° (blue) from top to bottom, obtained by dividing the spectrum (measured through a polarizer) of the device with buckles by that without buckles. (c) Enhancement ratios of TE-polarized light by buckles with the same information as (b). [68] Copyright 2011, Wiley-VCH.

within the solid circle (black) represents the escape zone to air mode, and that between the dotted and solid circle (black) indicates the glass mode. As the azimuthal angle of the SP vector increases, the polar and azimuthal angles of the outcoupled light increase and simultaneously the polarization conversion to the TE mode becomes strong. At an angle of 35°, below the azimuthal angle, the SP mode is outcoupled to the air mode by the grating, between 35°–55° it is trapped to the glass substrate, and above 55° it propagates into the ITO/organic layer with the highly TE-converted polarization. In such a restricted condition of a one-directional grating with a definite periodicity, this ITO/organic mode does not outcouple. However, a conical diffraction to air is expected to occur in our buckling structure over all the possible azimuthal angles, 0–360° because of the grating wavevector distributed over all azimuthal directions. Therefore, the enhancement of the TE-polarized light is observed as shown in Fig. 29(c). However, the greater enhancement of the TE-polarized light than that of the TM-polarized light for all polar angles indicates that more TE-polarized light must be outcoupled to the air mode through the diffraction by buckles, because of the polarization conversion to the TE mode being weak at low azimuthal angles below 35°. Considering the dimension of the emitting area (3 mm × 3 mm) and glass thickness (1.0 mm), most light propagating to the glass substrate cannot undergo reflection or scattering at the corrugated Al layer. Hence the scattering of the glass mode by buckles can be ignored. We believe that the TE-converted light propagating to the ITO/organic layer by the diffraction at an azimuthal angle above 55° can be coupled to the TE_0 leaky guided mode [75], which can then be outcoupled again by the diffraction through the grating vectors with different directions. The broad periodicity and random orientation of buckles contribute to the additional extraction of the TE-polarized light for all polar angles, thereby producing a higher enhancement of the TE-polarized light over all polar angles.

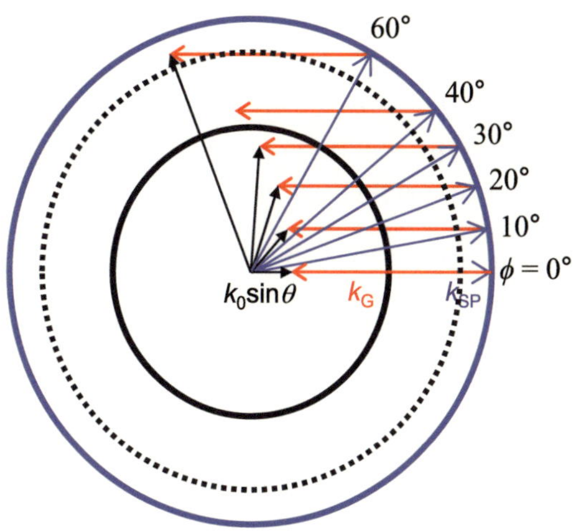

Figure 30. Momentum representations of SP mode (blue circle, k_{SP}), glass light-line (black dotted circle), air light-line (black solid line), grating wave vector (red arrow, k_G), and the outcoupled light to air mode (black arrow, $k_0\sin\theta$) for the emission wavelength of 600 nm. θ and ϕ represent the polar and azimuthal angle, respectively. Only one-directional grating vector from one-dimensional grating with a periodicity of 410 nm is assumed. [68] Copyright 2011, Wiley-VCH.

To confirm the polarization conversion by buckles on diffraction, a buckled resin layer on a glass substrate, coated with a 100-nm-thick Al layer, is irradiated using a linearly polarized He–Ne laser (632.8 nm) at an incident angle of 60° and the scattered light from the surface normal observed through a linear polarizer. We have found that the incident TM-polarized light is largely converted into the TE mode upon diffraction. The ratio of TE- to TM-polarized light intensities was around 0.7, irrespective of the incident azimuthal angle. This result is consistent with the enhancement of the TE-polarized light by buckles in the device structure shown in Fig. 29(c).

4. Summary

After summarizing methods for enhanced outcoupling in OLED devices, we reviewed enhancement methods using photonic structures at a surface. As explained in sections 2.3 and 2.4, corrugated EL devices with periodic or quasi-periodic nanostructures show enhanced light extraction performance compared with the reference flat device without such structures. The principle of light extraction is the same in both device configurations; however, periodically corrugated EL device shows higher light extraction only at a specific wavelength because of the well-defined corrugation pitch. This anisotropic angular dependence does not satisfy the Lambertian emission pattern, which is an important requirement in the lighting technologies. On the other hand, spontaneously formed buckling patterns on OLEDs towards air are effectively used as a quasi-periodic structure to extract light from the waveguide modes. The characteristics of the broad periodicity distribution and randomly oriented wave vectors of buckles provide an invaluable advantage of possible outcoupling of the waveguide light propagating along any direction with a wide spectral range. Namely a buckling device shows a Lambertian emission pattern with an increase in emission angle, which satisfies the requirement of OLED lighting. In particular, it enhances the outcoupling of waveguide modes at various wavelength ranges due to the broad distribution of the periodicity that can be applied to white OLEDs. We may conclude that the buckling device structure overcomes the limitation of periodic nanostructures and it can be used in the development of white OLEDs for lighting.

We have also demonstrated the generation of highly circularly polarized EL and its tunability using a liquid crystal phase retarder. A wide-band reflector has enabled us to obtain high ratio of brightness between R-CP-EL and L-CP-EL with the overall intensity ratio of about 10 and *g*-factor of about 1.6 over the whole emission band. Also, using a voltage dependent phase retarder, we have confirmed that there is no limitation for choosing emissive materials to obtain the tunable polarized EL light. We have also shown that the devices with buckles have double current and power efficiencies over the entire emission wavelengths and emission angles. This is achieved without any spectral changes even though ITO thickness is thin (40 nm) to outcouple the only SP (TM) mode which would otherwise be lost into the Al cathode layer of OLED devices. It is also found that the diffraction of the SP mode by buckles causes polarization conversion to the TE mode with a higher light intensity than in the TM mode, which occurs due to the random orientation of the buckling structure.

Author details

Soon Moon Jeong
Nano and Bio Research Division, Daegu Gyeongbuk Institute of Science and Technology, Sang-Ri, Hyeonpung-Myeon, Dalseong-Gun, Daegu, Republic of Korea

Hideo Takezoe*
Department of Organic and Polymeric Materials, Tokyo Institute of Technology, O-okayama, Meguro-ku, Tokyo, Japan

5. References

[1] Adachi. C, Baldo M. A, Thompson M. E, Forrest S. R. Nearly 100% Internal Phosphorescence Efficiency in an Organic Light-emitting Device. J. Appl. Phys. 2011;90: 5048-5051.
[2] Benisty H, Neve H. D, Weisbuch C. Impact of Planar Microcavity Effects on Light Extraction - Part I: Basic Concepts and Analytical Trends. IEEE J. Quantum Electron. 1998;34: 1612-1631.
[3] Benisty H, Neve H. D, Weisbuch C. Impact of Planar Microcavity Effects on Light Extraction - Part II: Selected Exact Simulations and Role of Photon Recycling. IEEE J. Quantum Electron. 1998;34: 1632-1643.
[4] Meerholz K, Muller D. C. Outsmarting Waveguide Losses in Thin-Film Light-Emitting Diodes. Adv. Funct. Mater. 2001;11: 251-253.
[5] Tsutsui T, Yahiro M, Yokogawa H, Kawano K, Yokoyama M. Doubling Coupling-Out Efficiency in Organic Light-Emitting Devices Using a Thin Silica Aerogel Layer. Adv. Mater. 2001;13: 1149-1152.
[6] Madigan C. F, Lu M. H, Sturm J. C. Improvement of Output Coupling Efficiency of Organic Light-emitting Diodes by Backside Substrate Modification. Appl. Phys. Lett. 2000;451: 1650-1652.
[7] A Chutinan, Ishihara K, Asano T, Fujita M, Noda S. Theoretical Analysis on Light-extraction Efficiency of Organic Light-emitting Diodes Using FDTD and Mode-expansion Methods. Org. Electronics. 2005;6: 3-9.
[8] Gu. G, Garbuzov D. Z, Burrows P. E, Venkatesh S, Forrest S. R, Thompson M. E. High-external-quantum Efficiency Organic Light-emitting Devices. Opt. Lett. 1997;22: 396-398.
[9] Greenham N. C, Friend R. H, Bradley D. D. C. Angular Dependence of the Emission from a Conjugated Polymer Light-Emitting Diode: Implications for Efficiency Calculations. Adv. Mater. 1994;6: 491-494.
[10] Windisch R., Heremans P., Knobloch A., Kiesel P., Dohler G. H., Dutta B., Borghs G. Light-emitting Diodes with 31% External Quantum Efficiency by Outcoupling of Lateral Waveguide Modes. Appl. Phys. Lett. 1999;74: 2256-2258.

* Corresponding Author

[11] Moller S, Forrest S. R. Improved Light Out-coupling in Organic Light Emitting Diodes Employing Ordered Microlens Arrays. J. Appl. Phys. 2002:91; 3324-3327.

[12] Sun J, Forrest S. R. Organic Light Emitting Devices with Enhanced Outcoupling via Microlenses Fabricated by Imprint Lithography. J. Appl. Phys. 2006;100: 073106.

[13] Yamasaki T., Sumioka K., and Tsutsui T. Organic Light-emitting Device with an Ordered Monolayer of Silica Microspheres as a Scattering Medium. Appl. Phys. Lett. 2000;76: 1243-1245.

[14] Jordan R. H., Rothberg L. J., Dodabalapur A., and Slusher R. E. Efficiency Enhancement of Microcavity Organic Light Emitting Diodes. Appl. Phys. Lett. 1996;69: 1997-1999.

[15] Dirr S, Wiese S, Johannes H, Kowalsky W. Organic Electro- and Photoluminescent Microcavity Devices. Adv. Mater. 1998;10: 167-171.

[16] Han S, Huang C, Lu Z. Color Tunable Metal-cavity Organic Light-emitting Diodes with Fullerene Layer. J. Appl. Phys. 2005;97: 093102.

[17] Lemmer U., Hennig R., Guss W., Ochse A., Pommerehne J., Sander R., Greiner A., Mahrt R. F., Bassler H., Feldmann J., Gobel E. O. Microcavity Effects in a Spin-coated Polymer Two-layer System. Appl. Phys. Lett. 1995;66: 1301-1303.

[18] Dodabalapur A., Rothberg L. J., Miller T. M., and Kwock E. W. Microcavity Effects in Organic Semiconductors. Appl. Phys. Lett. 1994;64: 2486-2488.

[19] Dodabalapur A, Rothberg L. J, Jordan R. H, Miller T. M, Slusher R. E, Phillips J. M. Physics and Applications of Organic Microcavity Light Emitting Diodes. J. Appl. Phys. 1996;80: 6954-6964.

[20] Tsutsui T., Takada N., Saito S., Ogino E. Appl. Phys. Lett. Sharply Directed Emission in Organic Electroluminescent Diodes with an Optical-microcavity Structure. 1994;65: 1868-1870.

[21] Matterson B. J, Matterson J, Lupton J. M, Safonov A. F, Salt M. G, Barnes W. L, Samuel I. D. W. Increased Efficiency and Controlled Light Output from a Microstructured Light-emitting Diode. Adv. Mater. 2001;13: 123-127.

[22] Lupton J. M., Matterson B. J., Samuel I. D. W., Jory M. J., Barnes W. L. Bragg Scattering from Periodically Microstructured Light Emitting Diodes. Appl. Phys. Lett. 2000;77: 3340-3342.

[23] Ziebarth J. M, Saafir A. K, Fan S, McGehee M. D. Extracting Light from Polymer Light-Emitting Diodes Using Stamped Bragg Gratings. Adv. Funct. Mater. 2004;14: 451-456.

[24] Ziebarth J. M, McGehee M. D. A Theoretical and Experimental Investigation of Light Extraction from Polymer Light-emitting Diodes. J. Appl. Phys. 2005;97: 064502.

[25] Hobson P. A, Wasey J. A. E, Sage I, Barnes W. L. The Role of Surface Plasmons in Organic Light-emitting Diodes. IEEE J. Sel. Top. Quantum Electron. 2002;8: 378-386.

[26] Ishihara K, Fujita M, Matsubara I, Asano T, Noda S, Ohata H, Hirasawa A. Nakada H., Shimoji N. Organic Light-emitting Diodes with Photonic Crystals on Glass Substrate Fabricated by Nanoimprint Lithography. Appl. Phys. Lett. 2007;90: 111114.

[27] Fujita M., Ueno T., Ishihara K., Asano T., Noda S., Ohata H., Tsuji T., Nakada H., and Shimoji N. Reduction of Operating Voltage in Organic Light-emitting Diode by Corrugated Photonic Crystal Structure. Appl. Phys. Lett. 2004;85: 5769-5771.

[28] Fujita M, Ishihara K, Ueno T, Asano T, Noda S, Ohata H, Tsuji T, Nakada H, Shimoji N. Optical and Electrical Characteristics of Organic Light-Emitting Diodes with Two-Dimensional Photonic Crystals in Organic/Electrode Layers. Jpn. J. Appl. Phys. 2005;44: 3669-3677.

[29] Schnitzer I., Yablonovitch E., Caneau C., Gmitter T. J., Scherer A. 30% External Quantum Efficiency from Surface Textured, Thin-film Light-emitting Diodes. Appl. Phys. Lett. 1993;63: 2174-2176.

[30] Lim J, Oh S. S, Kim D. Y, Cho S. H, Kim I. T, Han S. H, Takezoe H. Enhanced Out-coupling Factor of Microcavity Organic Light-emitting Devices with Irregular Microlens Array. Opt. Express 2006;14: 6564-6571.

[31] Schubert E. F., Wang Y. H., Cho A. Y., Tu L. W., Zydzik G. J. Resonant Cavity Light-emitting Diode. Appl. Phys. Lett. 1992;60: 921-923.

[32] Hunt N. E. J., Schubert E. F., Logan R. A., Zydzik G. J. Enhanced Spectral Power Density and Reduced Linewidth at 1.3 μm in an InGaAsP Quantum Well Resonant-cavity Light-emitting Diode. Appl. Phys. Lett. 1992;61: 2287-2289.

[33] Jeong S. M., Araoka F., Machida Y., Ishikawa K., Takezoe H., Nishimura S., Suzaki G. Enhancement of Normally Directed Light Outcoupling from Organic Light-emitting Diodes Using Nanoimprinted Low-refractive-index Layer. Appl. Phys. Lett. 2008;92: 083307.

[34] Jeong S. M., Ha N. Y., Araoka F., Ishikawa K., Takezoe H. Electrotunable Polarization of Surface-emitting Distributed Feedback Laser with Nematic Liquid Crystals. Appl. Phys. Lett. 2008;92: 171105.

[35] Jeong S. M, Araoka F, Machida Y, Takanishi Y, Ishikawa K, Takezoe H, Nishimura S, Suzaki G. Enhancement of Light Extraction from Organic Light-emitting Diodes with Two-Dimensional Hexagonally Nanoimprinted Periodic Structures Using Sequential Surface Relief Grating. Jpn. J Appl. Phys. 2008;47: 4566-4571.

[36] Koo W. H, Jeong S. M, Araoka F, Ishikawa K, Nishimura S, Toyooka T, Takezoe H. Light Extraction from Organic Light-emitting Diodes Enhanced by Spontaneously Formed Buckles. Nat. Photonics 2010;4: 222-226.

[37] Hubert C., Debuisschert C. F-., Hassiaoui I., Rocha L., Raimond P., Nunzi J.-M. Emission Properties of an Organic Light-emitting Diode Patterned by a Photoinduced Autostructuration Process. Appl. Phys. Lett. 2005;87: 191105.

[38] Lawrence J. R., Andrew P., Barnes W. L., Buck M., Turnbull G. A., Samuel I. D. W. Optical Properties of a Light-emitting Polymer Directly Patterned by Soft Lithography. Appl. Phys. Lett. 2002;81: 1955-1957.

[39] Jeong S. M, Takanishi Y, Ishikawa K, Nishimura S, Suzaki G, Takezoe H. Sharply Directed Emission in Microcavity Organic Light-emitting Diodes with a Cholesteric Liquid Crystal Film. Opt. Comm. 2007;273: 167-172.

[40] Burrows P. E, Shen Z, Bulovic V, McCarty D. M, Forrest S. R. Relationship Between Electroluminescence and Current Transport in Organic Heterojunction Light-emitting Devices. J. Appl. Phys. 1996;79: 7991-8006.

[41] Kalinowski J., Electroluminescence in organics. J. Phys. D 1999;32: R179-R250.

[42] Hobson P. A, Wedge S, Wasey J. A. E, Sage I, Barnes W. L. Surface Plasmon Mediated Emission from Organic Light-Emitting Diodes. Adv. Mater. 2002;14: 1393-1396.
[43] Bowden N, Brittain S, Evans A. G, Hutchinson J. W, Whitesides G. M. Spontaneous Formation of Ordered Structures in Thin Films of Metals Supported on an Elastomeric Polymer. Nature 1998;393: 146-149.
[44] Okayasu T, Zhang H. L, Bucknall D. G, Briggs G. A. Spontaneous Formation of Ordered Lateral Patterns in Polymer Thin-film Structures. Adv. Func. Mater. 2004;14: 1081-1088.
[45] Huck W. T. S, Bowden N, Onck P, Pardoen T, Hutchinson J. W, Whitesides G. M. Ordering of Spontaneously Formed Buckles on Planar Surfaces. Langmuir 2000;16: 3497-3501.
[46] Allen H. G., Analysis and Design of Structural Sandwich Panels (Pergamon, 1969).
[47] Cerda E, Mahadevan L. Geometry and Physics of Wrinkling. Phys. Rev. Lett. 2003;90: 074302.
[48] Bowden N., Huck W. T. S., Paul K. E., Whitesides G. M. The Controlled Formation of Ordered, Sinusoidal Structures by Plasma Oxidation of an Elastomeric Polymer. Appl. Phys. Lett. 1999;75: 2557-2559.
[49] Kalinowski J, Palilis L. C, Kim W. H, Kafafi Z. H. Determination of the Width of the Carrier Recombination Zone in Organic Light-emitting Diodes. J. Appl. Phys. 2003;94: 7764-7767.
[50] Tang C. W, VanSlyke S. A, Chen C. H. J. Electroluminescence of Doped Organic Thin Films. J. Appl. Phys. 1989;65: 3610-3616.
[51] Mochizuki H, Hasui T, Shiono T, Ikeda T, Adachi C, Taniguchi Y, Shirota Y. Emission Behavior of Molecularly Doped Electroluminescent Device Using Liquid-crystalline Matrix. Appl. Phys. Lett. 2000;77: 1587-1589.
[52] Furumi S, Sakka Y. Chiroptical Properties Induced in Chiral Photonic-Bandgap Liquid Crystals Leading to a Highly Efficient Laser-Feedback Effect. Adv. Mater. 2006;18: 775-780.
[53] O'Neill M, Kelly S. M. Liquid Crystals for Charge Transport, Luminescence, and Photonics. Adv. Mater. 2003;15: 1135-1146.
[54] Grell M, Knoll W, Lupo D, Meisel A, Miteva T, Neher D, Nothofer H, Scherf U, Yasuda A. Blue Polarized Electroluminescence from a Liquid Crystalline Polyfluorene. Adv. Mater. 1999;11: 671-675.
[55] Grell M, Bradley D. D. C. Polarized Luminescence from Oriented Molecular Materials. Adv. Mater. 1999;11: 895-905.
[56] Peeters E, Christians M. P. T, Janssen R.A. J, Schoo H. F. M, Dekkers H. P. J. M, Meijer E. W. Circularly Polarized electroluminescence from a Polymer Light-emitting Diode. J. Am. Chem. Soc. 1997;119: 9909-9910.
[57] Oda M, Nothofer H, Lieser G, Scherf U, Meskers S. C. J, Neher D. Circularly Polarized Electroluminescence from Liquid-Crystalline Chiral Polyfluorenes. Adv. Mater. 2000;12: 362-365.
[58] Grell M., Oda M., Whitehead K. S., Asimakis A, Neher D, Bradley D. D. C. A Compact Device for the Efficient, Electrically Driven Generation of Highly Circularly Polarized Light. Adv. Mater. 2001;13: 577-580.

[59] Belayev S. V., Schadt M., Barnik M. I., Funfschilling J., Malimoneko N. V., Schmitt K., Large Aperture Polarized Light Source and Novel Liquid Crystal Display Operating Modes. Jpn. J. Appl. Phys. 1990;29: L634-L637.

[60] Woon K. L, O'Neill M, Richards G. J, Aldred M. P, Kelly S. M, Fox A. M. Highly Circularly Polarized Photoluminescence over a Broad Spectral Range from a Calamitic, Hole-transporting, Chiral Nematic Glass and from an Indirectly Excited Dye. Adv. Mater. 2003;15: 1555-1558.

[61] Geng Y, Trajkovska A, S. Culligan W, Ou J. J, Chen H. M. P, Katsis D, Chen S. H. Origin of Strong Chiroptical Activities in Films of Nonafluorenes with a Varying Extent of Pendant Chirality. J. Am. Chem. Soc. 2003;125: 14032-14038.

[62] Jeong S. M., Ohtsuka Y., Ha N. Y., Takanishi Y., Ishikawa K., Takezoe H. Highly Circularly Polarized Electroluminescence from Organic Light-emitting Diodes with Wide-band Reflective Polymeric Cholesteric Liquid Crystal Films. Appl. Phys. Lett. 2007;90: 211106.

[63] Jeong S. M, Ha N. Y, Takezoe H, Nishimura S, Suzaki G. Polarization-tunable Electroluminescence Using Phase Retardation Based on Photonic Bandgap Liquid Crystal. J. Appl. Phys. 2008;103: 113101.

[64] Hwang J, Song M. H, Park B, Nishimura S, Toyooka T, Wu J. W, Takanishi Y, Ishikawa K, Takezoe H. Electro-tunable Optical Diode Based on Photonic Bandgap Liquid-crystal Heterojunctions. Nat. Mater. 2005;4: 383-387.

[65] Song M. H, Park B, Nishimura S, Toyooka T, Chung I. J, Takanishi Y, Ishikawa K, Takezoe H. Electrotunable Non-reciprocal Laser Emission from a Liquid-Crystal Photonic Device. Adv. Funct. Mater. 2006;16: 1793-1798.

[66] Song M. H, Park B, Shin K.-C, Ohta T, Tsunoda Y, Hoshi H, Takanishi Y, Ishikawa K, Watanabe J, Nishimura S, Toyooka T, Zhu Z, Swager T. M, Takezoe H. Effect of Phase Retardation on Defect-Mode Lasing in Polymeric Cholesteric Liquid Crystals. Adv. Mater. 2004;16: 779-783.

[67] Broer D. J, Mol G. N. Wide-band Reflective Polarizers from Cholesteric Polymer Networks with a Pitch Gradient. Nature 1995;378: 467-469.

[68] Koo W. H, Jeong S. M, Nishimura S, Araoka F, Ishikawa K, Toyooka T, Takezoe H. Polarization Conversion in Surface-Plasmon-Coupled Emission from Organic Light-Emitting Diodes Using Spontaneously Formed Buckles. Adv. Mater. 2011;23: 1003-1007.

[69] Inagaki T, Motosuga M, Yamamori K. Photo-acoustic Study of Plasmon Resonance-absorption in a Diffraction Grating. Phys. Rev. B. 1983;28: 1740-1744.

[70] Inagaki T, Goudonnet J. P, Arakawa E. T. Plasma Resonance Absorption in Conical Diffraction: Effects of Groove Depth. J. Opt. Soc. Am. B 1986;3: 992-995.

[71] Elston S. J, Bryan-Brown G. P, Sambles J. R. Polarization Conversion from Diffraction Gratings. Phys. Rev. B. 1991;44: 6393-6400.

[72] Bristow A. D., Astratov V. N., Shimada R., Culshaw I. S., Skolnick M. S., Whittaker D. M., Tahraoui A., Krauss T. F., Polarization Conversion in the Reflectivity Properties of Photonic Crystal Waveguides. IEEE J. Quantum. Elect. 2002;38: 880-884.

[73] Suyama T, Okuno Y, Matsuda T. Enhancement of TM-TE mode Conversion Caused by Excitation of Surface Plasmons on a Metal Grating and its Application for Refractive Index Measurement. Prog. Electromagn. Res. 2007;72: 91-103.

[74] Feng J, Okamoto T, Kawata S. Enhancement of Electroluminescence Through a Two-dimensional Corrugated Metal Film by Grating-induced Surface-plasmon Cross Coupling. Opt. Lett. 2005;30: 2302-2304.

[75] Reinke N. A, Ackermann C, Brütting W. Light Extraction via Leaky Modes in Organic Light Emitting Devices. Opt. Commun. 2006;266: 191-197.

Chapter 5

Photophysical Properties of Two New Donor-Acceptor Conjugated Copolymers and Their Model Compounds: Applications in Polymer Light Emitting Diodes (PLEDs) and Polymer Photovoltaic Cells (PPCs)

S. Ayachi, A. Mabrouk, M. Bouachrine and K. Alimi

Additional information is available at the end of the chapter

http://dx.doi.org/10.5772/54254

1. Introduction

Organic semiconductors hold the combined properties of inorganic semiconductors such as silicon and more desirable properties of plastics [1,2]. Since, the inception of the field of plastic electronics, various organic semiconductors including conjugated polymers and small molecules have been synthesized, studied, and applied to optoelectronic semiconductor device structures in order to improve efficiency, reduce cost or realize new applications that are difficult to achieve with silicon-based technology [3,4].

Recently, the exploitation of polymer as an active layer in organic electronic displays has received a particular attention. In this direction, greater efforts have been devoted to seek new possibilities for use in optoelectronic devices such as Polymer Light Emitting Diodes (PLEDs) [5-11], Polymer Photovoltaic Cells (PPCs) [12-23] and Polymer Field Effect Transistors (PFET) [24-33]. The field of PLEDs is still an active research area since the first conjugated conducting or semiconducting polymeric material, poly(p-phenylene-vinylene) (PPV), was reported by Burroughes et al. in 1990 [34]. In fact, only polymers can enable manufacturing of large-area light-emitting displays. These electronic devices need special polymers with specific and adapted properties. Since then, there have been increasing interests and research activities in synthesis and design of new polymeric materials for organic electronic devices. However, their properties and those of the related devices are still poorly understood.

One of the requirements for efficient PLEDs is balanced charge injection from the two electrodes and efficient transport of both holes and electrons within luminescent layer in the device structure [35]. More recently, much effort has been devoted to develop wide band gap conjugated polymers for application in light emitting diodes. Then, a number of conjugated polymers including poly(p-phenylene-vinylene) (PPV) [36,37], poly(p-phenylene) (PPP) [38-41], polythiophene (PT) [42,43] and polyfluorene (PF) [44] have been widely used as light-emitting materials in devices. However, one major problem with these polymers is that they are π-excessive in nature and hence are much better at accepting and transporting holes than electrons. Another series of polymers containing π-deficient heterocycles like pyridine [45] and oxadiazoles [46] show greater tendency to transport electrons than holes [47].

To tune the emission properties of PLEDs, sophisticated control of the polymer luminescence color, efficiency, and charge transport properties are required. The emission wavelength depends on the extent of conjugation/delocalization, and can be controlled by the modification of the configuration or conformation of the polymer and by interactions with the local environment [48,49]. This can be achieved by grafting functional moieties such as electron donor or acceptor groups, which allow the modulation of the electronic structure of the conjugated backbone [50,51]. Donor–acceptor (D–A) organic molecules are among the most important conjugated polymers, that produce low bad gap useful in technological fields novel materials, by adjusting the HOMO and LUMO levels [52-54]. The low optical band gaps of the compounds should result by alternating the electron-rich unit of donor segments and the strong electron-deficient unit of acceptor segments in the structure. Then, superior transport properties in organic materials can be achieved with planar and highly conjugated chains [55-57]. Many investigations have proven that conjugated D-A type polymers play important roles in their balanced charge transporting properties and show unique optical properties. The HOMO and LUMO energy levels of these systems are important for understanding charge injection processes in the luminescent devices [58-60].

On the other hand, due to their interesting electrical, optical and optoelectronic properties, conjugated oligomers represent a prominent class of compounds from the viewpoint of theory, synthesis, and applications in materials science [61-65]. Moreover, they are model compounds for the corresponding polymers [66,67]. In parallel to recent experimental work on oligomers, theoretical efforts have also begun complementing the experimental studies in the characterization of the nature and the properties of their ground- and lowest electronic excited states [68-73]. In addition, these approaches have provided significant insight into the electronic and optical properties of conjugated polymers. In the absence of structural information, the experimental measurement, in conjunction with molecular orbital theory, is a valuable tool in analyzing the electronic structure of polymers. This enables an estimate not only of the relative energies of the electronic levels but also of their detailed distribution over the whole molecule. The ionization Potential (IP), electron affinity (EA), molecular electronic structure of the ground and lowest excited states as well as the nature of absorption and photoluminescence obtained through quantum calculations are of great interest prior to fabricating organic devices.

In this context, two new alternating donor-acceptor conjugated copolymers, both of which may be used in organic electronics, are investigated here. The first one is a copolymer containing thienylene-dioctyloxyphenyle-thienylene (TBT) and bipyridine (BIPY) units as shown in Fig. 1 that can be used as an active layer in PLEDs. It is constructed with dioctyloxy substituted phenylene incorporated between two electron-rich-thiophene units, abbreviated as TBT unit, and a bipyridine (BIPY) unit (Fig. 1). It was obtained by the Stille reaction method and the detailed synthesis procedures and characterization have already been reported [74,75]. The soluble copolymer has a well-defined structure and exhibits excellent optical properties. The number average (Mn) and weight average (Mw) molecular weights of the copolymer, determined by gel permeation chromatography (GPC) using polystyrene as standard, are obtained as 3098 and 3477, respectively. The corresponding polymerization degree, DPn, is found to be 5 corresponding to 25 cycles of number. Photophysical properties of copolymer including Raman scattering, UV-Visible optical absorption and emission are studied.

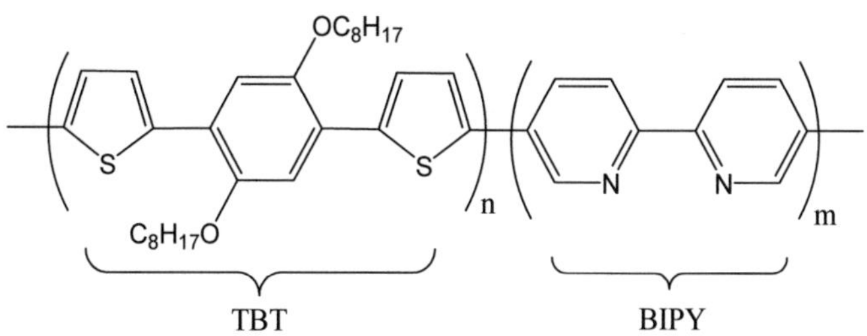

Figure 1. Chemical structure of TBT-BIPY copolymer.

Introducing long alkoxy pendants at 2 and 5 positions of the phenyl ring improves the solvent processability which is a prerequisite for fabricating organic light-emitting diodes (OLEDs) by the spin coating method.

The second part of this chapter concerns a composite based on Benzothiadiazole mixed with carbazole, or hexylthiophene that can be used for fabricating Polymer Solar Cells (PSCs). PSCs based on the bulk heterojunction (BHJ) structure have attracted broad attentions in recent years [76,77]. The requirements for the structure and properties of polymeric donors are low band gap, broad absorption range, high mobility and appropriate HOMO and LUMO levels [78]. Among the polymers tested for suitability as an active layer, poly(3-hexyl-thiophene) (P3HT) and poly(carbazole) (PCz) have emerged as promising candidates for applications in optoelectronic devices because of their exceptional properties [79,80]. However, alternative copolymers of [2,1,3]-benzothiadiazole (BT) acceptor units with various donor units have attracted particular attention for using them in high performance PSCs [81-83]. To optimize the material properties, conjugated polymers with alternating electron-rich and hole-rich units along their backbone have been extensively developed because their absorption spectra and band gap can be readily tuned by controlling the intra-molecular charge transfer (ICT) from donors to acceptors [84].

However, in these linear D-A polymers, the molecular interactions and packing orientation of the conjugating moieties need to be carefully controlled to ensure proper process ability and charge transporting properties [85]. A fundamental understanding of the ultimate relations between structure and properties of these materials is necessary for using them in photovoltaic cells. A number of studies demonstrate that the interplay between theory and experiment is very important in providing useful insights in understanding the molecular electronic structure of the ground and excited states as well as the nature of absorption and photoluminescence [86]. To rationalize our theoretical results, the simulated data are compared with the available experimental data [87].

In what follows, we elucidate the photophysical properties of the benzothiadiazole derivative compounds with structures as shown in Fig. 2 (a,b). These two D-A polymers provide a basis for a more comprehensive study of the backbone ring, heteroatom and fused ring effects on polymer properties. Therefore, it is of practical significance to extend our previous work to a comprehensive theoretical investigation on these two types of BTD-based derivatives. Moreover, poly(3-hexyl-thiophene) (P3HT) units have relative higher charge mobility in comparison with other conjugated polymers and have been widely used as π-conjugating spacers [88,89]. Its insertion in the polymer backbone serves the dual purpose of transporting carriers and providing sites for exciton dissociation [90]. Moreover, the incorporation of electron-withdrawing moieties (3HT) as side chains leads to some useful properties which can further widen the absorption spectrum.

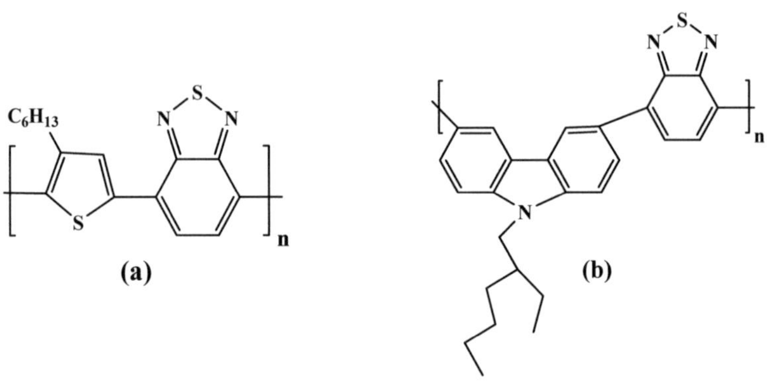

Figure 2. Chemical structure of compounds under study: (a): P3HTBT, (b): PCzBT.

Recently, the conjugated P3HT2BTCz compound, built as carbazole-thiophene-benzothiadiazole, has been copolymerized onto the backbone of the copolymer as shown in Fig. 3. This compound has been synthesized and experimentally characterized, using only photoluminescence and optical absorption spectroscopy. Their related intense and broad absorption bands as well as favorable excited-state energy levels make them good candidates for fabricating PSCs. Thus, if P3HT2BTCz compound is blended with [6,6]-phenyl-C_{61}-bytric acid methyl ester (PCBM) fullerene derivative into BHJ photovoltaic devices [87], then the conversion efficiency may be increased.

Here further investigations of geometrical parameters, electronic structures, photo-physical and vibrational properties of these compounds are carried out, on the basis of quantum-chemical calculations, providing a reasonable interpretation of the experimental results and better understanding of the relationship between the structure and resulting properties. Finally, the parameters that influence the photovoltaic efficiency are elucidated. We think that the presented study of structural, electronic, optical, and charge transfer properties for this compound will help the design more efficient functional photovoltaic copolymers.

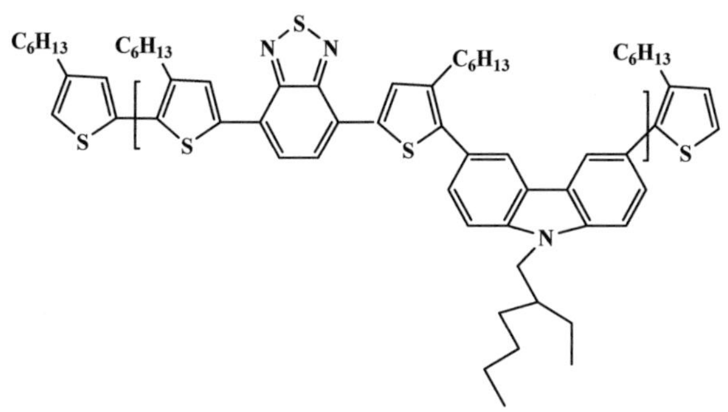

Figure 3. Chemical structure of P3HT2BTCz compound.

The objective of the presented result here is not to develop or optimize any applications, but to understand why and how the combined theoretical and experimental studies on copolymers can be conducted in developing optimized Polymer Light Emitting Diodes (PLEDs) and Photovoltaic Cells (PPCs).

2. Theoretical methodology

All molecular calculations are performed in the gas phase using Density Functional Theory (DFT) implemented in the GAUSSIAN (03) program [91]. We have used the B3LYP (Becke three-parameter Lee-Yang-Parr) exchange correlation functional [92,93] with 3-21G* and 6-31G* as basis sets. In the first part, the calculation of conformational characteristics has been done by varying the torsion angle in steps of 20° from $\theta = 0°$ to $\theta = 180°$. For each increment, the dihedral angle is held fixed while the remainder of the molecule is optimized. The energy differences in electronic states are always calculated relative to the corresponding absolute minimum conformation and then the relative potential energy surfaces are drawn.

In the optimization procedure of these compounds, the alkyl chains at the N-9 positions of carbazole (Cz) motifs and dioctyloxy groups in TBT-BIPY copolymer are replaced by methyl and methoxy groups, respectively. This has been proven that the presence of alkyl/alkoxy groups does not significantly affect the equilibrium geometry and hence the electronic and the optical properties [94]. Hexyl groups in 3HT motifs are then replaced by methyl groups. The optimization of the composite (P3HT2BTCz: PCBM) is done in two steps. First optimization with PM3 semi-empirical method was carried out, then the resulting structure was re-optimized by DFT/B3LYP/3-21G* to find the equilibrium geometrical structure.

Optical absorption spectra are calculated using the Time-Dependant Density Functional Theory (TDDFT) [95] based on optimized ground state geometries [96]. Theoretically the transition energies and their respective intensities in a given configuration interaction (CI) expansion of singly excited determinants are determined [97]. The electronic configurations for the lowest 50 singlet-singlet transitions are obtained using the same basis set. Then, the obtained data are transformed using the SWizard program [98] into simulated spectra as described in the literature [99]. Finally, the nature and the energy of vertical electron transitions (the main singlet-singlet electron transitions with highest oscillator strengths) of molecular orbital wave functions are presented. The photoluminescence (PL) spectrum has been derived from CIS/TDDFT calculation [100]. Similar procedures are applied on TBT-BIPY model compound on the basis of ground and lowest singlet excited-states, but with two additional methods (CIS/3-21G* and the semi-empirical quantum-chemical ZINDO levels) for absorption and emission properties [101]. The vibrational properties as well as force constants are also examined through results derived from the Molecular Orbital Package (MOPAC 2000) [102].

3. Part I: TBT-BIPY copolymer for Light Emitting Diodes (PLEDs)

3.1. Raman scattering spectroscopy

The Raman spectrum recorded for the excitation line of 1064 nm is presented in Fig. 4a. We have found that the Raman spectrum is dominated by bands originating from the thiophene, the di-alkoxy-substituted phenylene and pyridine rings vibration. According to the literature [103,104], the major band in the spectrum can be attributed to C_α=C_β stretching vibration of the thienyl ring at roughly 1444 cm^{-1} and the relatively weaker band at about 1604 cm^{-1} can be assigned to the C=C stretching vibration of the phenylene ring. The 1302 cm^{-1} can be attributed to the interring $C_{thienyl}$-C_{phenyl} vibration. In addition, we notice a strong asymmetry in intensity of the dominant triplet, occurring at 1444, 1543 and 1604 cm^{-1}, resulting from the short conjugation length of the material [105].

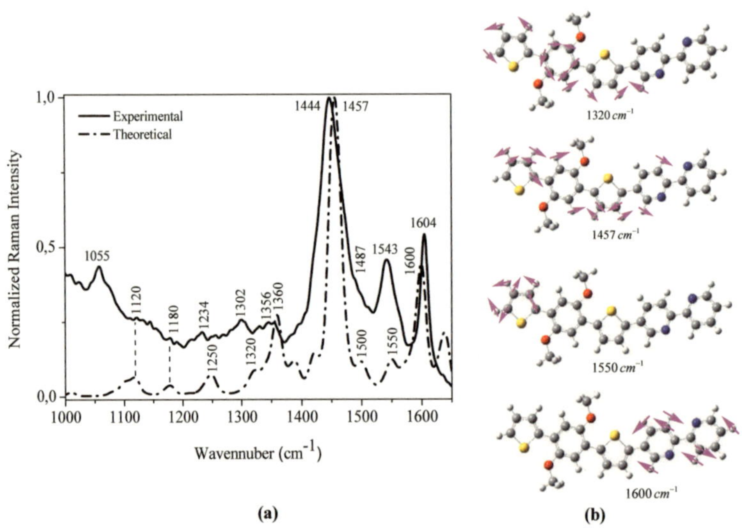

Figure 4. (a) Experimental and theoretical normalized Raman spectra and (b) Selected Raman vibrational modes of the calculated frequencies of TBT-BIPY copolymer.

3.2. Optical absorption and emission properties

The optical properties of the copolymer were studied, in chloroform solution and recorded at ambient temperature, by using UV-Vis and fluorescence emission spectroscopies (Fig. 5). The TBT-BIPY solution showed a sharp peak absorption maximum at 436 nm corresponding to the $\pi\rightarrow\pi^*$ electronic transition in the polymer backbone. This band appears at 517 nm for the polymer film. Obviously, the red shift of about 79 nm in the film state is due to the π-π^* stacking effect [106]. The optical band gap, defined by the onset absorption of the polymer in the chloroform solution state is 2.43 eV. The polymer showed low band gap when compared to that of TBT-BIPH (2.48 eV). This may be due to the strong interaction between electron acceptor (TBT) and strong electron acceptor segments (BIPY) in the polymer backbone. Then, this optical band gap of the copolymer could be attributed to the D-A structure of polymer matrix.

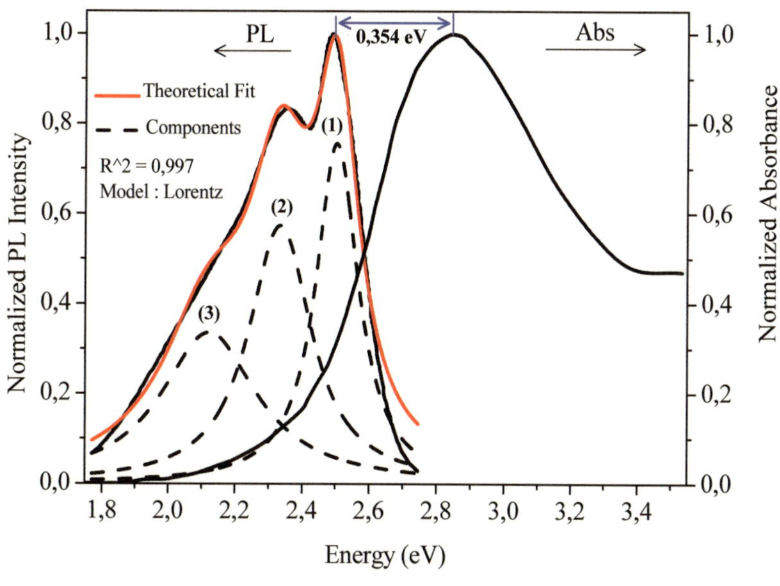

Figure 5. Normalized optical absorption and photoluminescence spectra of TBT-BIPY copolymer. The PL deconvolution spectrum was given in the same figure.

Fig. 5 includes also the fluorescence spectrum of copolymer that gives a bright blue-greenish fluorescence with the maximum emission wavelength of 498 nm with the excitation wavelength at 450 nm in chloroform solution state. This emission is corresponding to the onset of $\pi\rightarrow\pi^*$ transition of the electronic absorption spectra. The band gap of the polymer (2.43 eV) estimated from the onset position of the absorption (510 nm) essentially agrees with the λ_{max} value (498 nm, 2.48 eV) of the main fluorescence peak, indicating that the fluorescence takes places by migration of electrons in the conducting band to the valence band. It is worthy to note that the PL spectrum of the compound shows well-resolved structural features with maxima at 498, 527 and shoulder at about 580 nm assigned to the 0–0, 0–1, and 0–2 intra-chain singlet transition, respectively (the 0–0 transition, the most intense) [107]. The stocks shift was found to be 62 nm (0.35 eV). This shift points to large structural differences between the ground and excited states in the material. In addition, from PL deconvolution spectrum, it should be noted that the energy difference (~ 0.18 eV) agrees well with that of the most intense Raman vibration modes at around 1450 cm^{-1}.

3.3. Theoretical part

3.3.1. *Conformational analysis*

In the absence of structural information, we have assumed that the oligomer tends to be planar because of two reasons: (i) interchain interactions (packing force) tend to significantly reduce the torsion angles between adjacent units in the solid state and (ii) electronic and optical properties are weakly affected by small changes in torsional angels. To determine the minimum energy configuration, we perform fully geometrical optimizations on TBT-BIPY with B3LYP/3-21G*. Since there is only one type of substitution on the phenyl ring (substitution 2 is equivalent to the site 5), three different conformation types can occur in TBT-BIPY copolymer structure. The potential energy surface (PES) of copolymer is obtained by partial optimization as shown in Fig. 6.

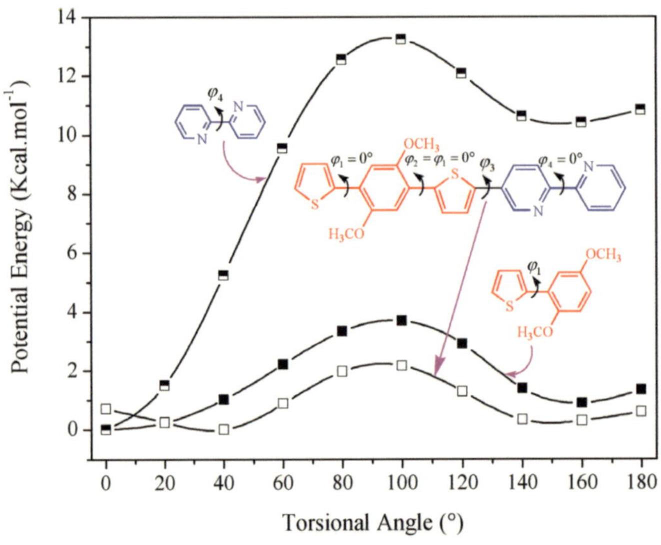

Figure 6. Potential energy curves of thienylene-2,5-di-methoxy-phenylene, BIPY and TBT-BIPY obtained from DFT/B3LYP/3-21G* level of theory.

As these structures show flexibility in the molecule, first of all, individual torsion potentials for the two structures of thiophene-di-methoxy-phenylene (TDMP) and bipyridine (BIPY) are obtained for each molecule as a function of the inter-ring C-C dihedral angle ϕ_1 (torsional angle between the thiophene and di-methoxy-phenylene rings) and ϕ_4 (torsional angle between the two pyridine rings) by varying them from 0° (syn-planar) to 180° (anti-planar) in steps of 20°. Therefore, to construct the potential energy curve for TBT-BIPY copolymer, ϕ_1 and ϕ_4 are held fixed and the torsional angle ϕ_3 (dihedral angle between the thiophene and pyridine rings) is calculated in the same way by varying the torsional angles (ϕ_1 and ϕ_4) as described above. From the conformational analysis of TDMP and BIPY, it is found that both show a minimum at the torsional angle around 0°, and they adopt co-planar conformations. However, when BIPY is connected to TBT unit, molecules get twisted out of the planarity with an angle $\phi 3 = 40°$. Accordingly, all the inter-ring dihedral angles are kept constant at $\phi_1 = \phi_2 = \phi_4 = 0°$ and $\phi_3 = 40°$ during the geometry optimizations.

3.3.2. *Ground- and excited-state structures*

The optimized structure of TBT-BIPY optimized using DFT//B3LYP/3-21G* is shown Fig. 7. The selected bond lengths and twist angles are collected in Table 1.

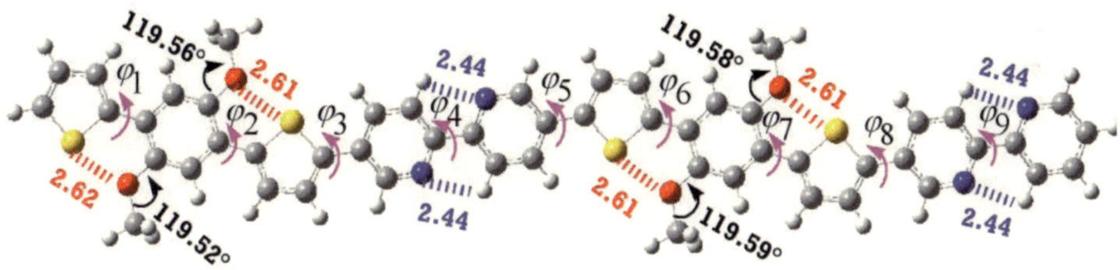

Figure 7. Ground state B3LYP/3-21G* optimized structure of 2-TBT-BIPY copolymer. The values written in red (blue) color represent S--O (N--H) distances in Angstrom. φ_n (n=1-9) represents the dihedral angle between rings and the values expressed in degree are the C-O-C angles.

As shown in Table 1, the two TBT and bipyrdine units of 2-TBT-BIPY adopt planar conformations with dihedral angles inferior to 1°. Whereas, the dihedral angles φ_3, φ_5 and φ_8 for 2-TBT-BIPY are twisted out of plane of ~24°. In addition, the C-O-C angles are not affected along the polymer chains and are evaluated to be 119.5°.

Dihedral Angle (°)	Ground State	Excited State
φ_1	-0.28	-0.21
φ_2	-0.39	-0.31
φ_3	23.61	20.31
φ_4	0.007	-0.72
φ_5	-24.52	-2.42
φ_6	-0.20	-0.34
φ_7	-0.55	-0.41
φ_8	23.86	1.85
φ_9	-0.088	-0.091

Table 1. Calculated dihedral angles in their ground- and excited-states of 2-TBT-BIPY copolymer.

It is worth noting that the interaction forces between the oxygen atom (negatively charged) and the sulfur atom (positively charged) in the TBT unit are attractive [108-110]. Similar results are found for the Bipyridine unit; in which intra-molecular interaction occurs between non-bonded nitrogen and hydrogen atoms (the atomic charges are listed in Table 2 referred to the individual atoms in the numbering sequence shown Fig. 8). In fact, the calculated bond lengths of S--O (N--H) bonds are found to be ~2.62 Å (~2.44 Å), which correspond to ~79% (~92%) of the sum of their Van der Waals radii, fall inside the Van der Waals contact distance of the S--O (3.32 Å) and N--H (2.64 Å) and outside of their covalent contacts of 1.70 Å for S-O and 0.91 Å for N-H. In this case, the planar conformations are stabilized by the non-bonded S--O and N--H interactions [111].

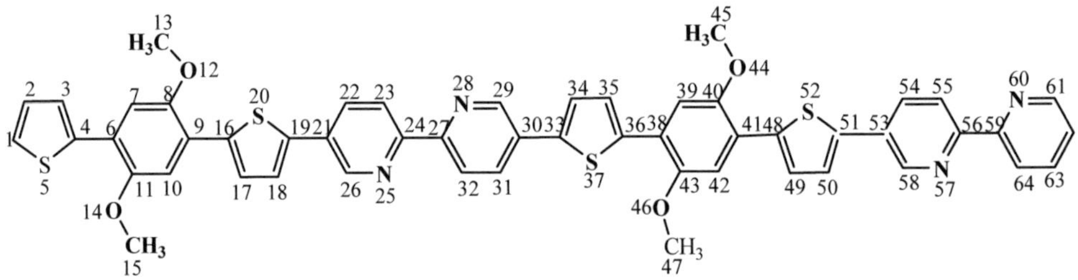

Figure 8. 2-TBT-BIPY copolymer structure with individual atoms in the numbering sequence.

	Atomic charges (e)	
Atoms	Ground State	Excited State
S_5/O_{14}	0,459/-0,565	0,513/-0,757
S_{20}/O_{12}	0,492/-0,566	0,560/-0,759
N_{25}/H	-0,619/0,220	-0,764/0,287
N_{28}/H	-0,619/0,219	-0,767/0,287
S_{37}/O_{46}	0,494/-0,565	0,568/-0,760
S_{52}/O_{44}	0,494/-0,565	0,569/-0,760
N_{57}/H	-0,618/0,220	-0,767/0,286
N_{60}/H	-0,605/0,221	-0,753/0,288

Table 2. Atomic charges of sulfur, oxygen, nitrogen and hydrogen atoms in S--O and N--H intra-molecular interactions.

Further, the highest occupied molecular orbital (HOMO), lowest unoccupied molecular orbital (LUMO) as well as the HOMO-LUMO energy gap ($\Delta_{LUMO-HOMO}$) are studied. Accordingly, for 2-TBT-BIPY, the HOMO is at -4.922 eV, LUMO at -2,152 eV and the energy difference between these levels is thus 2.77 eV. To further understand the optical property changes, Fig.9 illustrates the three highest occupied and three lowest unoccupied orbital levels for the 2-TBT-BIPY copolymer.

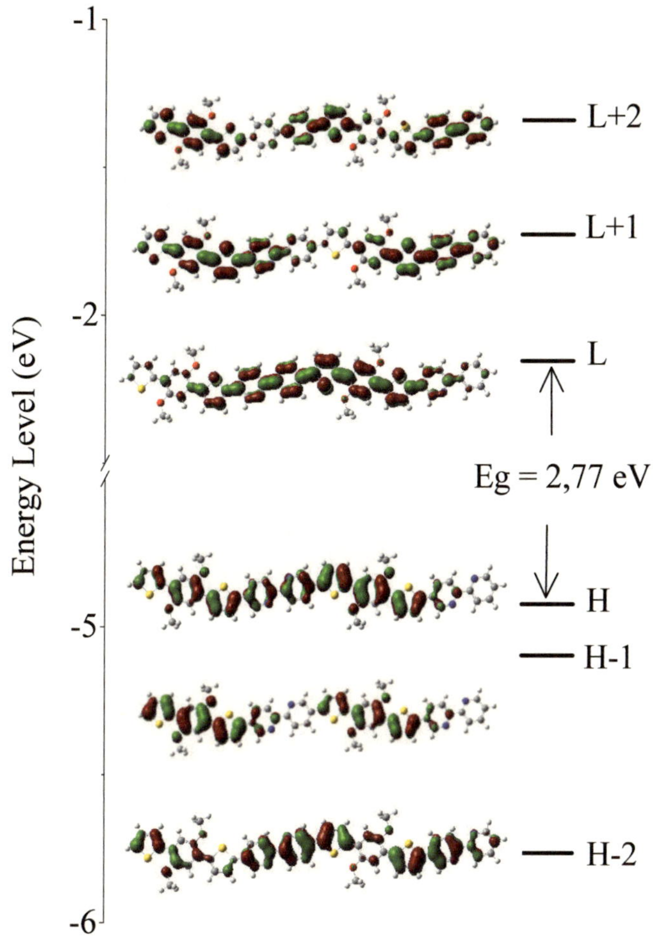

Figure 9. The DFT//B3LYP/3-21G* calculated energy levels for 2-TBT-BIPY copolymer.

The vibrational Raman frequencies are calculated using the same method on geometry-optimized structure and are directly compared to those obtained from the Raman spectroscopy measurements. In Fig. 4a, we have plotted the normalized theoretical and experimental Raman spectra of the TBT-BIPY copolymer compound. It is relevant to note here that the vibrational spectrum calculated by DFT methodology agree satisfactorily with the experimental spectrum both in relative intensities and peak positions. The deviation between the measured Raman scattering and theoretically vibrational frequencies are less than 30 cm^{-1}. Moreover, it was found that there were no negative vibrational frequencies, which indicate that optimized structure was at the energy minimum. This implies that the theoretically determined structure of copolymer is the most accurate description of the electronic structure. Accordingly, the experimental and calculated Raman bands at 1444 and 1457 cm^{-1}, respectively, assigned to the thiophene ring vibrations [103-104], are strongly resonant with the $\pi \rightarrow \pi^*$ electronic transition of compound. The most important Raman vibrational modes are shown in Fig.4b.

By combining the experimental data (optical band gap and Raman frequencies) with DFT calculations, two units of TBT-BIPY copolymer were considered as model structure for predicting the optical and emission properties.

For better understanding of the optical and emission processes, we have firstly computed the bond lengths of the ground and excited states, where the changes of bond lengths can be compared. The values of bond lengths for the 2-TBT-BIPY copolymer, in their ground- and excited-states are shown in Fig. 10. It can be seen that some bond lengths increase and some decrease in the excited state. Furthermore, we find that all the bond lengths of two bipyridine as well as those of C-O-C are shortened. Whereas, in the left TBT unit, the C-C single bond of thiophene rings as well as that connecting the thiophene ring to phenylene and bipyridine rings increase. In addition, in the second TBT unit, double bonds of thiophene rings and single/double bonds of substituted phenylene rings also increase.

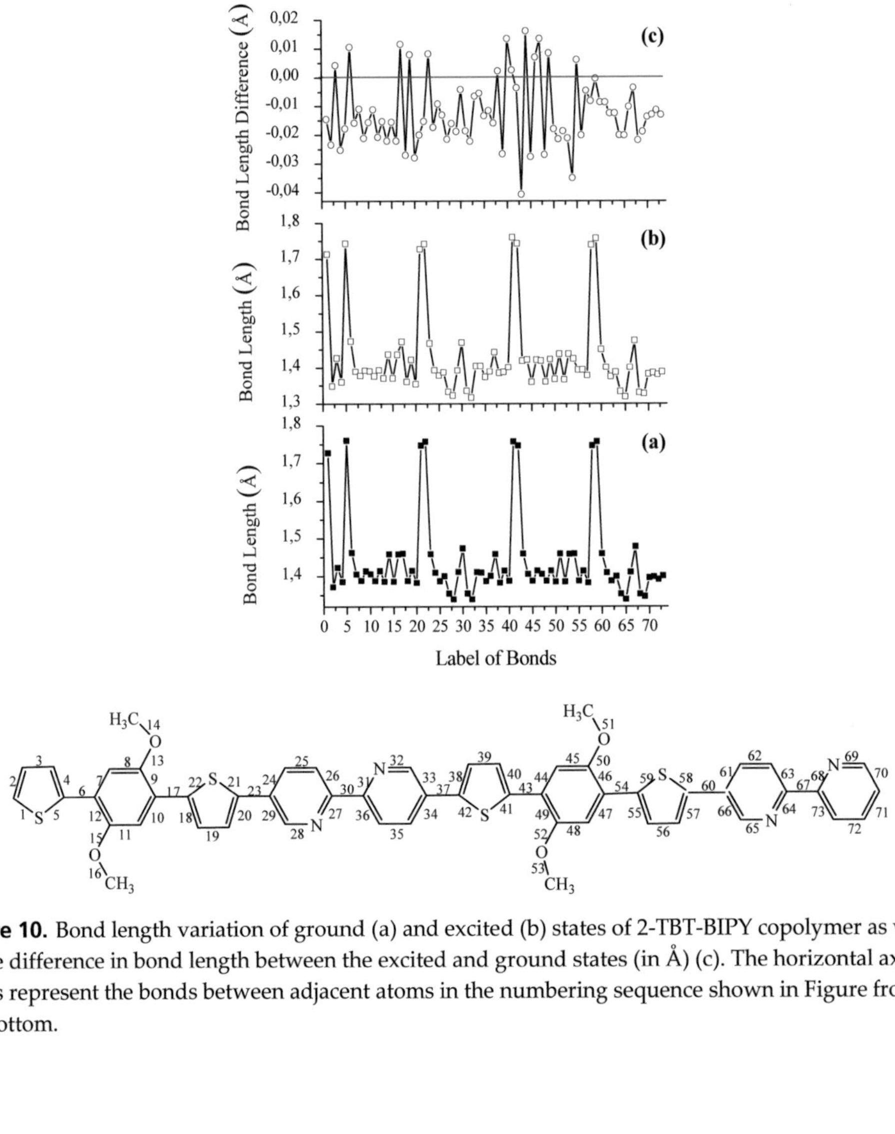

Figure 10. Bond length variation of ground (a) and excited (b) states of 2-TBT-BIPY copolymer as well as the difference in bond length between the excited and ground states (in Å) (c). The horizontal axis labels represent the bonds between adjacent atoms in the numbering sequence shown in Figure from the bottom.

On the other hand and whatever the state is, the non-bonded S--O and N--H contacts were found to be considerably shorter than the sum of their Van der Waals radii. These distances vary from ~2.62 Å to ~2.64 Å (S--O) and from ~2.43 Å to ~2.46 Å (N--H), when excited from the ground to excited states, which confirm the occurrence of non-covalent intra-molecular interactions. We believe that attractive interaction forces can modify the C-O-C angles in the excited state. This indicates that the singlet excited state should be much more planar than their ground state.

3.3.3. *Electronic transitions*

We have applied a variety of theoretical approaches, including CIS/3-21G*, TD-B3LYP/3-21G* and ZINDO methods to study the optical and emission properties of TBT-BIPY copolymers. The theoretical results thus obtained are compared with the experimental ones. All the energy levels calculated using the Time Dependent Density Functional Theory (TD-DFT), the CIS/3-21G* and the semi-empirical quantum-chemical ZINDO are used to predict the optical absorption and emission spectra of the ground (S_0) and first excited (S_1) optimized structures. The assignment of electronic transitions and their oscillator strengths are also calculated using these three methods.

From theoretical calculations, the wavelength of transitions from the ground to the first excited state ($S_0 \rightarrow S_1$) and from the first excited state to ground state ($S_1 \rightarrow S_0$) having the largest oscillator strength as well as their corresponding molecular orbital character for 2-TBT-BIPY are listed in Table 3. The corresponding experimental optical absorption and emission wavelengths measured in TBT-BIPY copolymer in chloroform solution are also listed in the same table. Clarke et al [112] suggest that the importance of the HOMO-LUMO transition may be easily understood from the spectral distribution of molecular orbitals. Accordingly, to a first approximation, a significant overlap found between HOMO and LUMO implies an intense transition between HOMO to LUMO and vice-versa. Here, the vertical $S_0 \rightarrow S_1$ transition dominates the H→L excitation by 60-81%.

Method of calculation	Optical absorption properties of ground state ($S_0 \rightarrow S_1$) λmax (nm)	f	MO/Character	Coefficient (%)	Emission properties of excited state ($S_1 \rightarrow S_0$) λmax (nm)	f	MO/Character	Coefficient (%)	Stokes shift (nm/eV)
CIS	357.2	4.364	H→L	60	398.7	3.055	L→H	80	41.5
			H-1→L+1	22			L+2→H-2	5	(0.36)
			H-2→L+2	9					
TD-DFT	481.6	2.943	H→L	78	508.8	3.224	L→H	82	27.2 (0.13)
ZINDO	439.4	3.781	H→L	81	519.2	4.352	L→H	89	79.8
			H-1→L+1	5					(0.43)
Exp	436 nm				498-527 nm				62 (0.35)

(H=HOMO, L=LUMO, L+1=LUMO+1, etc.), f: Oscillator strength

Table 3. The vertical transition energies (nm) and their oscillator strengths of absorption from the ground to the first excited state ($S_0 \rightarrow S_1$) and emission from the first excited to ground ($S_1 \rightarrow S_0$) states of TBT-BIPY copolymers calculated by CIS/3-21G*, TD//B3LYP/3-21G* and ZINDO methods.

3.3.4. Frontier molecular orbitals (HOMO and LUMO)

To gain insight into the excitation properties and the ability of electron or hole transport, we have shown in Fig. 11 HOMO and LUMO together known as frontier molecular orbitals which contribute significantly to the electronic transitions between the ground and excited states in 2-TBT-BIPY.

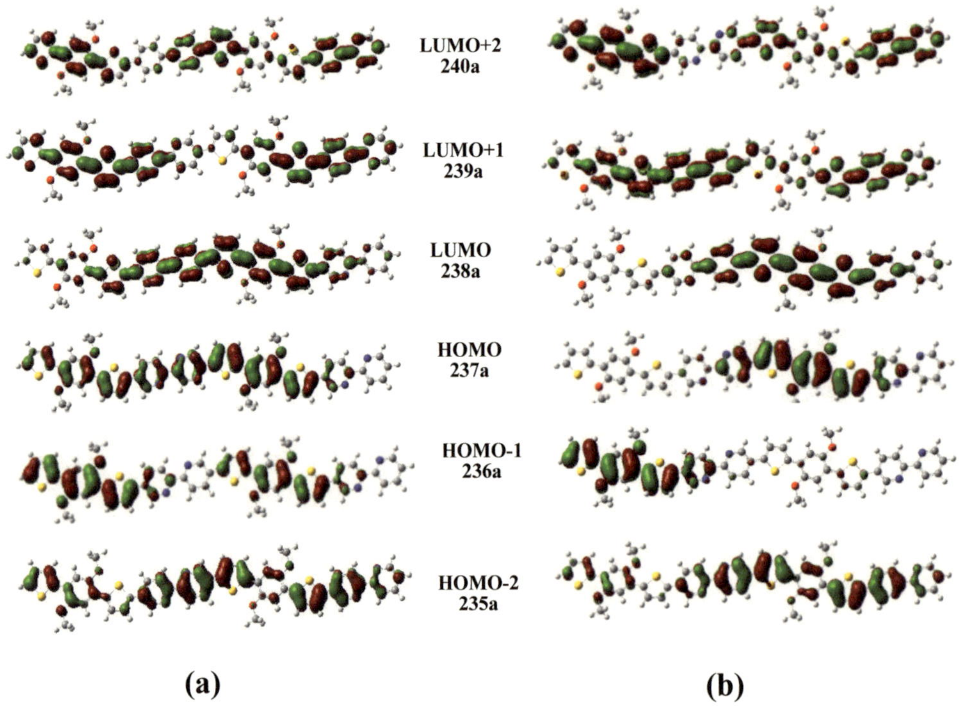

Figure 11. Contour plots for the HOMO and LUMO molecular orbitals which contribute significantly to the electronic transitions in 2-TBT-BIPY copolymers: (a) absorption from ground to excited and (b) emission from excited-to ground states.

We have examined and found that the presence of methoxy side chain does not have a significant effect on the molecular orbital distribution. In the HOMO, the C=C segments are π-bonding and have anti-bonding character with respect to their neighboring C=C units. Whereas, in the case of LUMO, the C=C units are anti-bonding and bonding in the bridge single bond. In general, excitation of a π-electron from HOMO to LUMO leads to increase the localization of electron density on the acceptor part of the molecule. Here, the promotion of one electron from HOMO to LUMO is explained by the frontier molecular orbital. For the TBT-BIPY copolymer, the LUMO favors the inter-ring mobility of electrons, while the HOMO only promotes the intra-ring mobility of electrons [116]. As outlined before, in the excited state of copolymer, both the HOMO and LUMO frontier molecular orbitals topology are significantly affected, particularly in the left TBT unit indicating their contribution to the excitation processes. In fact, in the ground state, the spatial distribution of the molecular orbitals is rather delocalized over the molecule. Changing to the excited state geometry, they become more localized.

3.3.5. Mulliken charge distribution for TBT-BIPY

A schematic representation for the intra-molecular charge transfer (CT) in the ground and excited states of 2-TBT-BIPYcopolymer, calculated as the average of the summation of Mulliken charge distribution of the TBT and BIPY units, is displayed in Fig.12. In general, intra-molecular charge transfer is generated through the alternating donor-acceptor conjugated systems [117]. From this figure, we think that the alternating TBT (positively charged) and BIPY (negatively charged) can be used as donor and acceptor, respectively. We have separately examined their HOMO and LUMO levels, which indicates that for the TBT unit, the HOMO is at -4.29 eV and the LUMO at -1,29 eV and for bipyridine unit we get -6.52 eV for the HOMO and -1.33 eV for the LUMO. Although the LUMO levels for both are quite similar, a weak intra-molecular charge transfer in these molecules can established. Based on the comparison between ground and excited-state geometries for 2-TBT-BIPY, we deduce that the charge distributions are predominantly restricted to the substituted phenylene and thiophene units.

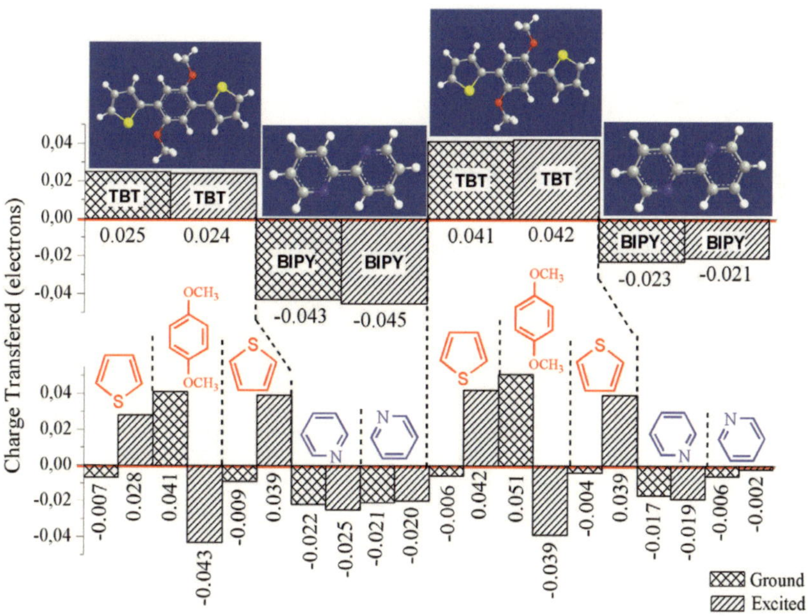

Figure 12. Illustration of the 2-TBT-BIPY copolymer structure with Mulliken charges distributions for TBT and BIPY units at the ground and excited states. All segments presented with dotted line separate the sub-units involved in the copolymer structure.

We can also predict the geometrical structure changes between the ground (S_0) and singlet excited (S_1) from the molecular orbitals. Therefore, to better understand the excitation process in TBT-BIPY copolymer, we have investigated the molecular orbitals involved in the electronic transition.

3.3.6. Simulated optical and emission spectra for TBT-BIPY copolymer

In Fig.13, we have depicted the simulated results of the optical absorption and emission spectra for 2-TBT-BIPY copolymer using the above three methods. To select the accurate method for predicting these optical properties, we show the potential energy surface (PES) of the ground (U_g) and excited (U_e) states along with their normal coordinates for 2-TBT-BIPY copolymer in Fig. 14 and Table 4. In Fig. 14, the two potential energies surfaces (PES) are plotted along with their normal coordinate q and the absorption and fluorescence spectra obtained from the transition between these two PES, using CIS/3-21G* and TD-DFT, respectively. The optical absorption energy (E_{VA}), emission energy (E_{VE}) and the relaxation energy (E_R^{GS}, E_R^{ES}) are presented in Table 5. The Stokes shift (SS), which is defined as the difference between the absorption and emission energies (E_{VA}-E_{VE}), is usually related with the band widths of both the absorption and emission bands [118] and it is a measure of the energy loss due to the molecular relaxation. It can be expressed as: $SS = E_R^{GS} + E_R^{ES} = E_{VA} - E_{VE}$. From the results given in table 5, we show that the SS calculated by CIS/3-21 G* is about two times higher than that calculated by TD-DFT. Accordingly, due to the neglect of the effects of electron correlation and higher order excitations, the geometrical relaxation after the excitation contributes much to the Stokes shift calculated by CIS/3-21 G*. It is well known that the absorption energy (E_{VA}) is usually considered to be maximum in the absorption spectrum, but it must be corrected for the zero-point vibrational energy (ZPE). In our case, compared with the results given in Table 4, SS energies calculated by CIS/3-21G* and TD-DFT methods as given in Table 5 deviate only by 0.084 eV and 0.088 eV, respectively. This difference of about 0.08 eV probably represents the value that needs to be used to correct the theoretical data. By such correction to the experimental value an excellent agreement is obtained with the result calculated by ZINDO method as shown in Table 4 for 2-TBT-BIPY copolymer.

	CIS/3-21G*	TD//B3LYP/3-21G*
E_{VA} (eV)	7.613	2.899
E_{VE} (eV)	7.169	2.681
E_R^{GS}(eV)	0.25	0.099
E_R^{ES}(eV)	0.194	0.119
SS (eV)	0.444	0.218

Table 4. The optical absorption energy (E_{VA}), emission energy (E_{VE}), relaxation energy (E_R^{GS}, E_R^{ES}) and Stokes shift (SS) calculated by CIS/3-21G* and TD//B3LYP/3-21G* methods for 2-TBT-BIPY.

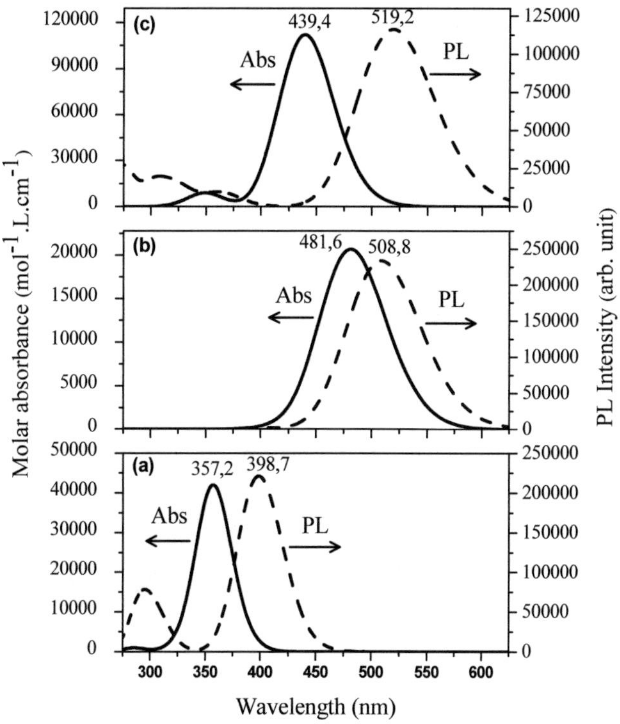

Figure 13. The simulated optical absorption and emission spectra of 2-TBT-BIPY copolymer with CIS/3-21G* (a), TD-B3LYP/3-21G* (b) and ZINDO (c) methods.

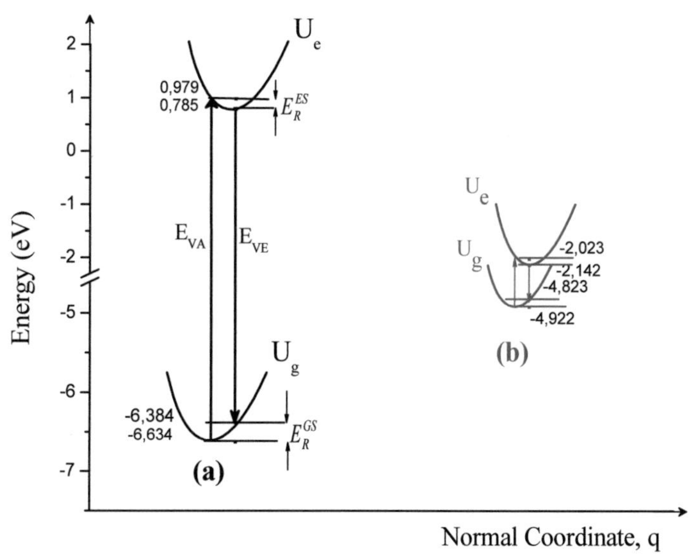

Figure 14. Schematic representation of the potential energy surface (PES) of the ground (U_g) and excited (U_e) states along with their normal mode coordinates of 2-TBT-BIPY copolymer calculated by CIS/3-21G* (a) and TD//B3LYP/3-21G* (b) methods. The parameters indicated are the absorption energy (E_{VA}), emission energy (E_{VE}) and relaxation energy (E_R^{GS}, E_R^{ES}).

For understanding better the results optical absorption and emission spectra calculated by ZINDO method experimental results are presented in Fig. 15. All curves are normalized to unity at their respective maximum. Prior to comparing the results calculated by ZINDO with those of obtained from experiments, it may be noted that no solvent effects have been taken into account in the ZINDO calculation. Keeping this in mind and comparing the spectra shapes, we believe that ZINDO results are in agreement with the experimental ones.

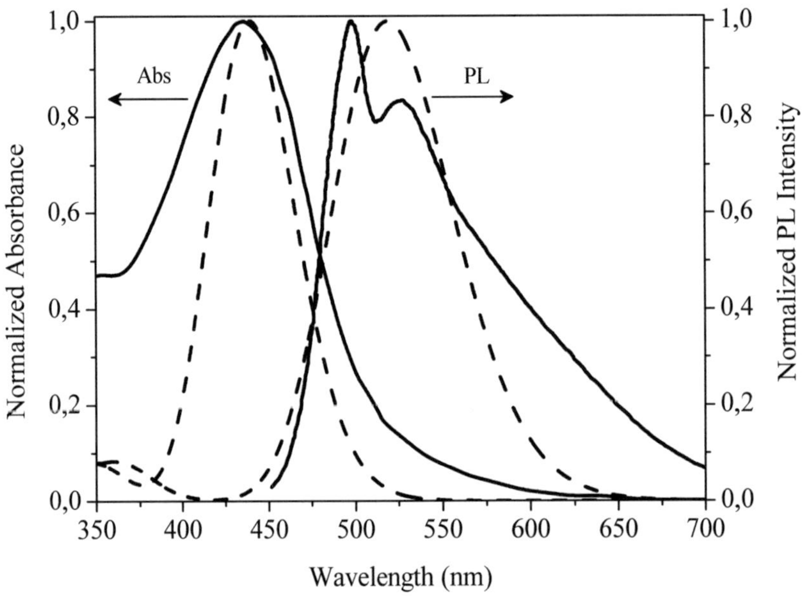

Figure 15. Normalized experimental optical absorption and photoluminescence spectra of TBT-BIPY copolymer (–) and those calculated by ZINDO method for 2-TBT-BIPY copolymer (- - -).

3.4. PLEDs architecture

In general, conjugated organic materials have smaller hole injection barriers than electron injection barriers due to the electron richness in a π-conjugated system, leading to poor electron transport ability in these materials. There are two possible approaches to improve this poor electron-transporting ability in organic materials used for fabricating LEDs. The most straightforward modification is to deposit a low work function (WF) metals such as Mg or Ca as cathode by high vacuum sublimation. However, the sensitivity of these metals towards oxygen and moisture limits their practical applications. The other more practical approach is to design or invent a material with lower LUMO energy by increasing its electron affinity, so that LUMO is to WF of the cathode material.

The electron injection energy barrier (ΔE_e) is determined by the electron affinity (EA) or by the difference between LUMO and WF of the cathode (ϕc), while the hole injection energy barrier (ΔE_h) is determined by the difference between IP or HOMO and WF of the anode (ϕa). In the most simple case, a single organic layer OLED, the organic layer is sandwiched between two electrodes of different work functions, one of which has to be transparent to light. For this electrode ITO coated glass substrates are frequently used. As for the counter electrode aluminum is used mostly.

The energy barriers between the emitting polymer and electrodes can be estimated by comparing the work function of the electrodes with HOMO and LUMO energy levels of emitting polymer. Thus, the hole-injection barrier is $\Delta Eh = E_{HOMO}-4.8$ eV, where 4.8 eV is the work function of the ITO anode and the electron-injection barrier is $\Delta E_e = \phi_X - E_{LUMO}$, where ϕ_X is the work function of cathode. The difference between the electron- and hole-injection barriers (ΔE_e-ΔE_h) is a useful parameter to evaluate the balance in electron and hole injection. Lower the (ΔE_e-ΔE_h) better the injection balance of electrons and holes from the cathode and anode, respectively. For TBT-BIPY copolymer, we have shown in Fig.16 the energy level diagrams of a single-layer PLED.

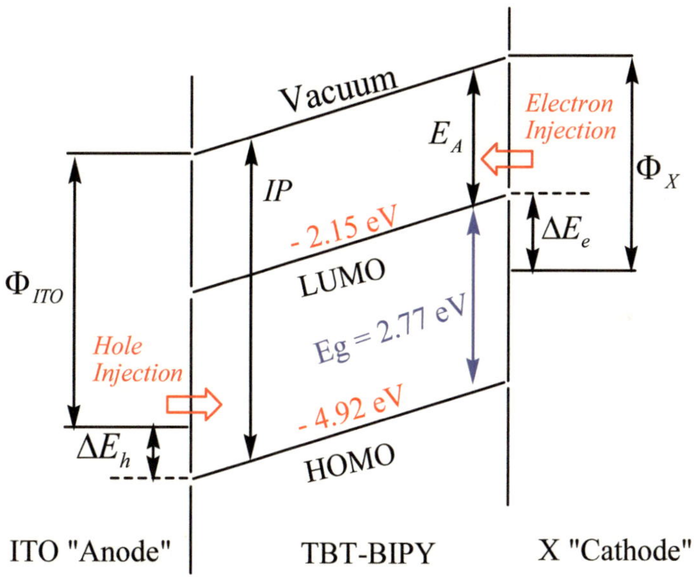

Figure 16. Energy-level diagrams of a single-layer PLED (ITO/TBT-BIPY/AL, Mg or Ca).

The ionization potentials (IP) and electron affinity (EA) are calculated by DFT/B3LYP/3-21G*on the geometry of the neutral, cationic and anionic states to estimate the energy barrier for the injection of both holes and electrons into TBT-BIPY copolymer. The calculated values are obtained as 5.62 eV and 1.35 eV, respectively. From Table 5, we showed that low work function metals such as Mg or Ca are typically used to minimize the barrier and then to provide for an ohmic contact.

X	ϕ_X (eV)	ΔE_h (eV)	ΔE_e (eV)	ΔE_e-ΔE_h (eV)
Al	4.2	0.12	2.05	1.93
Mg	3.6	0.12	1.45	1.33
Ca	2.8	0.12	0.65	0.53

Table 5. Parameters to evaluating the balance in electron and hole injections in PLED.

4. Part II: Donor-acceptor polymers for photovoltaic cell devices

4.1. Results and discussion

4.1.1. Conformational study

As first step, an accurate representation of the bond rotations in the chain is extremely important, since the properties of such polymers depend strongly on the conformational statistics of polymer chains [119]. Besides, the geometries obtained for the most stable conformations are used as input data for full optimization calculations. DFT/B3LYP calculations are performed on the following three model compounds, poly(3-hexylthiophene)-benzothiadiazole (P3HTBT), poly (carbazole-benzothiadiazole) (PCzBT) and poly(3-hexylthiophene)- di-benzothiadiazole-carbazole (P3HT2BTCz).

In conformational part, two basis sets 3-21G* and 6-31G* have been used for the sake of comparison. We note that the results derived from these two basis sets are almost similar. The relative energy for the first model (Fig. 17) shows two local minima in both sides of the spectrum (0° and 180°) and a maximum at about 90°.

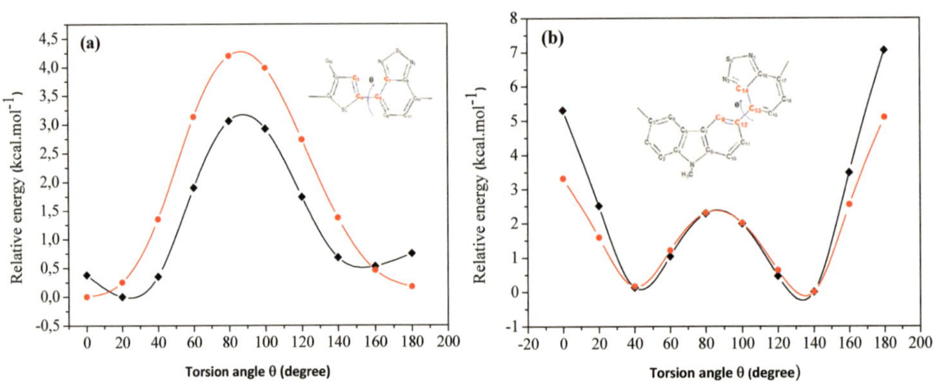

Figure 17. Potential energy curves of: (a) P3HTBT and (b) PCzBT monomer simulated at DFT/B3LYP level with (-•-) 3-21G* and (-♦-) 6-31G* basis sets.

The results indicate that the P3HTBT is completely planar with the inter-ring torsion angle 0°. It's obvious that this planarity is caused by intra-molecular repulsion between sulphur atoms in the main polymer backbone. In the case of benzothiadiazole copolymerized with carbazole, the conformational behaviour is completely different. The twisted conformations have two torsion angles θ_1 and θ_2 at around 40° and 140°, respectively. The latter conformation (140°) is slightly more stable by about 0.5 kcal.mol^{-1}.

4.1.2. Structural and characteristic parameters

The fully optimized structures with DFT/B3LYP/3-21G* method, with the respect to the torsion angles of both P3HTBT and PCzBT copolymers are shown in Fig. 18.

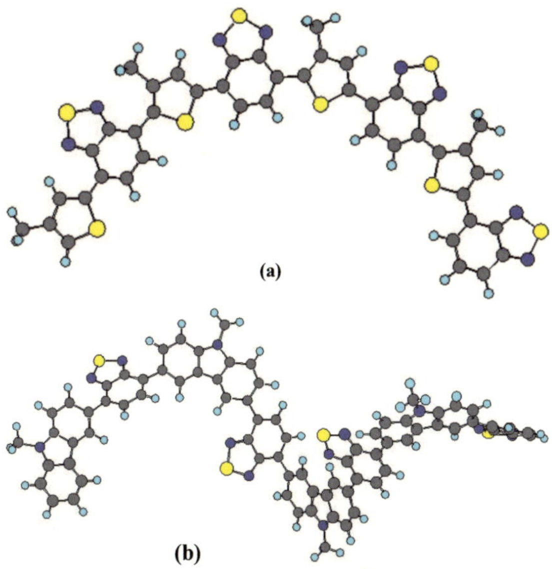

Figure 18. DFT/B3LYP/3-21G* optimized structure of: (a) $(P3HTBT)_4$ copolymer and (b) $(PCzBT)_4$ copolymer.

Based on these optimized structures, the principal physico-chemical parameters of the two copolymers are collected in Table 6. Along with the torsional angle (θ) (the deviation from co-planarity between the donor and acceptor units), intra-molecular charge transfer (D_{CT}) (the summation of all charges for the donor unit 3-hexyl thiophene (3HT) and carbazole (Cz)), bridge length (L_B) (the bond length between the donor and acceptor) are summarized.

	L_B (Å)	D_{CT} (e)	θ (°)
P3HTBT	1.458	0.038	0.0
PCzBT	1.482	0.063	147.0

Table 6. The optimum characteristic parameters (L_B, D_{CT}, and θ) of P3HTBT and PCzBT model compounds.

Considering the most stable conformation, we can deduce that the optimized structure of the PCzBT appears under a twisted configuration with a large torsional angle (θ = 147.0°). This suggests that a strong steric hindrance effect exists between the donor and acceptor moieties, whereas the P3HTBT structure has perfectly a planar structure. Moreover, compared to PCzBT, the order of the L_B of P3HTBT remains smaller indicating the formation of the mesomeric structures induced by intra-molecular charge transfer, that is, $D{-}A \rightarrow\leftarrow D^{+}{=}A^{-}$. The large intra-molecular charge (D_{CT}) of PCzBT copolymer backbone is probably originating from the nitrogen atoms with high electronegativity in the main backbone of Cz donor group. DCT significantly enhances the π-electron delocalization which is largely dependent on θ rather than on the acceptor strength.

The optimized structure of the resulting P3HT2BTCz composite and its main geometrical parameters (torsion angle and interring bond length) are illustrated in Fig. 19. The inspection of these data reveals that the resulting composite shows an almost non planar conformation which is more underlined on both sides of carbazole units to reach the values of 45° and 49°. Moreover, compared to those of P3HTBT and PCzBT, the central bonds connecting the two neighbouring central rings are slightly shorter, showing that this compound is more conjugated to extend the delocalization on all the chain backbone.

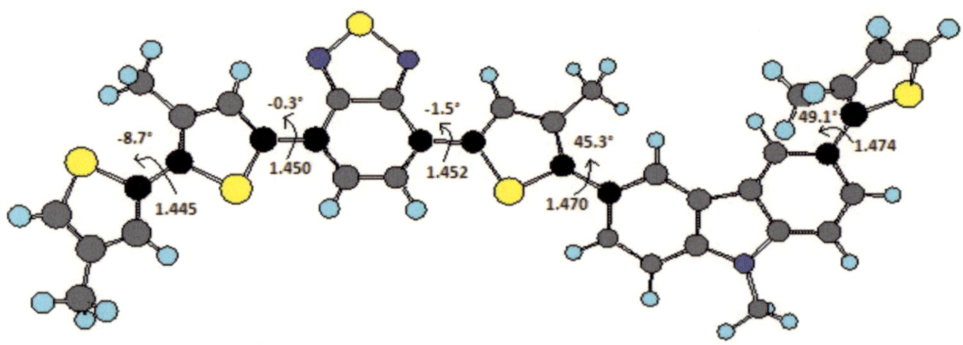

Figure 19. DFT/B3LYP/3-21G* optimized geometric structure of the resulting P3HT2BTCz composite.

4.1.3. Optical properties and electronic structures

As firstly discussed, the oligomer geometries and their corresponding band gap are calculated using DFT/B3LYP method with 3-21G* and 6-31G* basis sets. The band gap is estimated as the difference between the HOMO and LUMO energies. In our case, the band gap of $(P3HTBT)_n$ and $(PCzBT)_n$ (n = 1-4) oligomers are listed in Table 7.

Polymer	Number of monomer	Band gap energy (E_g) (eV)	
		B3LYP/3-21G*	B3LYP/6-31G*
P3HTBT	1	3.23	3.26
	2	2.67	2.45
	3	2.11	2.12
	4	1.94	1.96
	∞	1.61	1.55
PCzBT	1	3.10	3.09
	2	2.80	2.82
	3	2.71	2.63
	4	2.68	2.61
	∞	2.52	2.44

Table 7. Band gap energy E_g of $(P3HTBT)_n$ and $(PCzBT)_n$ (n: from 1 to 4 units).

By using the linear extrapolation technique [120], it can be seen from Fig. 20 that this value decreases with increasing the chain length from monomer to quatermer. Moreover, the theoretical data resulting from the two considered basis sets are very close and no significant changes are noticed when going from 3-21G* to 6-31G* basis set calculations.

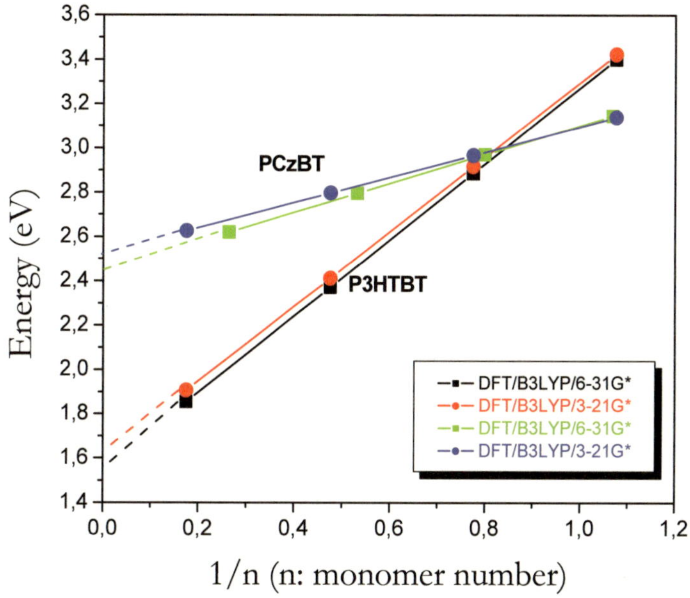

Figure 20. Representation of the band gap energy (E_g) as function of inverse chain length (1/n) for P3HTBT and PCzBT calculated by DFT/B3LYP with 6-31G* and 3-21G*basis sets.

The band gap of P3HTBT is found to be around 1.55 and 1.61 eV with 6-31G* and 3-21G* basis sets, respectively. These values are lower than that of pristine P3HT (1.90 eV) [121], due to the presence of benzothiadiazole in the main backbone copolymer. In parallel, a wide band gap for PCzBT is estimated to be 2.44-2.52 eV (Fig. 20). Nevertheless, the band gap of resulting composite P3HT2BTCz is found to be 2.31 eV which is in agreement with the experimental values $E_g \approx 1.97$ eV (derived from the UV-visible absorption spectrum in chloroform solution) [87]. These results are in close agreement with the experimental data by taking into account the packing effects (interchain interaction) in the solid state [122]. The HOMO level energy is estimated to be - 4.9 eV making this copolymer photo-chemically stable.

The TDDFT method was applied on the basis of the ground state optimized geometry of different compounds under study. As shown in Fig. 21, the absorption spectrum of the P3HT2BTCz composite seems to be the superposition of the two absorption spectra of P3HTBT and PCzBT copolymers. Compared to PCzBT and P3HTBT polymers, the absorption spectra is broader due the red shifted absorption, which may be attributed to the much better conjugation along the polymer backbone. Besides, the simulated absorption spectra show that the P3HT2BTCz compound absorbs from the UV at a wavelength of 600 nm, with two main absorption peaks centred at 478 and 319 and a weak peak at 260 nm. The band located at 319 nm arises from the delocalized π→π* transition in the polymer and the visible absorption peak located at longer wavelength centered at 478 nm could be assigned to the intra-molecular charge transfer transition between the Cz donor moiety and the BT acceptor unit [123].

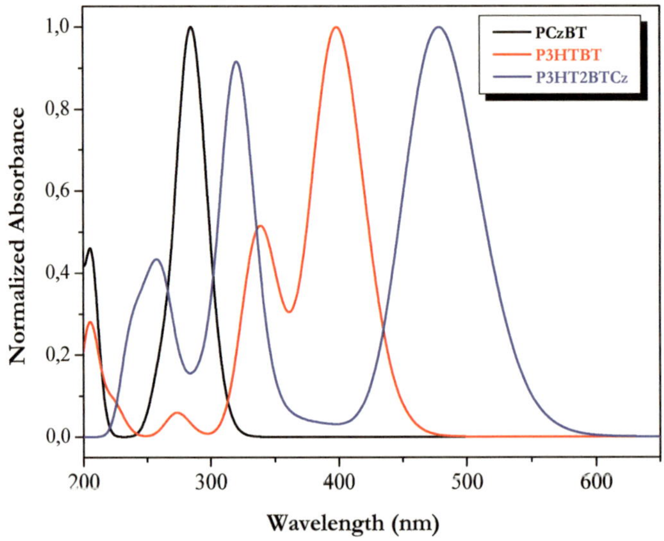

Figure 21. TD/B3LYP/3-21G* simulated UV-Visible optical absorption spectra: of PCzBT, P3HTBT and P3HT2BTCz.

The vertical excitation energy and their corresponding oscillator strength along the main excitation configuration are listed in Table 8. The first optically allowed electronic transition of P3HT2BTCz populates the HOMO→LUMO excitation with high oscillator strength (f = 1.0898). The two other transitions are mainly assigned respectively to HOMO→LUMO+1 and HOMO-1→LUMO+3 excitations. All intermediate states with low oscillator strength, so-called dark states, have intra-molecular charge transfer (ICT) character. Through this study, it is found that the calculated results reproduce very well the corresponding experimental data [87].

Electronic transition	Wavelength (nm)	Oscillator Strength (f)	Main MO/character	Coefficient	Experimental value (nm)
$S_0 \rightarrow S_1$	478	1.0898	HOMO→LUMO	80%	504[a] 518[b]
$S_0 \rightarrow S_2$	319	0.6912	HOMO→LUMO+1	51%	327[a] 338[b]
$S_0 \rightarrow S_3$	260	0.1905	HOMO-1→LUMO+3	54%	----

[a]in chloroform solution [87]
[b]in solid film [87]

Table 8. Main electronic transitions in P3HT2BTCz composites and their assignments.

In order to study the emission properties of P3HT2BTCz compounds, the TD/B3LYP method was applied to the geometry of the lowest singlet excited state optimized at the CIS level with 3-21G* basis set [124]. The normalized photoluminescence (PL) spectrum of P3HT2BTCz (Fig. 22) shows a maximum at 649 nm with strongest intensity (f = 0.8415), compared to 630 nm in experimental spectrum as indicated in Table 9. This may be regarded as an electronic transition reverse of the absorption corresponding mainly from LUMO to HOMO. Moreover, the observed red-shifted emission in the PL spectra is found to be in reasonable agreement with the experimental one by taking into account the packing effects (inter-chain interaction) in the solid state (0.49 eV (124 nm)) [87].

Electronic transition	Emission wavelength (nm)	Emission energy (cm^{-1})	Oscillator Strength (f)	MO/character	Coefficient	Experimental value (nm)
$S_1 \rightarrow S_0$	649	15400	0.8415	HOMO→LUMO	75%	630[a]

[a]in chloroform solution [87]

Table 9. Emission energy of P3HT2BTCz obtained by the TDDFT/B3LYP/3-21G* method.

We also find relatively high values of Stokes Shift (SS) in P3HT2BTCz (0.62 eV (172 nm)) (Fig. 22).

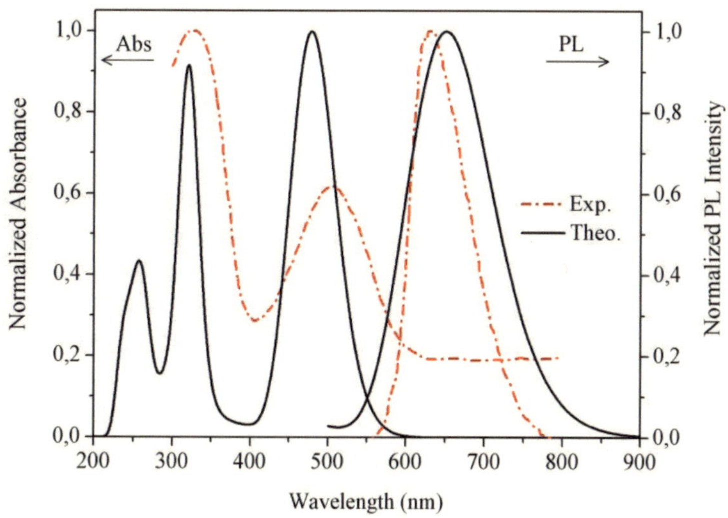

Figure 22. Experimental and TD-DFT calculated normalized absorption and emission spectra of P3HT2BTCz.

Based on the above results, the energy band structures are plotted in Fig. 23. When carbazole (Cz) is replaced by 3-hexylthiophene (3HT), the energy of HOMO level increases, while that of LUMO decreases. This change on the electronic structure facilitates both the hole and electron-transporting ability. The electronic structure differs greatly from one model to another, showing the effect of donor units in D-A architecture polymer and it results from the coupling behaviour of 3HT, Cz and BT in the main backbone. Moreover, further insights are obtained comparing the DFT calculated density of states (DOS) of the P3HTBT and PCzBT with that of P3HT2BTCz composite. This comparison is showed in Fig. 23 (at the right). Two striking things immerge from DOS diagram: 1) the ground state interaction between the donor and acceptor units and 2) this interaction induces intra-gap charge transfer states lying inside the gap of the PCzBT. As a result, P3HT2BTCz composite orbitals are shifted towards higher energies compared to the isolated PCzBT orbitals and towards lower energies compared to the isolated P3HTBT orbitals.

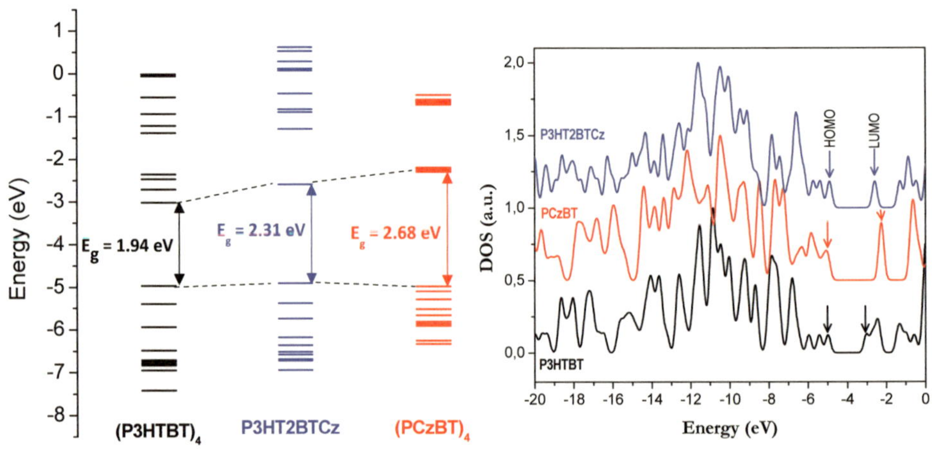

Figure 23. Electronic structure and DOS diagram of P3HTBT, PCzBT and P3HT2BTCz, simulated using DFT/B3LYP/3-21G* method.

The electron density iso-contours of HOMO and LUMO of P3HT2BTCz compound are plotted in Fig. 24. It can be seen that an asymmetric character within the rings and between subunits prevails for the HOMO orbital of this copolymer. Moreover, the localization of electronic charge lies mainly in the side part of HOMO orbital, which is typically expected due to the chain-end effects, which changes the shape of LUMO orbital. Due to the non-planarity observed for the P3HT2BTCz compound geometry, in its ground state, electrons are mainly localized on the benzothiadiazole units, as result of the weak interactions between the two building blocks. This fact is particularly noticeable in the LUMO orbital with a symmetric character between the subunits.

According to our calculations, electron densities in the first excited state namely LUMO and LUMO+1 are delocalized on BT and P3HT units with a symmetric character. Whereas, for higher energy levels, e.g., LUMO+2 levels take part in electron transitions on the P3HT and Cz units. Yet, the charge density of HOMO, HOMO-1 shows that the charge density spreads over the main chain of the compound to become much concentrated around the P3HT and Cz units in HOMO-2.

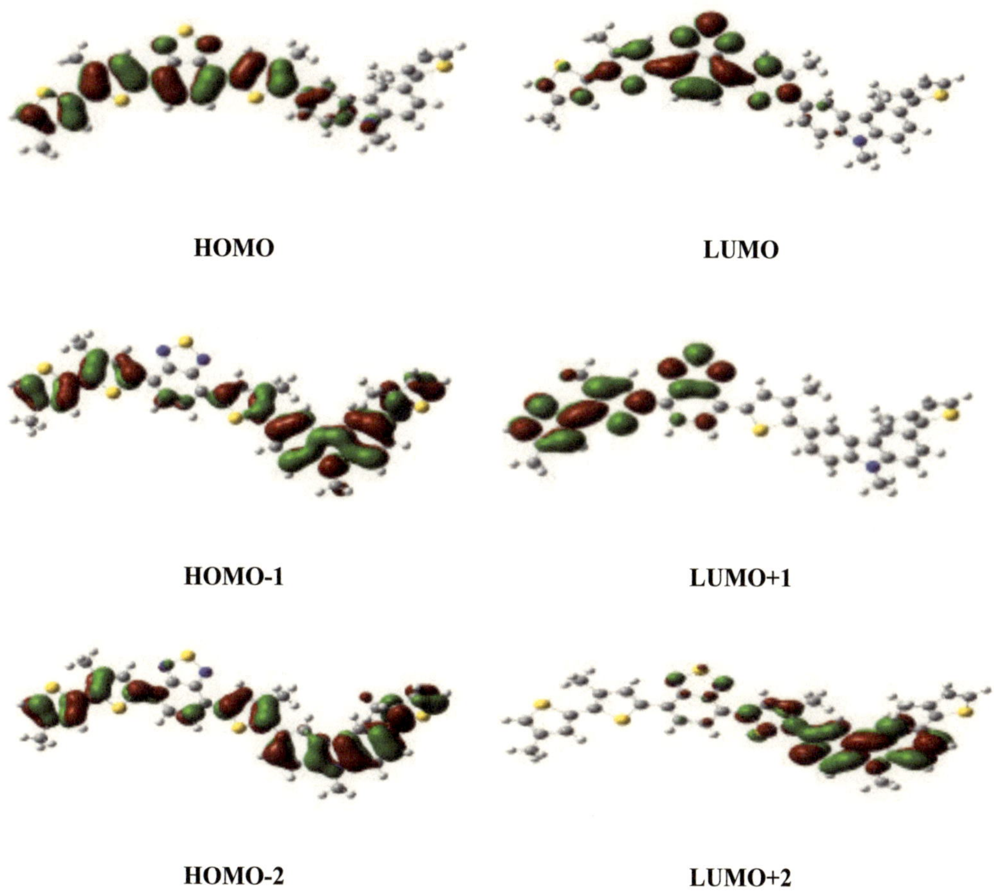

Figure 24. Contour plots for the main HOMO and LUMO molecular orbitals of P3HT2BTCz compound.

4.1.4. Vibrational study and force constant analysis

The most intense calculated bands of the infrared absorption (IR) of these compounds, shown in Fig. 25 are collected in Table 10 together with their corresponding assignments.

PBT		**P3HTBT**		**PCzBT**		**P3HT2BTCz**		**Assignments**
υ (cm^{-1})	**I**	**υ (cm^{-1})**	**I**	**υ (cm^{-1})**	**I**	**υ (cm^{-1})**	**I**	
707	m	725	w	728	w	-	-	Out of plan C-H wagging (BT + Cz).
-	-	957	w	-	-	-	-	Ring breathing (BT and P3HT).
956	s	961	w	960	vw	947	vw	S-N Scissoring (BT).
1080	s	1024	s	1037	w	1037	vw	S-N stretching (BT) + Rocking CH_3 (P3HT and Cz).

1234	vw	1234	w	1231	w	1224	w	C-H Rocking (3HT + Cz) + C-H wagging (BT)
-	-	1350	w	-	-	1347	w	CH_3 Wagging (3HT).
-	-	-	-	1372	w	1372	-	CH_3 Scissoring (Cz).
		-	-	1432	w	1429	w	C=C stretching (Cz).
1445	s	1445	vw	-	-	-	-	Ring vibration.
1514	vw	1516	m	-	-	1516	m	Ring vibration + C-H Rocking.
-	-	-	-	1582	m	-	-	Aromatic C-H and C-C stretching (Cz and BT).
-	-	1607	m	-	-	-	-	C=C ring stretch (P3HT and BT).
1621	w	-	-	-	-	-	-	C-C Bending vibration (BT).
-	-	-	-	1646	m	1652	s	C=C bending (Cz).
-	-	1675	w	-	-	1652	s	Antisymmetric C=C stretching (3HT).
-	-	1690	w	-	-	1689	w	Symmetric C=C stretching (3HT).
1792	m	1787	m	-	-	-	-	C-C Bending (BT).
		-	-	1792	s	1791	s	Symmetric C-C stretching (Cz).
-	-	-	-	1841	m	1841	w	C-N Scissoring and C-C stretching (Cz).
3330	vs	3310	S	3312	s	-	-	C-H stretching (BT).

Table 10. Main selected infrared modes of PBT, P3HTBT, PCzBT and P3HT2BTCz and their corresponding assignments (υ: frequency, I: intensity, s: strong, vs: very strong, m: medium, w: weak, vw: very weak).

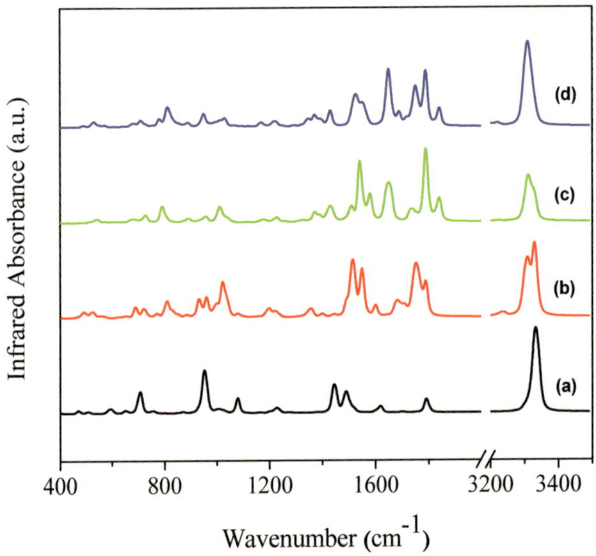

Figure 25. Theoretical infrared spectra of: (a) PBT, (b) P3HTBT, (c) PCzBT and (d) P3HT2BTCz.

A large number of bands appear with very high peaks due to an induced strong dipolar moment. All characteristics of infrared bands in P3HT [125], PCz and BT vibration modes are observed.

The inspection of these spectra shows that after combining the two copolymers to obtain the P3HT2BTCz composite, some bands undergo slight changes in their positions and intensities. The main vibrational modes of PBT persist following the addition of 3HT, Cz groups in the P3HT2BTCz composite. Firstly, a down shift of the band assigned to C-H stretching in benzothiadiazole unit is observed at high frequencies with strong intensity located at 3330 cm^{-1} in $(P3HTBT)_4$ and $(PCzBT)_4$.. The band at 1785 cm^{-1} assigned to the anti-symmetric C=C stretching mode becomes clear in the other PCzBT and P3HT2BTCz compounds and the C-C bending vibrational mode located at 1621 cm^{-1} in PBT becomes clearly pronounced in PCzBT and P3HT2BTCz with a high energy shift of about 20 cm^{-1}.

The signal attributed to the S-N stretching at 1073 cm^{-1} completely disappears in the functionalized composite following a significant interaction of different groups. This effect is also confirmed by the shift (from 1486 to 1550 cm^{-1} and from 1443 to 1513 cm^{-1}) observed in IR bands ascribed to symmetric and anti-symmetric C=C stretching, respectively, as a consequence of the presence of more conjugated backbone. The band at 707 cm^{-1}, ascribed as out-of-plane C-H wagging of PBT polymer, decreases in intensity in the first two copolymers and disappears completely in the case of the P3HT2BTCz composite. This effect is due to a significant interaction between donor Cz as donor and BT as acceptor acceptor groups. Thus, this analysis highlights the effective charge transfer in the main backbone of these compounds targeted for photovoltaic applications.

In order to support the above discussed results further, the force constant analysis of benzothiadiazole unit in P3HTBT, PCzBT copolymers and P3HT2BTCz composite, have been investigated as shown in Fig. 26.

Generally, the bond stretch depends on two main parameters, the bonding energy (E_0) and the force constant k. The latter represents the potential energy surface (PES) curvature near the minimum $k = (\frac{\partial^2 E}{\partial R^2})_{R=R_0}$. The force constant (k) is proportional to the strength of the covalent bond [126]. Considering the BT unit as shown in Fig. 26, one can deduce that the bond length variation marks important changes in benzothiadiazole bonding, depending on the electronic configuration and hence force constant. One can also notice that the inter-ring force constants (F_1 and F_{12}) increase dramatically with a significant decrease in their corresponding bond length, leading to a more conjugated composite compared to P3HTBT and PCzBT. It clearly shows that the intra-ring delocalization is larger for P3HT2BTCz than for P3HTBT or PCzBT. In addition, the modifications on benzene moiety are clearly seen through the force constants F_5 and F_7. On the thiadiazole part, force constants such as F_8, F_9, F_{10} and F_{11} are very similar but a slight variation can be noticed in the case of the composite because of the presence of Cz and BT together in the main backbone structure with 3HT as spacer. This configuration can enhance the charge transfer between donor and acceptor units. These observations are consistent with the above discussed properties.

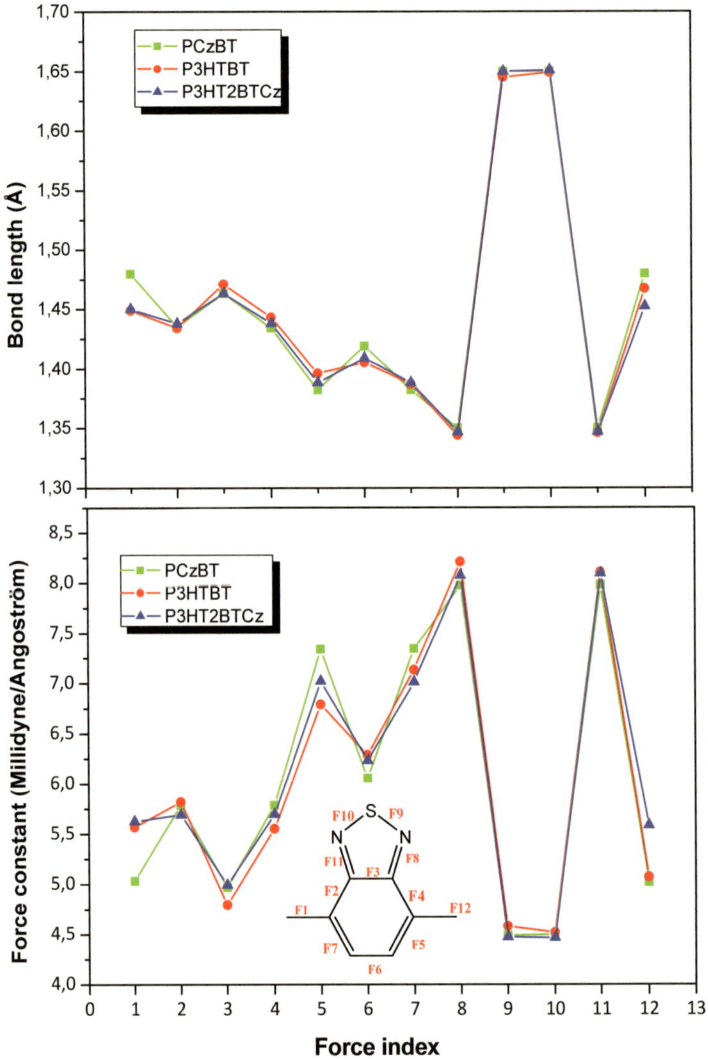

Figure 26. Main force constants and bond lengths of equivalents benzothiadiazole sites.

4.2. Photovoltaic properties

Favorable values of HOMO and LUMO levels, band gaps, and strong absorptions in the visible region suggest that the P3HT2BTCz may be used as active layer in PSCs devices when blended with [6,6]-phenyl-C_{61}-butyric acid methyl ester (PCBM), which is the most broadly used acceptor in solar cell devices [127-128]. After an optimization procedure via the semi-empirical PM3 method, the resulting blend geometrical structure of the composite (P3HT2BTCz:PCBM) in weight ratio of 1:1 is then re-optimized by DFT/B3LYP/3-21G* as shown in Fig. 27, where the substituents in the case of PCBM play the role of the spacers between the donor and C_{60} acceptor units. This stable configuration is governed by the interaction of oxygen of the PCBM with the sulphur atom of thiophene units on both sides of carbazole motives.

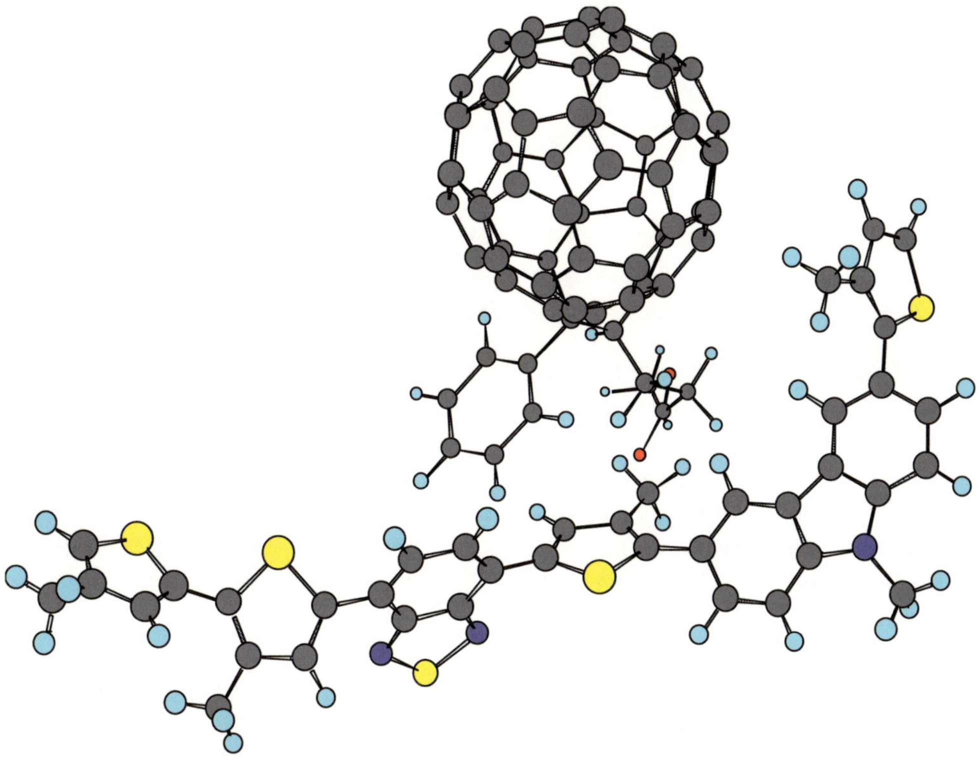

Figure 27. Optimized geometric structure of P3HT2BTCz:PCBM (1:1) simulated by DFT/B3LYP/3-21G* method.

Based on the comparison between the donor and the acceptor compounds, the resulting composite shows some interesting electronic properties, such as a low band gap of 1.93 eV and a lower HOMO energy level of -5.32 eV which indicate that this composite can be used as an active layer in photovoltaic cells. The corresponding structure of a photovoltaic device is schematically presented in Fig. 27. The difference in the LUMO energy levels of P3HT2BTCz and PCBM is close to 1.0 eV, suggesting that the photo-excited electron transfer from P3HT2BTCz to PCBM may be sufficiently efficient in photovoltaic devices [129-130]. Energetically, in comparing different anode and cathode metals as shown in Fig. 28 one can notice that the transparent ITO (Indium Tin Oxide) anode and Al, Ag or Mg (Aluminium, Silver or Magnesium) as cathode are the most suitable metals for for effective charge collection on two electrodes.

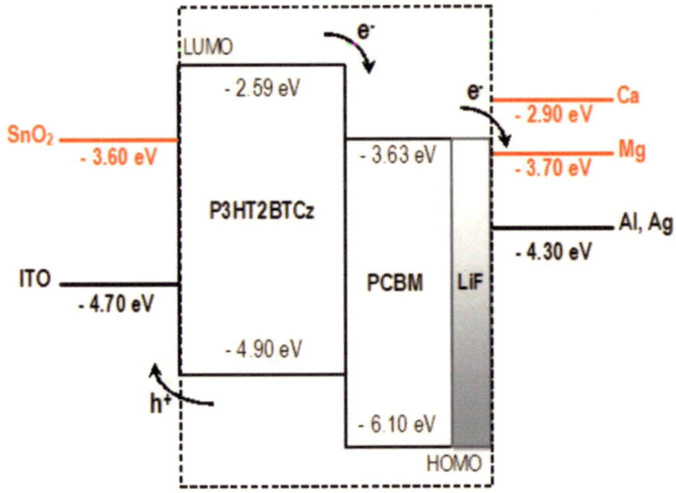

Figure 28. Schematic energy diagram of the proposed bulk heterojunction solar cell.

The photovoltaic efficiency performance data of the photovoltaic cell (power conversion efficiency (PCE) values, including the open circuit voltage (V_{oc}), short circuit current (J_{sc}), fill factor (FF) and incident-light power (Pin), are derived from the following equation: PCE = V_{oc}. J_{sc}. FF/P_{in}. The maximum open circuit voltage (V_{oc}) of the BHJ solar cell is related to the difference between HOMO of the electron donor and LUMO of the electron acceptor, taking into account the energy lost during the photo-charge generation [131-133]. The theoretical values of open-circuit voltage V_{oc} have been calculated from the following expression [134]:

$$V_{OC} = |E(HOMO)^{donor}| - |E(LUMO)^{acceptor}| - 0.3$$

Based on this formula, it can be seen that the V_{oc} value of P3HT2BTCz: PCBM is about 0.97 V but it depends on the difference of the output of the electrodes [135]. Starting from the above results, P3HT2BTCz composite seems to be a good candidate for photovoltaic application due to its high V_{oc} and wider absorption range broader than the range of absorption of other copolymers. Based on Scharber model [136], the maximum power-conversion efficiency of the photovoltaic solar cell, with P3HT2BTCz:PC_{61}BM (1:1) composite as active layer can be up to 5%.

5. Conclusion

The aim of this chapter is to combine experimental analyses and theoretical calculations to present a comprehensive study of the structural and optical properties of organic electronic devices. Based on model compounds, Highest Occupied Molecular Orbital (HOMO), Lowest Unoccupied Molecular Orbital (LUMO) levels, Ionization Potential (IP), Electron Affinity (EA) as well as electronic structures for two samples are examined. The optoelectronic parameters studied here are essential for better understanding of the exchange between polymer and electrodes in PLEDs and PPCs. The experimental and computational results are compared and discussed.

The first part of this chapter shows how important it is to combine thienylene, dialkoxyphenylene and bipyridine fragments to obtain compounds with a strong electronic delocalization. As a result, analysis of the results obtained in the gas phase has allowed us to understand the crucial role played by the intra-molecular S--O and N--H interactions in determining the planarity of the compound. This leads to the formation of a donor–acceptor type of arrangement within the polymer backbone and an intra-molecular charge transfer for the TBT-BIPY copolymer model compound. In addition, we have presented the optical and emission properties of these compounds by studying the ground and first excited states of copolymer models.

In the second part, we have used the density functional theory DFT/B3LYP to investigate the photo-physical properties of some copolymers in alternate donor-acceptor structure. In fact, the modification of chemical structures can greatly modulate and improve the electronic and optical properties of pristine copolymers. Hence, added to benzothiadiazole units, the introduction of carbazole motives in the copolymer backbone results in a better overlap of the absorption spectrum with the solar spectrum. In addition, the hexylthiophene linkage is found not only as a conjugated bridge but also it reduces the steric interaction between aromatic rings and thus enhances the effective charge transfer between donor and acceptor units.

In fact, the obtained theoretical data derived from DFT/B3LYP/3-21G* method are in good agreement with the available experimental data. The resulting optimized BHJ active layer shows a π-stacking configuration governed by a Wander walls interaction. A model energy band diagram is introduced, simulating the energy behaviour of this active layer. Based on this design concept, the PSC using the blend of P3HT2BTCz with fullerene derivatives, exhibit a promising performance with a PCE up to 5%. This approach provides great flexibility in fine-tuning of the absorption spectra and energy levels of the resultant polymers for achieving high device performance.

Finally, these results clearly indicate that these new compounds with alternating donor-acceptor structures are promising materials for application in optoelectronic devices. Devices fabrication and characterization are in progress and will be published elsewhere.

Author details

S. Ayachi, A. Mabrouk and K. Alimi*

Research Unit: New Materials and Organic Electronic Devices (UR 11ES55), Faculty of Sciences of Monastir, University of Monastir, Tunisia

M. Bouachrine

UMIM, Polydisciplinary Faculty of Taza, University Sidi Mohamed Ben Abdellah, Taza, Morocco

Acknowledgement

This work has been supported by the cooperation Tunisio-Marocain (CMPTM No. 11/TM/72) and Tunisian-French cooperative action (CMCU/07G1309).

---

* Corresponding Author

6. References

[1] Chiang C.K, Fincher C.R. Jr, Park Y.W, Heeger A.J, Shirakawa H, Louis E.J, Gau S.C, MacDiarmid, A.G, Phys. Rev. Lett. 39, 1098 (1977).

[2] Chiang C.K, Druy M.A, Gau S.C, Heeger A.J, Louis E.J, MacDiarmid A.G, Park Y.W, Shirakawa H, J. Am. Chem. Soc., 100, 1013 (1978).

[3] Balan A, Baran D, Gunbas G, Durmus A, Ozyurt F, Toppare L, Chem. Commun. 44, 6768 (2009).

[4] Thompson, B.C, Kim, Y-G, McCarley T.D, Reynolds J.R, J. Am. Chem. Soc. 128, 12714 (2006).

[5] Kraft A, Grimsdale AC, Holmes A.B, Ange. Chem. Intern. Ed. 37, 402 (1998).

[6] Evans N.R, Devi L.S, Mak C.S.K, Watkins, S.E, Pascu S.I, Kohler A, Friend R.H, Williams C.K, Holmes A.B, J. Am. Chem. Soc., 128, 6647 (2006). Huang Q, Evmenenko G.A, Dutta P, Lee P, Armstrong N.R, Marks T.J, J. Am. Chem. Soc. 127, 10227 (2005).

[8] Tsuji H, Mitsui C, Sato Y, Nakamura E, Adv. Mater. 21, 3776 (2009).

[9] Zhen C.G, Chen Z.K, Liu. Q.D, Dai Y.F, Shin R.Y.C, Chang, S-Y, Kieffer J, Adv. Mater. 21, 2425 (2009).

[10] Huang J, Qiao X, Xia Y, Zhu X, Ma D, Cao Y, Roncali J, Adv. Mater. 20, 4172 (2008).

[11] Song M.H, Kabra D, Wenger B, Friend R.H, Snaith H, J. Adv. Funct. Mater. 19, 2130 (2009).

[12] Huang F, Chen K.S, Yip H.L, Hau S.K, Acton O, Zhang Y, Luo J, Jen A.K.Y, J. Am. Chem. Soc.131, 13886 (2009).

[13] Bijleveld J.C, Zoombelt A.P, Mathijssen S.G.J, Wienk M.M, Turbiez M, de Leeuw, D.M, Janssen R.A, J. Am. Chem. Soc. 131, 16616 (2009).

[14] Westenhoff S, Howard I.A, Hodgkiss J.M, Kirov K.R, Bronstein H.A, Williams C.K, Greenham N.C, Friend R.H, J. Am. Chem. Soc. 130, 13653 (2008).

[15] Beek W.J.E, Sloof L.H, Wienk M.M, Kroon J.M, Janssen R.A.J, Adv. Funct. Mater. 15, 1703 (2005).

[16] Coakley K.M, McGehee M.D, Chem. Mater. 16, 4533 (2004).

[17] Hoppe H, Sariciftci N.S, J. Mater. Res. 19, 1924 (2004).

[18] Brabec C.J, Sariciftci N.S, Hummelen J.C, Adv. Funct. Mater. 11, 15 (2001).

[19] Günes, S, Baran D, Günbas G, Özyurt F, Fuchsbauer A, Sariciftci N.S, Toppare L, Sol. Energy Mater. Sol. Cells. 92, 1162 (2008).

[20] Zoombelt A.P, Fonrodona M, Wienk M.M, Sieval A.B, Hummelen J.C, Janssen R.A, Org Lett. 11, 903 (2009).

[21] Colladet K, Fourier S, Cleij T.J, Lutsen L, Gelan J, Vanderzande D, Nguyen L.H, Neugebauer H, Sariciftci S, Aguirre A, Janssen G, Goovaerts E, Macromol. 40, 65 (2007).

[22] Sang G, Zhou E, Huang Y, Zou Y, Zhao G, Li Y, J. Phys. Chem. C.113, 5879 (2009).

[23] Friend R.H. Gymer R.W, Holmes A.B, Buffoughes J.H, Marks R.N, Taliani C, Bradley D.D.C, dos Santos D.A, Bradas J.L, Logdlund M, Salaneck W.R, Nature. 397, 121 (1999).

[24] Fong H.H, Pozdin V.A, Amassian A, Malliaras G.G, Smilgies D.M, He M, Gasper S, Zhang F, Sorensen M, J. Am. Chem. Soc. 130, 13202 (2008).

[25] Lin C-J, Lee W-Y, Lu C, Lin H-W, Chen W-C, Macromol. 44, 9565 (2011).

[26] Beaujuge, P.M, Pisula W, Tsao H.N, Ellinger S, Mullen K, Reynolds J.R, J. Am. Chem. Soc. 131, 7514 (2009).

[27] Pang S, Tsao H.N, Feng X, Mullen K, Adv. Mater. 21, 3488 (2009).
[28] Cho S, Lee K, Heeger A.J, Adv. Mater. 21, 1941 (2009).
[29] Gao P, Beckmann D, Tsao H.N, Feng X, Enkelmann V, Baumgarten M, Pisula W, Müllen K, Adv. Mater. 21, 213 (2009).
[30] McCulloch I, Heeney M, Bailey C, Genevicius K, MacDonald I, Shkunov M, Sparrowe D, Tierney S, Wagner R, Zhang W, Chabinyc M.L, Kline R.J, McGehee M.D, Toney M.F, Nat. Mater. 5, 328 (2006).
[31] Chua L.L, Zaumseil J, Chang J.F, Ou E.C.W, Ho P.K.H, Sirringhaus H, Friend R.H, Nature. 434, 194 (2005).
[32] Irringhaus H, Brown P.J, Friend R.H, Nielsen M.M, Bechgaard K, Langeveld B.M.W, Spiering A.J.H, Janssen R.A.J, Meijer E.W, Herwig P, de Leeuw D.M, Nature. 401, 685 (1999).
[33] Horowitz, G, Adv. Mater. 10, 365 (1998).
[34] Burroughes J.H, Bradley D.D.C, Brown A.R, Marks R.N, Mackay K, Friend R.H, Bum P.L, Holmes A.B, Nature. 347, 539 (1990).
[35] Lin Y-J, Chou W-Y, Lin S-T, Appl. Phys. Lett. 88, 071108 (2006).
[36] Curtis M.D, Macromol. 34, 7905 (2001).
[37] Lee D.W, Kwon K.Y, Jin J.I, Park Y, Kim Y.R, Wang I.W, Chem. Mater. 13, 565 (2001).
[38] Benincori T, Consinni V, Gramatica P, Pilati T, Rizzo S, Sannicolo F, Odeschini R.T, Zotti G, Chem. Mater. 13, 665 (2001).
[39] Yang Y, Pei Q, Heeger A.J, Synth. Met. 78, 263 (1996).
[40] Grem. G, Leditzky. G, Leising. G, Adv. Mater. 4, 36 (1992).
[41] Grem. G, Leising. G, Synth. Met. 57, 4105 (1993).
[42] Berggren M, Inganas O, Gustafsson G, Rasmusson J, Anderson M.R, Hjertberg T, Wennerström O, Nature. 372, 444 (1994).
[43] Pei J, Yu W-L, Ni J, Lai L-H, Huang W, Heeger A.J, Macromol. 34, 7241 (2001).
[44] Pei Q, Yang Y, J. Am. Chem. Soc. 118, 7416 (1996).
[45] Wang C, Kilitziraki M, MacBride J.A.H, Bryce M.R, Horsburgh L.E, Sheridan A.K, Monkman A.P, Samuel I.D.W, Adv. Mater. 12, 217 (2000).
[46] Murali M.G, Naveen P, Udayakumar D, Yadav V, Srivastava R, Tetrahed. Lett. 53, 157 (2012).
[47] Wang L-H, Kang E-T, Huang W, Thin Solid Films 417, 151 (2002).
[48] Sarrazin P, Beneventi D, Denneulin A, Stephan O, Chaussy D, Int. J. Polym. Sci. 2010, 1 (2010).
[49] Nguyen T-Q, Doan V, Schwartz B.J, J. Chem. Phys. 110, 4068 (1999).
[50] Catellani M, Luzzati S, Lupsac N-O, Mendichi R, Consonni R, Famulari A, Meille S.V, F Giacalone, Segura J.L, Martín N, J. Mater. Chem. 14, 67 (2004).
[51] Champion R.D, Cheng K.F, Pai C-L, Chen W-C, Jenekhe S.A, Macromol. Rapid. Commun. 26, 1835 (2005).
[52] Yu W.L, Meng H, Pei J, Huang W, J. Am. Chem. Soc.120, 11808 (1998).
[53] Huang W, Meng H, Pei J, Chen Z, Lai Y, Macromol. 32, 118 (1999).
[54] Ng S.C., Ding M, Chan H.S.O, Yu W-L, Macromol. Chem. Phys. 202, 8 (2001).
[55] Tanese M.C, Farinola G.M, Pignataro B, Valli L, Giotta L, Conoci S, Lang P, Colangiuli D, Babudri F, Naso F, Sabbatini L, Zambonin P.G, Torsi L, Chem. Mater. 18, 778 (2006).

[56] Santos Silva H, Nogueira S.L, Manzoli J.E, Barbosa Neto N.M, Marletta A, Serein-Spirau F, Lère-Porte J.-P Sandrine Lois, Silva, R.A, J. Phys. Chem. A. 115, 8288 (2011).
[57] Lois S, Florès J-C, Lère-Porte J-P, Spirau F.S, Moreau J.J.E, Miqueu K, Sotiropoulos J-M, Baylère P, Tillard M, Belin C, Eur. J. Org. Chem. 2007, 4019 (2007).
[58] Watkins N.J, Mäkinen A.J, Gao Y, Uchida M, Kafafi Z.H (2004) "Direct observation of the evolution of both the HOMO and LUMO energy levels of a silole derivative at a magnesium/silole interface". Organic Light-Emitting Materials and Devices VII, edited by Zakya H. Kafafi, Paul A. Lane, Proceedings of SPIE Vol. 5214 (SPIE, Bellingham, WA, Doi: 10.1117/12.515300
[59] Yan L, Gao Y, Thin Solid Films 417, 101 (2002).
[60] J. C. Bernède, A. Godoy, L. Cattin, F. R. Diaz, M. Morsli and M. A. del Valle (2010) in Solar Energy, Book edited by: Radu D. Rugescu, , pp. 432, INTECH, Croatia, downloaded from SCIYO.COM, Organic Solar Cells Performances Improvement Induced by Interface Buffer Layers, ISBN 978-953-307-052-0.
[61] Herbert M, Angew. Chem. Int. Ed. 44, 2482 (2005).
[62] Kanibolotsky A.L, Perepichka I.F, Skabara P.J, Chem. Soc. Rev., 39, 2695 (2010).
[63] Wallace J.U, Chen S.H, Adv. Polym. Sci. 212, 145 (2008).
[64] Chen P, Lalancette R.A, Jäkle, F, J. Am. Chem. Soc. 133, 8802 (2011).
[65] Lehnherr D, Tykwinski R.R, Aust. J. Chem. 64, 919 (2011).
[66] Pai C-L, Liu C-L, Chen W-C, Jenekhe S.A, Polym. 47, 699 (2006).
[67] Ayachi S, Ghomrasni S, Bouachrine M, Hamidi M, Alimi K, J. Mol. Str (2012, In Press)
[68] Yang L, Liao Y, Feng J-K, Ren A-M, J. Phys. Chem. A. 109, 7764 (2005).
[69] Yang L, Feng J-K, Ren A-M, J. Org. Chem. 70, 5987 (2005).
[70] Yang L, Ren A-M, Feng J-K, Wang J-F, J. Org. Chem. 70, 3009 (2005).
[71] Yang L, Feng J.K, Ren A.M, Sun C.C, Polym. 47, 3229 (2006).
[72] Yang G, Su T, Shi S, Su Z, Zhang H, Wang Y, J. Phys. Chem. A. 111, 2739 (2007).
[73] Telesca R, Bolink H, Yunoki S. Hadziioannou G. Van Duijnen P. Th, Snijders J.G, Jonkman H.T, Sawatzky G.A, Phys. Rev. B 63, 155112 (2001).
[74] Ayachi S, Alimi K, Bouachrine M, Hamidi M, Mevellec J.Y Lère-Porte J.P, Synth. Met. 156, 318 (2006).
[75] Bouachrine M, Bouzakraoui S, Hamidi M, Ayachi S, Alimi K, Lère-Porte J-P, Moreau J, Synth. Met. 145, 237 (2004).
[76] Brabec CJ, Dyakonov V, Parisi J, Sariciftci NS (2003) Organic Photovoltaics: Concepts and Realization Eds. 60, Springer, Berlin Heidelberg, pp. 169-177.
[77] Dennler G, Scharber M.C, Brabec C, Adv. Mater. 21, 1323 (2009).
[78] Wu Z, Fan B, Xue F, Adachi C, Ouyang J, Solar Energy Materials & Solar Cells 94, 2230 (2010).
[79] Al-Ibrahim M, Roth H.K, Schroedner M, Konkin A, Zhokhavets U, Gobsch G, Scharff P, Sensfuss S, Org. Electron. 6, 65 (2005).
[80] Chemek M, Wéry J, Bouachrine M, Paris M, Lefrant S, Alimi K, Synth. Met. 160, 2306 (2010).
[81] Yang Y, Zhou Y, He Q, He C, Yang C, Bai F, Li Y, J. Phys. Chem. B 113, 7745 (2009).
[82] Soci C, Hwang I, Moses D, Zhu Z, Waller D, Gaudiana R, Brabec C. J, Heeger A. J, Adv. Func. Mat. 17, 632 (2007).
[83] Huang J, Li C, Xia Y, Zhu X, Peng J, Cao Y, J. Org. Chem. 72, 8580 (2007).

[84] Colladet K, Fourier S, Cleij T.J, Lutsen L, Gelan J, Vanderzande D, Macromol. 40, 65 (2007).
[85] Roncali J, Leriche P, Cravino A, Adv. Mater. 19, 2045 (2007).
[86] Morvillo P, Bobeico E, Sol. Ener. Mat. Sol. Cells 92, 1192 (2008).
[87] Li J.C, Meng Q.B, Kim J.S, Lee Y.S, Bull. Korean Chem. Soc. 30, 951 (2009).
[88] Hou Q, Xu Y, Yang W, Yuan M, Peng J, Cao Y, J. Mater. Chem. 12, 2887 (2002).
[89] Svensson M, Zhang F, Veenstra S.C, Verhees W.J.H, Hummelen J.C, Kroon J.M, Inganäs O, Andersson M.R, Adv. Mater. 15, 988 (2003).
[90] Wu M.C, Lin Y.Y, Chen S, Liao H.C, Wu Y.J, Chen C.W, Chen Y.F, Su W.F, Chem. Phys. Lett. 468, 64 (2009).
[91] Frisch M.J, Trucks G.W, Schlegel H.B, Scuseria G.E, Robb M.A, Cheeseman J.R, Montgomery J.A, Vreven T, Kudin K.N, Burant J.C, Millam J.M, Iyengar S.S, Tomasi J, Barone V, Mennucci B, Cossi M, Scalmani G, Rega N, Petersson G.A, Nakatsuji H, Hada M, Ehara M, Toyota K, Fukuda R, Hasegawa J, Ishida M, Nakajima T, Honda Y, Kitao O, Nakai H, Klene M, Li X, Knox J.E, Hratchian H.P, Cross J.B, Adamo C, Jaramillo J, Gomperts R, Stratmann RE, Yazyev O, Austin A.J, Cammi R, Pomelli C, Ochterski J.W, Ayala P.Y, Morokuma K, Voth G.A, Salvador P, Dannenberg J.J, Zakrzewski V.G, Dapprich S, Daniels A.D, Strain M.C, Farkas O, Malick D.K, Rabuck A.D, Raghavachari K, Foresman J.B, Ortiz J.V, Cui Q, Baboul A.G, Clifford S, Cioslowski J, Stefanov B.B, Liu G, Liashenko A, Piskorz P, Komaromi I, Martin R.L, Fox D.J, Keith T, Al-Laham M.A, Peng C.Y, Nanayakkara A, Challacombe M, Gill P.M.W, Johnson B, Chen W, Wong M.W, Gonzalez C, Pople J.A, Gaussian, Inc., Pittsburgh PA, 2003.
[92] Becke AD, J. Chem. Phys. 98, 1372 (1993).
[93] Lee C, Yang W, Parr R.G, Phys. Rev. B. 37, 785 (1988).
[94] Belletête M, Beaypré S, Bouchard J, Blondin P, Leclerc M, Durocher G, J. Phys. Chem. B. 104, 9118 (2000).
[95] Matsuzawa N.N, Ishitani A, Doxon D.A, Uda T, J. Phys. Chem. A. 105, 4953 (2001).
[96] Broo A, Zerner MC, Chem. Phys. Lett. 227, 551 (1994).
[97] Sun L, Bai F.Q, Zhao Z.X, Zhang H.X, Sol. Ener. Mat & Sol. Cell. 95, 1800 (2011).
[98] Gorelsky S.I, SWizard program, University of Ottawa, Ottawa, Canada, 2009. http://www.sg-chem.net/
[99] Gorelsky S.I, Lever A.B.P, J. Organometal. Chem. 635, 187 (2001).
[100] Yang L, Ren A.M, Feng J.K, Liu X.J, Ma Y.G, Zhang M, Liu X.D, Shen J.C and Zhang HX, J. Phys. Chem. 108, 6797 (2004).
[101] Ayachi S, Ghomrasni S, Alimi K, J. App. Polym. Sci. 123, 2684 (2012).
[102] Stewart J.J.P, MOPAC 2000 Manual; Fujitsu Ltd.: Tokyo, Japan, 1999.
[103] Botta C, Destri S, Porzio W, Rossi L, Tubino R, Synth. Met. 95, 53 (1998).
[104] Lin V, Grasselli J.G, Colthup N.B (1989) Handbook of Infrared and Raman Characteristics Frequencies of Organic Molecules; Academic Press, Harcourt Brace Jovanovich Publishers: New York.
[105] Ip J, Nguen T.P, Le Rendu P, Tran V.H, Synth. Met. 122, 45 (2001).
[106] Murali M.G, Naveen P, Udayakumar D, Yadav V, Srivastava R, Tetrahed. Lett. 53, 157 (2012).
[107] Fukuda M, Sawada K, Yoshino K, J. Polym. Sci., Part A: Polym. Chem. 31, 2465 (1993).
[108] Meille S.V, Farina A, Bezziccheri F, Gallazzi M.C, Adv. Mat. 6, 848 (1994).

[109] DiCésare N, Belletête M, Leclerc M, Durocher G, J. Mol Struct: THEOCHEM. 467, 259 (1999).
[110] Leriche P, Frère P, Roncali J, J. Mater. Chem. 15, 3473 (2005).
[111] Savitha G, Hergué N, Guilmet E, Allain M, Frère P, Tetrahed. Lett. 52, 1288 (2011).
[112] Clarke T. M, Gordon K. C, Officer D. L, Hall S. B, Collis G. E, Burrell A. K, J. Phys. Chem. A. 107, 11505 (2003).
[113] Yang L, Feng J.K, Ren A.M, Polym. 46, 10970 (2005).
[114] Yang L, Feng J.K, Liao Y, Ren A.M, Opt. Mat. 29, 642 (2007).
[115] Irving D.L, Devine B.D, Sinnott S.B, J. Lumin. 126, 278 (2007).
[116] Casanovas J, Zanuy D, Aleman C, Polym. 46, 9452 (2005).
[117] Wu P-T, Kim, F.S, Champion R.D, Jenekhe S.A, Macromol. 41, 7021 (2008).
[118] May, V, Kühn O (2000) Charge and energy transfer dynamics in molecular system, Wiley-VCH, Berlin.
[119] Bouzakraoui S, Bouzzine S.M, Bouachrine M, Hamidi M, J. Mol. Struct. (THEOCHEM) 725, 39 (2005).
[120] Cao H, Ma J, Zhang G.L, Jiang Y.S, Macromol. 38, 1123 (2005).
[121] Lee Y, Russell T.P, Jo W.H, Org. Elec. 11, 846 (2010).
[122] Yang L, Feng J.K, Ren A.M, Sun C.C, Polym. 47, 3229 (2006).
[123] Koster L.J.A, Mihailetchi V.D, Blom P.W, Appl. Phys. Lett. 88, 093511 (2006).
[124] Kato S, Matsumoto T, Shigeiwa M, Gorohmaru H, Maeda S, Ishi-i T, Mataka S, Chem. Eur. J. 12, 2303 (2006).
[125] Wei H, Scudiero L, Eilers H, Appl. Surf. Sci. 255, 8593 (2009).
[126] Mabrouk A, Alimi K, Hamidi M, Bouachrine M, Molinie P, Polym. 46, 9928 (2005).
[127] Chen J.W, Cao Y, Acc. Chem. Res. 42, 1709 (2009).
[128] Rong Z.C, Jiang L.Z, Hong C.Y, Shan C.H, Zhi W.Y, Hua Y.L, J. Mol. Str. THEOCHEM 899, 86 (2009).
[129] He Y, Chen H.Y, Hou J.H, Li Y.F, J. Am. Chem. Soc. 132, 1377 (2010).
[130] Kooistra F.B, Knol J, Kastenberg F, Popescu L.M, Verhees W.J.H, Kroon J.M, Hummelen J.C, Org. Lett. 9, 551 (2007).
[131] Gadisa A, Svensson M, Andersson M.R, Inganas O, Appl. Phys. Lett. 84, 1609 (2004).
[132] Mühlbacher D, Scharber M, Morana M, Zhu Z, Waller D, Gaudiana R, Brabec C, Adv. Mater. 18, 2884 (2006).
[133] Brabec C.J, Cravino A, Meissner D, Sariciftci N.S, Fromherz T, Rispens M.T, Sanchez L, Hummelen J.C, Adv. Funct. Mater. 11, 374 (2001).
[134] Gûnes S, Neugebauer H, Sariciftci N.S, Chem. Rev. 107, 1324 (2007).
[135] Liu J, Shi Y, Yang Y, Adv. Funct. Mat. 11, 420 (2001).
[136] Scharber M.C, Mühlbacher D, Koppe M, Denk P, Waldauf C, Heeger A.J, Barbec C.J, Adv. Mater. 18,789 (2006).

Chapter 6

The Advanced Charge Injection Techniques Towards the Fabrication of High-Power Organic Light Emitting Diodes

Dashan Qin and Jidong Zhang

Additional information is available at the end of the chapter

http://dx.doi.org/10.5772/52446

1. Introduction

1.1. General descriptions of OLEDs

In general, an organic light emitting diode (OLED) contains a system of thin organic layers sandwiched between two electrode layers (anode and cathode). When a voltage is applied, light is generated within the system of organic layers and emerges through one of the transparent electrodes. The OLED fabricated onto a glass substrate is considered as a surface light source less than 2 millimeters (mm) thick. In contrast to conventional light sources, OLEDs emit wide-area light with high-quality color rendering, which is very pleasant for the human eye, and also needs no reflectors to reduce glare. As a result, OLEDs are one of the hottest research areas being intensely developed in the photonic industry.

The thin organic layers of an OLED can be prepared either via the vacuum deposition or solution casting, with a total thickness of approx. 200-400 nanometers (nm). If the solution casting is employed to make organic or polymeric thin films, the under layer must be free from the destruction brought by the solvent of the upper layer, therefore, solution-processed OLEDs often have only one or two organic layers. However, if the vacuum deposition is adapted to make organic small-molecule thin films, the upper layer causes no harm to the under ones. Thus, vacuum-deposited OLEDs are always multi-layered. Compared to the solution casting, the vacuum deposition enables OLEDs to have much higher injection efficiency, luminance, and lifetime.

1.2. Hole injection in OLEDs

The light emission from an electrically driven OLED occurs due to the recombination of positive and negative charge carriers, hereafter called as holes and electrons, respectively

[1]. The number of holes (or the hole current) is limited by the hole injection from anode to organic layer, which is controlled by the Schottky barrier height at the anode/organic interface. If pristine based organic hole transporters are used, for example, N,N'-bis-(1-naphthl)-diphenyl-1,1'-biphenyl-4,4'-diamine (NPB), 4,4'-N,N'-dicarbazole-biphenyl (CBP), 4,4',4''-tris(N-3-methylphenyl-N-phenylamino)triphenylamine (m-MTDATA), because the Fermi level of the high-work function anode is pinned at about 0.5-0.6 eV above the energy level of highest occupied molecular orbital (HOMO) of organic hole transporters [2, 3], the injection barrier is always no less than 0.5 eV, leading to inefficient hole injection and thereby high driving voltage in such OLEDs. However, if the p-doped materials are utilized, an efficient hole injection can be achieved. In this case, athough the barrier height of hole injection remains almost the same and unchanged by the intervention of the p-doped layer, holes can easily tunnel into organic layer of an OLED through the very thin depletion zone formed at the interface of p-doped material and anode even at very low driving voltage [4]. In this chapter, we introduce a method of efficient hole current generation at the interface of electron donor and acceptor, in clear contrast to the above-mentioned hole injection technologies, and discuss its potential applications in OLEDs [5-7].

1.3. Electron injection in inverted OLEDs

The active matrix displays based on OLEDs have been successfully applied to portable electronic devices, e.g., mobile phones and music players. In order to facilitate the large-scale commercialization for active matrix OLED displays, it is of great necessity to reduce their fabrication cost. Hence, the n-channel amorphous silicon thin-film transistor technology may preferably be utilized to drive the light emitting elements, but requires the OLED structure with inverted layer sequence [8, 9]. Compared to the regular OLEDs, higher driving voltages are always obtained in the inverted ones due to the poor electron injection as a consequence of the inefficient metal penetration into organics [10,11]. Thus, the n-doped electron transport layers (n-ETLs) are adopted to enhance the electron current in inverted OLEDs [4], e.g., lithium doped bathcuproine (BCP). However, the electron injection from the cathode into n-ETL in the inverted OLED is found always less efficient than that in the regular OLED. Thomschke et al. [12] improved the electron injection in a n-i-p OLED via inserting an interlayer of Bphen double-doped with cesium and silver between the silver cathode and cesium-doped Bphen, followed by the thermal annealing of the whole device. Though this method led to almost the same electrical properties in the inverted OLEDs as those in the regular OLEDs, it is relatively complicated and therefore unsuitable for use in the mass production of the active matrix OLED displays. Here, we introduce the increased electron injection in inverted bottom-emission OLEDs (IBOLEDs) via using the combination of two n-ETLs [13-16].

2. Experimental methods

100-nm-thick indium tin oxide (ITO) thin film coated glass substrates were commercially bought with a sheet resistance of 10-30 Ω per square, used as the anode in the conventional OLEDs and as the cathode in the inverted OLEDs. After being cleaned in acetone, alcohol, and de-ionized water sequentially by an ultrasonic horn, the patterned ITO substrates were blown dry by nitrogen gun and then treated in UV-ozone for 15 min. The base vacuum pressures of the thin film and device fabrications were 1×10^{-5} to 4×10^{-4} Pa.

Copper phthalocyanine (CuPc, electron donor) and 3, 4, 9, 10 perylenetetracarboxylic dianhydride (PTCDA, electron acceptor) were used to form the interfaces of generating holes and electrons. Molybdenum oxide (MoO_3) and NPB were adopted as hole injection and transport materials, respectively. Tris(8-quinolinolato) aluminum (Alq3) and BCP were chosen as the emissive and electron transport materials, respectively. The lithium carbonate was selected as an n-dopant to the PTCDA and BCP. 1,4,5,8-naphthalene-tetracarboxylic-dianhydride (NTCDA) was also chosen to form an n-doped material with ITO.

The current versus voltage (I-V) characteristics of the devices were measured by a programmable Keithley 2400 sourcemeter or a Keithley electrometer 617, and the device luminance was recorded by an ST-86LA spot photometer. The optical absorption spectra of organic thin films were obtained using a Cary 300 spectrophotometer or an UV-3100 spectrophotometer. The X-ray diffraction (XRD) measurements were performed on an X-ray diffractometer (D/max-RB).

3. Results and discussion

3.1. Hole generation at the interfaces of electron donor (CuPc) and acceptor (PTCDA) in OLEDs

Pairs of organic donors and acceptors have been applied in organic solar cells to realize the efficient photo-to-electricity conversion since 1986 [17]. The underlying mechanism for organic photovoltaic devices is the photoinduced electron transfer from donor to acceptor, generating free holes in donor and electrons in acceptor. The donors and acceptors in organic photovoltaic devices constitute either planar interfaces or mixed interfaces. Recently, the planar donor/acceptor interfaces have been successfully used to realize the ambipolar transport in organic field effect transistors [18, 19]. The recent work shows that the planar and mixed interfaces of donor and acceptor can generate very efficient hole current in OLEDs [5-7].

3.1.1. The hole generation at the planar CuPc/PTCDA interface

We firstly fabricated the following light emitting devices:
Device 1: ITO/ CuPc 5 nm/ NPB 75 nm/ Alq3 60 nm/ Mg:Ag/ Ag;
Device 2: ITO/ PTCDA 10 nm/ NPB 70 nm/ Alq3 60 nm/ Mg:Ag/ Ag;
Device 3: ITO/ PTCDA 20 nm/ NPB 60 nm/ Alq3 60 nm/ Mg:Ag/ Ag;
Device 4: ITO/ PTCDA 10 nm/ CuPc 5 nm/ NPB 65 nm/ Alq3 60 nm/ Mg:Ag/ Ag;
Device 5: ITO/ PTCDA 20 nm/ CuPc 5 nm/ NPB 55 nm/ Alq3 60 nm/ Mg:Ag/ Ag;

Note that, the devices 4 and 5 are fabricated with planar PTCDA/ CuPc interfaces for hole injection.

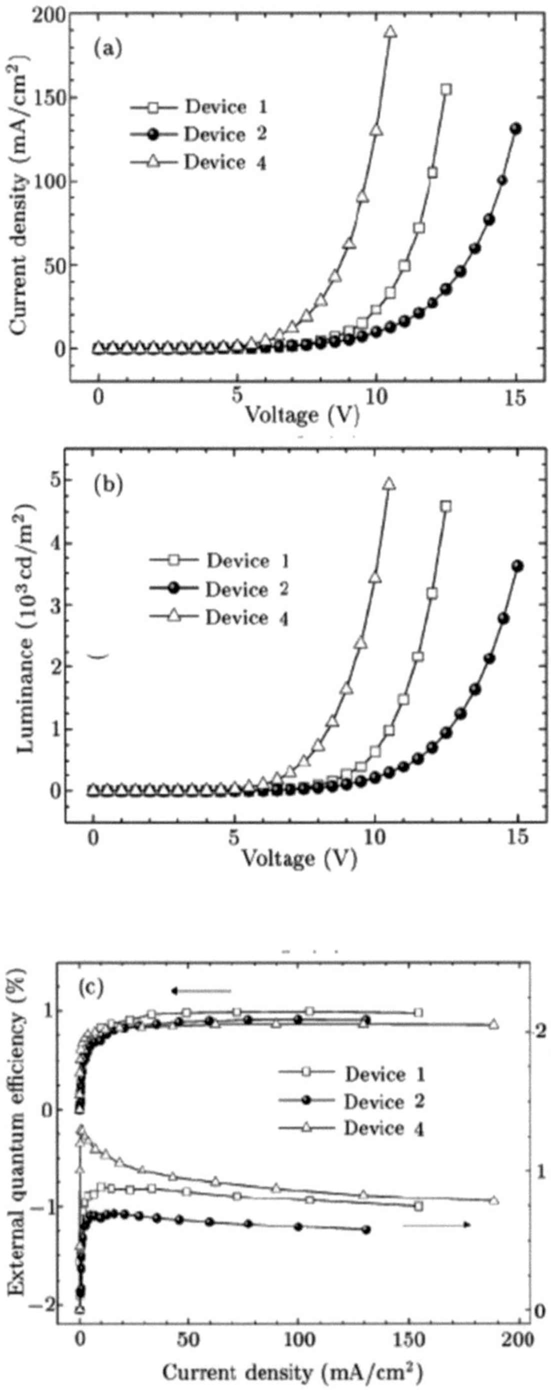

Figure 1. Characteristics of (a) I-V, (b) luminance-voltage (L-V), and (c) efficiency-current density of devices 1, 2, and 4. The external quantum efficiencies are calculated, provided that all of the devices are assumed to function as Lambertian sources.

Fig. 1 shows the properties of devices 1, 2, and 4. As shown in Fig. 1(a), device 4 reaches a current density of 130 mA/cm² at a voltage of 10 V, which is 4.6 times higher than that of 23.2 mA/cm² for device 1 and 12.3 times higher than that of 9.8 mA/cm² for device 2. There is a decrease in the operating voltage at a given current density in the order of device 2 > device 1 > device 4, clearly demonstrating the structure of ITO/ PTCDA 10 nm/ CuPc 5 nm can generate much enhanced hole current than the structures of ITO/ CuPc 5 nm and ITO/ PTCDA 10 nm. Likewise, Fig. 1(b) shows that device 4 also exhibits much higher luminance than devices 1 and 2. As seen in Fig. 1(c), there is a decrease in the external quantum efficiency (EQE) in the order of device 1 > device 2 > device 4, mostly resulting from some absorption in CuPc and PTCDA over the Alq3 emission; nevertheless, the power efficiency at a certain current density decreases in an order of device 4 > device 1 > device 2. It may thus,be concluded that the structure of ITO/ PTCDA 10 nm/ CuPc 5 nm outperforms the structures of ITO/ CuPc 5 nm and ITO/ PTCDA 10 nm in OLEDs.

In order to figure out the hole generation in OLEDs with the ITO/ PTCDA (10 and 20) nm/ CuPc structure, the influence of the PTCDA layer thickness on the device current has been studied. We have shown the observed I-V characteristics of devices 2, 3, 4 and 5 in Fig. 2. It shows that the device 2 (10 nm) has got nearly the same I-V characteristics as device 3 (20 nm), indicating that the hole generation in the ITO/ PTCDA (10 and 20) nm structure is weakly dependent on the PTCDA thickness [20], while device 4 gives much increased current density than device 5, implying that the hole generation in the ITO/ PTCDA (10 and 20) nm/ CuPc structure is strongly subject to the PTCDA layer thickness. It may therefore be concluded that besides a minor process of ITO injecting holes directly into the PTCDA layer [20, 21], devices 4 (10 nm) and 5 (20 nm) possess a major hole-generating process which devices 2 and 3 do not have.

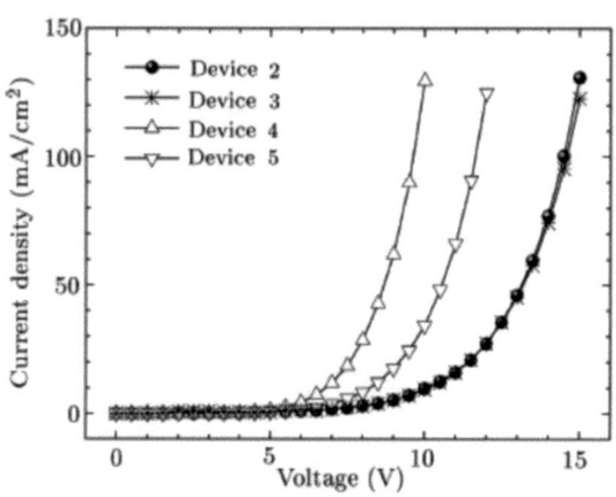

Figure 2. The I-V characteristics of devices 2 (10 nm PTCDA), 3 (20 nm PTCDA), 4 (10nm PTCDA/ CuPc) and 5 (20 nm PTCDA/ CuPc).

Fig. 3 depicts the major hole generation process in the ITO/ PTCDA (20) nm/ CuPc structure. Under the forward bias, due to the influence of holes accumulated at the ITO/ PTCDA interface [22], the PTCDA/ CuPc interface may become polarized, that is, half electrons from the highest occupied molecular orbital (HOMO) of CuPc are transferred into the lowest unoccupied molecular orbital (LUMO) of PTCDA, which is possibly favored by the following three reasons: Firstly, the PTCDA/ CuPc interface can effectively dissociate the photo-generated excitons into hole and electrons and then prevent them from recombining [23]; secondly, there is a strong Coulomb force between PTCDA and CuPc [7], thereby making the overlap between the HOMO of CuPc and LUMO of PTCDA significant, and furthermore the electric field can assist electrons in the HOMO of CuPc overcoming 0.6 eV energy barrier to get into the LUMO of PTCDA; thirdly, the interfacial dipole at the PTCDA/ CuPc interface can facilitate the generation of hole-electron pairs [19] instead of excitons. After electrons get transferred into the LUMO of PTCDA, the vacancies (holes) are generated in the HOMO of CuPc, and those transferred into the PTCDA layer are rapidly transported into the ITO anode, generating very efficient hole current. For the ITO/ PTCDA (10 and 20) nm/ CuPc structure, when the thickness of the PTCDA layer increases, the electrostatic inducement effect of holes accumulated at the ITO/ PTCDA interface on the PTCDA/ CuPc interface, presumably considered proportional to the inverse of the square of the PTCDA layer thickness, decreases, and accordingly the charge generation at the PTCDA/ CuPc interface becomes less efficient, accounting for the decreased current density in device 3 as compared to that in device 4.

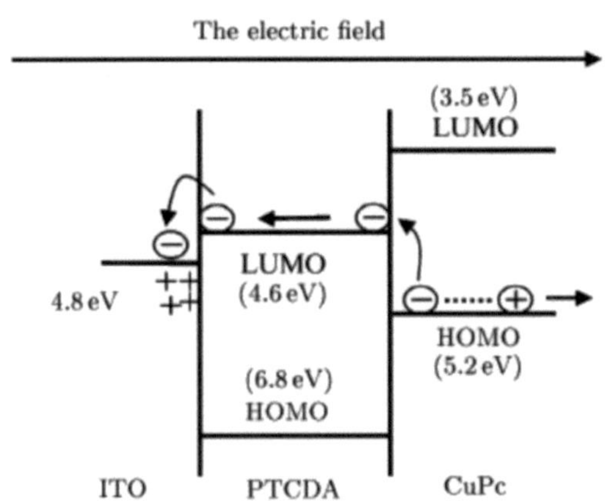

Figure 3. Schematic flat-band energy level diagram for description of the major hole-generating process in devices 4 and 5. The energy level alignments at the PTCDA/ CuPc interface have been ignored for simplicity.

3.1.2. Characterization of OLEDs with ITO/ CuPc and ITO/PTCDA:CuPc anodes

We compare the performance of two devices with structures of ITO/CuPc 5 nm/NPB 75 nm/Alq3 60 nm/Mg:Ag and ITO/PTCDA:CuPc (1:2 in mass ratio) 10 nm/NPB 70 nm/Alq3 60 nm/Mg:Ag, respectively. As shown in Fig. 4(a), the device with the ITO/ PTCDA:CuPc anode has higher current density and luminance at the same driving voltage compared to the device with the ITO/CuPc anode. In addition, as shown in Fig. 4(b), the latter has better current and power efficiencies than the device with the ITO/CuPc anode. For example, at a brightness of 2000 cd/m2, the device with the ITO/PTCDA:CuPc anode has a current efficiency of 2.7 cd/A and a power efficiency of 1 lm/W, whereas the device with the ITO/CuPc anode has 2.3 cd/A and 0.7 lm/W, respectively. Therefore, the ITO/PTCDA:CuPc anode structure is more efficient than the ITO/CuPc anode structure.

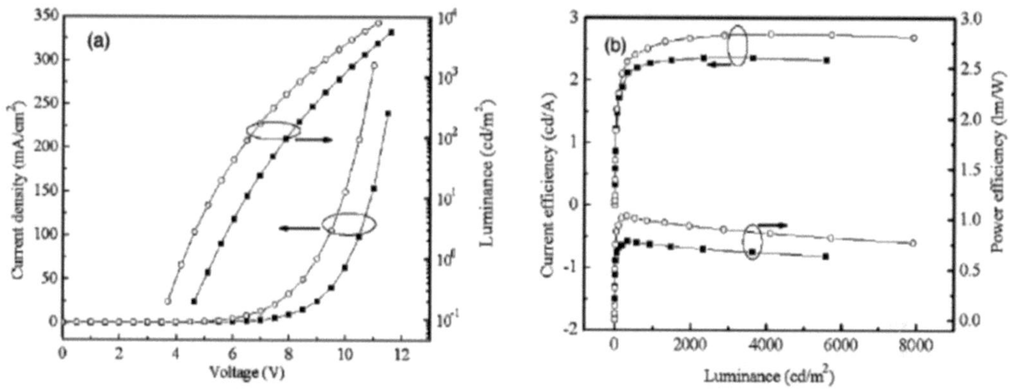

Figure 4. (a) I-V-L and (b) efficiency-luminance characteristics of two OLEDs with structures of ITO/CuPc 5 nm/NPB 75 nm/Alq3 60 nm/Mg:Ag (filled squares) and ITO/PTCDA:CuPc (1:2 in mass ratio) 10 nm/NPB 70 nm/Alq3 60 nm/Mg:Ag (open circles), respectively.

The influence of PTCDA:CuPc thickness on the device performance was studied. Fig. 5(a) shows that the maximum current efficiencies of the devices with 10 and 20 nm PTCDA:CuPc are slightly different and higher than those of the devices with 5 and 30 nm PTCDA:CuPc. The driving voltage for the devices with 10, 20, and 30 nm PTCDA:CuPc, such as at I=50 mA/cm2 or I=150 mA/cm2, is very close and all are lower than that for the device with 5 nm PTCDA:CuPc. Fig. 5(b) shows that the power efficiency of the devices with 10 and 20 nm PTCDA:CuPc is nearly the same and both are higher than those of the devices with 5 and 30 nm PTCDA:CuPc. It may therefore be concluded that the ITO/ PTCDA:CuPc anode structure is better when the thickness of PTCDA:CuPc is changed from 10 to 20 nm.

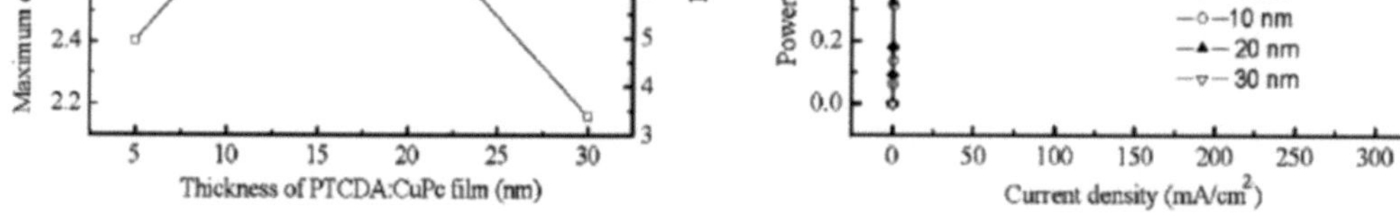

Figure 5. (a) Maximum current efficiency and driving voltage vs PTCDA:CuPc thickness and (b) plots of power efficiency as a function of current density for OLEDs with structure of ITO/1:2 PTCDA:CuPc x nm/NPB 80-x nm/Alq3 60 nm/Mg:Ag (open circles), where the PTCDA:CuPc thicknesses are 5, 10, 20, and 30 nm, respectively.

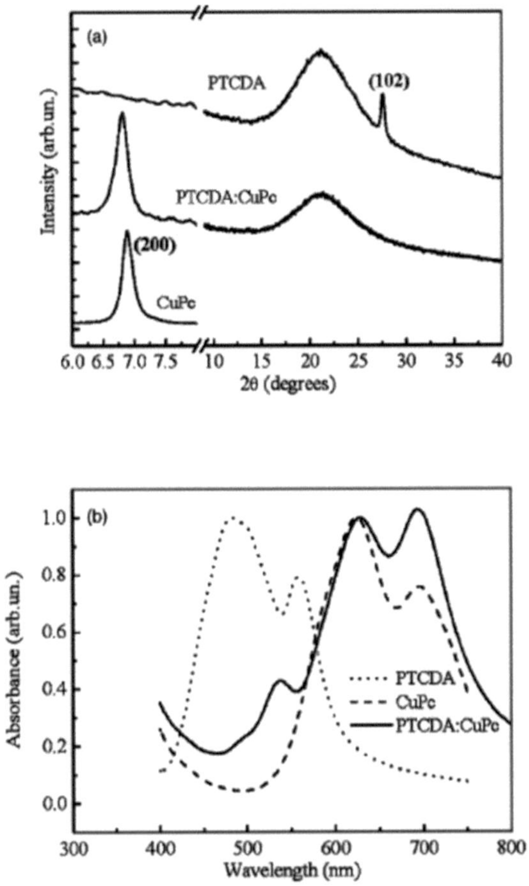

Figure 6. (a) XRD patterns and (b) absorption spectra for 100 nm CuPc, PTCDA, and 1:2 PTCDA:CuPc thin films deposited on quartz glasses.

The structural and optical properties of 1:2 PTCDA:CuPc composite have also been studied. As shown in Fig. 6(a), for the mixed PTCDA:CuPc thin film, there is a prominent peak present at 2θ =6.8°, corresponding to diffraction from the (200) plane of the α-CuPc phase [18], and a broad peak present at 2θ =21.0°, assigned to the reflection from the quartz substrate. This clearly demonstrates that some CuPc aggregates are crystalline, while almost all PTCDA aggregates are amorphous in the 1:2 PTCDA:CuPc film. Intermolecular hydrogen bonding is possibly formed between the outer ring of CuPc and carboxylic dianhydride of PTCDA [25]. Thus, during the codeposition, CuPc could be bonded to PTCDA and accordingly destroy PTCDA (102) aggregates, but PTCDA does not affect all of the CuPc (200) stacks, presumably ascribed to the smaller amount of PTCDA than CuPc. Fig. 6(b) displays the electronic absorption spectra of CuPc, PTCDA, and PTCDA:CuPc films. In the low-energy Q band of CuPc, as compared to CuPc film, PTCDA:CuPc film shows a decreased relative intensity of the CuPc dimeric peak at λ=629 nm to the CuPc monomeric peak at λ=693 nm, consistent with the earlier analyses of XRD data [25]. In the high-energy zone, there is a prominent absorption peak at λ=536 nm for PTCDA:CuPc film; the broad absorption feature at λ=539 nm for neat PTCDA film [26] is strongly suppressed in PTCDA:CuPc film .

Hole generation in the ITO/PTCDA:CuPc anode cannot be simply described as holes being directly injected from ITO to PTCDA:CuPc, as the hole injection barrier from ITO to PTCDA:CuPc is not reduced, as compared to that from ITO to CuPc [5]. In order to figure out the working mechanism of the ITO/PTCDA:CuPc hole injection structure, the dependence of the device performance on NPB spacer thickness was studied. As shown in Fig. 7(a), with increasing NPB spacer thickness, the device current decreases significantly; but when the NPB spacer thickness is greater than or equal to 20 nm, the devices exhibit nearly identical I–V characteristics. Fig. 7(b) shows that the device efficiency decreases with increasing NPB spacer thickness.

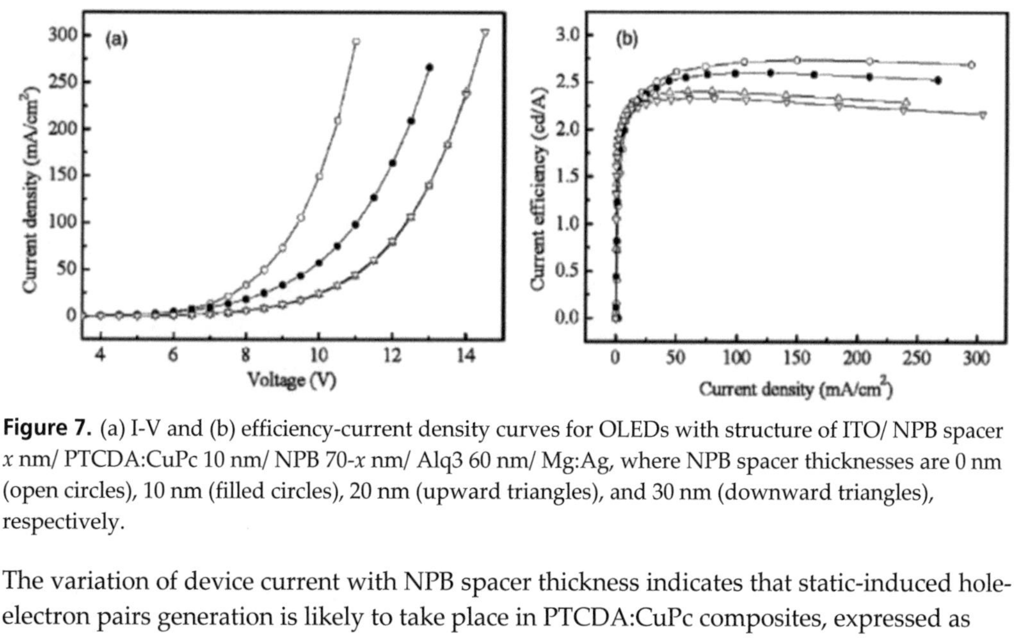

Figure 7. (a) I-V and (b) efficiency-current density curves for OLEDs with structure of ITO/ NPB spacer x nm/ PTCDA:CuPc 10 nm/ NPB 70-x nm/ Alq3 60 nm/ Mg:Ag, where NPB spacer thicknesses are 0 nm (open circles), 10 nm (filled circles), 20 nm (upward triangles), and 30 nm (downward triangles), respectively.

The variation of device current with NPB spacer thickness indicates that static-induced hole-electron pairs generation is likely to take place in PTCDA:CuPc composites, expressed as

$$\mathrm{CuPc + PTCDA} \xrightarrow{\mathrm{SI}} \mathrm{CuPc^{+} + PTCDA^{-}}, \tag{1}$$

where SI stands for static inducement. One paired electron in the HOMO of one CuPc molecule can be transferred to the LUMO of one PTCDA molecule in proximity, favored by (i) the PTCDA/CuPc interface can effectively dissociate the photogenerated excitons into holes and electrons and prevent them from recombining [23] and (ii) there is intermolecular hydrogen bonding between PTCDA and CuPc, and consequently the overlap between the HOMO of CuPc and the LUMO of PTCDA becomes significant. Thus, via process (1) holes and electrons can be efficiently generated in the HOMO of CuPc and in the LUMO of PTCDA, respectively, in PTCDA:CuPc composites. Note that the static inducement effect decreases rapidly with increasing NPB spacer thickness and almost vanishes when the NPB spacer thickness is greater than or equal to about 20 nm, as shown in Fig. 7(a). The working mechanism of the ITO/PTCDA:CuPc anode is described as follows: when the ITO is biased positive, holes confined at the ITO/PTCDA:CuPc interface may effectively polarize PTCDA:CuPc pairs, efficiently generating holes in CuPc and electrons in PTCDA, and then holes are transferred into the NPB layer by CuPc aggregates and electrons into the ITO by PTCDA aggregates, thus generating a very efficient hole current. The static-inducement for very efficient hole-electron pair generation in PTCDA:CuPc is the reason why the ITO/PTCDA:CuPc anode structure is superior to the ITO/CuPc anode structure. For the ITO/ PTCDA:CuPc anode structure, if the thickness of the PTCDA: CuPc composite (such as 5 nm) is smaller than the optimal thickness of 10–20 nm, the number of PTCDA:CuPc pairs involved in hole-electron pair generation is smaller, hence generating less efficient hole current and resulting in reduced device performance. If the thickness of the PTCDA:CuPc composite is larger (e.g. 30 nm) than the optimal thickness of 10–20 nm, there is no more enhancement of hole-electron pairs generation, but the absorption of PTCDA:CuPc composite over Alq3 emission becomes remarkable, causing the degradation in device performance.

3.1.3. The significance and future development of the donor/ acceptor interface in OLEDs

Very recently, the planar and mixed donor/acceptor interfaces have also been used as very efficient interconnecting structures for tandem OLEDs [27], showing the versatility of the donor/acceptor pairs. However, it should be noted that all the donors and acceptors currently used to form the hole-generating interfaces in OLEDs exhibit significant absorption in the visible-light range, thus reducing the light out-coupling from the devices. Therefore, it is of great importance to develop wide-bandgap donors and acceptors. In addition, in order to increase the amplitude of hole current at the donor/acceptor interfaces, it is of much necessity to narrow down the offset between the HOMO of donor and LUMO of acceptor.

3.2. Increasing electron current in IBOLEDs via the combination of two n-doped layers

The n-doped organic electron acceptors, e.g., n-NTCDA, n-PTCDA, n-C_{60}, possess markedly higher conductivities but markedly lower capabilities of injecting electrons into electron transport layer (such as BCP, Alq3, etc.), as compared to the frequently used n-doped materials (such as n-BCP, n-Alq3, etc.) in OLEDs. However, recent work [13-16] shows that the combination of the above two classes of n-doped materals may increase current in IBOLEDs as described below.

3.2.1. Inverted OLEDs incorporated with the combination of Li_2CO_3:PTCDA and Li_2CO_3:BCP

Here first we will describe the optical and structural properties of n-doped materials and then the prperties of OLEDs fabricated from these materials.

3.2.1.1. Optical and structural properties of Li_2CO_3:PTCDA and Li_2CO_3:BCP

The optical properties of 1:2 Li_2CO_3:PTCDA and 1:4 Li_2CO_3:BCP are studied here and presented in Fig. 8. As seen in Fig. 8(a), compared to the intrinsic PTCDA, Li_2CO_3:PTCDA exhibits much lower intermolecular optical absorption at 558 nm [26], indicating severe distortion of the ordered π-π stack of PTCDA molecules caused by the doping with Li_2CO_3. In addition, a new sub-gap absorption centered at a wavelength of 686 nm appears, indicating the formation of the charge transfer state between the matrix and dopant in Li_2CO_3:PTCDA composite [28]. Fig. 8(b) shows that the edge of the optical absorption for 1:4 Li_2CO_3:BCP was slightly blue-shifted relative to that of intrinsic BCP, which implies the occurrence of electron transfer from the O^{2-} bonded to Li to BCP 1:4 Li_2CO_3:BCP composite [29].

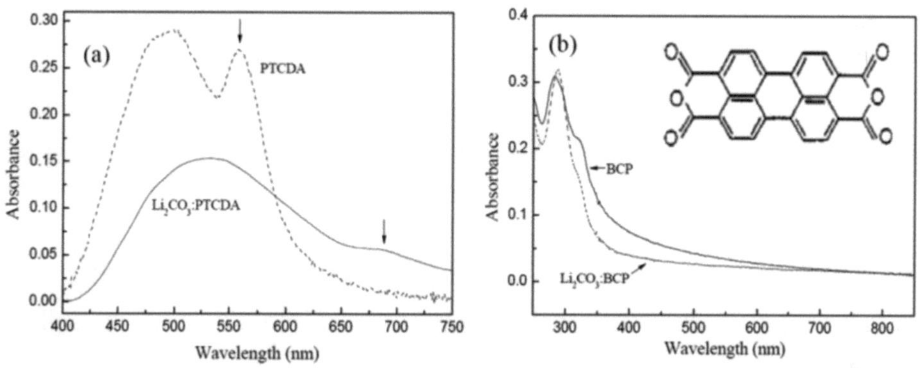

Figure 8. UV-vis absorption spectra for the 20 nm PTCDA and 1:2 Li_2CO_3:PTCDA thin films (a), and the 20 nm BCP and 1:2 Li_2CO_3:PTCDA thin films (b). The inset in (b) is the molecular structure of PTCDA.

3.2.1.2. Electrical properties of Li_2CO_3:PTCDA and Li_2CO_3:BCP

For studying the electrical properties we have fabricated the following two devices:
Device 6: ITO (anode)/ 1:4 Li_2CO_3:BCP 10 nm/ 1:2 Li_2CO_3:PTCDA 5 nm/ Alq3 65 nm/ Al (cathode);
Device 7: ITO (anode)/ 1:4 Li_2CO_3:BCP 5 nm/ 1:2 Li_2CO_3:PTCDA 10 nm/ Alq3 65 nm/ Al (cathode).

In device 6 with the structure of ITO (anode)/1:4 Li_2CO_3:BCP x nm/1:2 Li_2CO_3:PTCDA 15-x nm/Alq3 65 nm/Al (cathode), the electron current passes through Alq3 to 1:2 Li_2CO_3:PTCDA and to 1:4 Li_2CO_3:BCP connected in series. Therefore, the relative conductivity of 1:2 Li_2CO_3:PTCDA to 1:4 Li_2CO_3:BCP can be detected by observing the influence of 1:4 Li_2CO_3:BCP thickness on the device current. Fig. 9 exhibits that the current of device 6 is much lower than that of device 7 at a given driving voltage, demonstrating that 1:2 Li_2CO_3:PTCDA is much more conductive than 1:4 Li_2CO_3:BCP. This can be due to the following two factors: firstly, the charge carriers are more efficiently generated in 1:2 Li_2CO_3:PTCDA than in 1:4 Li_2CO_3:BCP due to the lower LUMO level of PTCDA than that of BCP. Secondly, the electron mobility of intrinsic PTCDA (3×10^{-6} $cm^2 \cdot V^{-1} \cdot s^{-1}$) is much higher than that of BCP (6×10^{-7} $cm^2 \cdot V^{-1} \cdot s^{-1}$). Therefore, it might be feasible to improve the performance of IBOLED via the introduction of the Li_2CO_3:PTCDA composite onto the ITO cathode for enhancing current conduction.

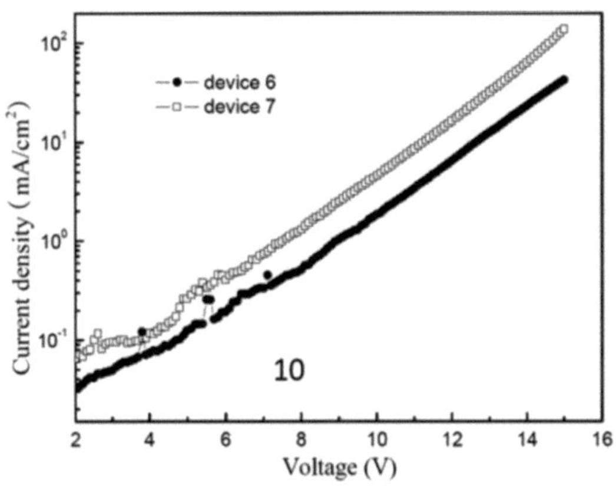

Figure 9. I-V characteristics for devices 6 and 7. As there is no Alq3 emission observed from the two devices in the measurement range, the current in both the two devices is considered to be comprised of electrons only.

3.2.1.3. Inverted OLED using the combination of Li_2CO_3:PTCDA and Li_2CO_3:BCP

We have fabricated the following three inverted devices and studied their characteristics.
Device 8: ITO (cathode)/ 1:2 Li_2CO_3:PTCDA 5 nm/ 1:4 Li_2CO_3:BCP 5 nm/ Alq3 40 nm/ NPB 40 nm/ MoO_3 10 nm/Al (anode);
Device 9: ITO (cathode)/ 1:4 Li_2CO_3:BCP 10 nm/ Alq3 40 nm/ NPB 40 nm/ MoO_3 10 nm/Al (anode);
Device 10: ITO (cathode)/ 1:2 Li_2CO_3:PTCDA 5 nm/ BCP 5 nm/ Alq3 40 nm/ NPB 40 nm/ MoO_3 10 nm/Al (anode).

Fig.10 shows the performance of devices 8 and 9. It can be seen in Fig. 10(a) that the driving voltage of device 8 for achieving a given current was reduced than that of device 9. To generate a current density of 100 mA/cm², device 8 needs a driving voltage of 7.1 V, smaller than that needed by device 9 (8.3 V). Device 8 also gives higher luminance than device 9 as shown in Fig. 10(b). At a driving voltage of 8 V, the luminance of device 8 is 6983 cd/m², in contrast to that of 1354 cd/m² for device 9. Fig. 10(c) shows that the current efficiency of device 8 is slightly lower than that of device 9, mostly because 1:2 Li_2CO_3:PTCDA exhibits some absorption over Alq3 emission, while 1:4 Li_2CO_3:BCP is nearly transparent in the visible-light range. Nevertheless, due to the marked reduction of driving voltage, device 8 provides higher power efficiency than device 9. It can be concluded that the two-layer combination of 1:2 Li_2CO_3:PTCDA and 1:4 Li_2CO_3:BCP outperforms the single 1:4 Li_2CO_3:BCP with regard to promoting the current conduction for the IBOLED.

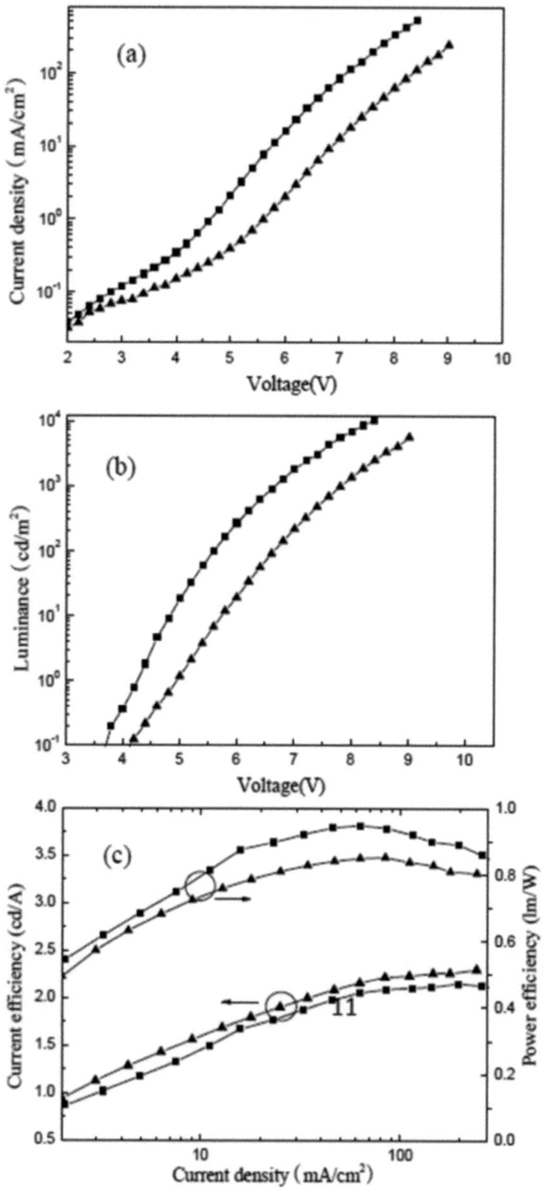

Figure 10. (a) I-V characteristics, (b) L-V characteristics, (c) current and power efficiencies versus current density of devices 8 (filled squares) and 9 (filled triangles). Both the devices give nearly the same Alq3 emission.

Figs. 10 (a) and (b) show that device 8 performs better that device 9, which may be attributed to the enhanced electron current due to the following : (i) In device 8, the electron injection is realized not only via tunneling, but also via thermionic emission due to relatively smaller energy barrier from ITO into the LUMO of PTCDA. (ii) The conductivity of 1:2 Li_2CO_3:PTCDA is higher than that of 1:4 Li_2CO_3:BCP, leading to reduced ohmic loss of current conduction. Fig. 10(c) shows that the current efficiency of device 8 is slightly lower than that of device 9, mostly because 1:2 Li2CO3:PTCDA exhibits some absorption over Alq3 emission, while 1:4 Li2CO3:BCP is nearly transparent in the visible-light range. Nevertheless, due to the marked reduction of driving voltage, device 8 provides higher power efficiency than device 9. In terms of device 8, the issue of the efficient electron transfer from 1:2 Li_2CO_3:PTCDA into 1:4 Li_2CO_3:BCP needs to be addressed, since the LUMO level (4.6 eV) of PTCDA is 1.6 eV lower than that (3.0 eV) of BCP. In quest of the mechanism of electron transport over such a big Schottky barrier, device 10 is fabricated and characterized. As seen in Fig. 11, device 10 shows poor current-voltage and luminance-voltage characteristics. At a driving voltage of 10 V, it exhibits a current of 2.3 mA/cm^2 and a luminance of 4.1 cd/m^2, in clear contrary to 531 mA/cm^2 and 10492 cd/m^2 achieved at a driving voltage of 8.4 V in device 8. This greatly improved performance of device 8 relative to device 10 implies that the energy barrier for electron injection from PTCDA into BCP can be significantly reduced by fulfilling the double-sided n-doping to the Schottky interface.

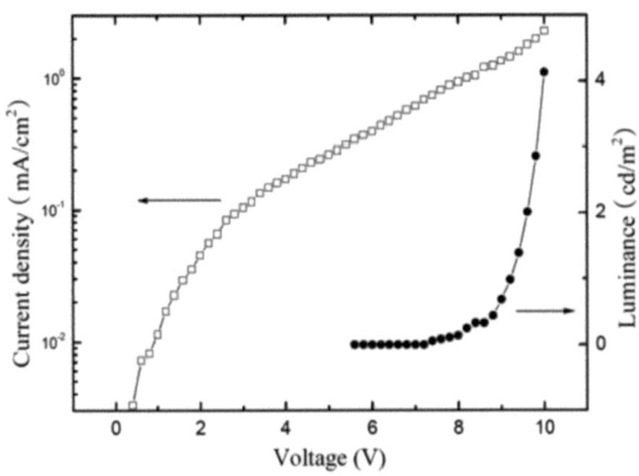

Figure 11. The I-V and L-V characteristics for device 10.

3.2.1.4. The working mechanism of the combination of Li_2CO_3:PTCDA and Li_2CO_3:BCP

The electronic structures of the interfaces of 1:2 Li_2CO_3:PTCDA/BCP and 1:2 Li_2CO_3:PTCDA/1:4 Li_2CO_3:BCP are schematically shown in Fig. 5. The LUMO-LUMO offset at organic heterojunction can be varied by the interfacial dipole Δ given by [30]:

$$\Delta = \left(1 - \frac{1}{2}\left(\frac{1}{\varepsilon_1} + \frac{1}{\varepsilon_2}\right)\right)(CNL_1 - CNL_2)_{initial}, \tag{2}$$

where ε_1 and ε_2 are the low frequency dielectric constants for the two organic materials, CNL_1 and CNL_2 represent their charge neutrality levels. In the case of 1:2 Li_2CO_3:PTCDA/BCP heterojunction, considering that the dielectric constant of n-doped PTCDA is larger compared to that of the undoped PTCDA (1.9) and taking ε_{BCP}~1.4, $(1/\varepsilon_1+1/\varepsilon_2)$ is estimated to be ~0.7. Provided that the CNL for n-doped PTCDA is roughly equal to that of undoped PTCDA(-4.8 eV) due to the Fermi level pinning [22] and CNL of BCP~-3.8 eV, Eq. (2) yields an interface dipole Δ_1=-0.65 eV, resulting a 0.65 eV downwards shift of the vacuum level (VL) on the BCP side LUMO as shown in Fig. 12(i). Thus, the electron injection barrier from the LUMO of 1:2 Li_2CO_3:PTCDA to hat of BCP reduces roughly to 0.95 eV. This shows that the electron transport through the 1:2 Li_2CO_3:PTCDA/BCP interface can be small in device 10. For the 1:2 Li_2CO_3:PTCDA/1:4 Li_2CO_3:BCP heterojunction, the term $(1/\varepsilon_1+1/\varepsilon_2)$ is assumed to be ~0.2, because the low-frequency dielectric constants of the doped materials are larger compared to the undoped materials. Provided the CNL for n-doped PTCDA~-4.8 eV and the CNL for n-doped BCP~3.2 eV [31], Eq. (2) yields an interface dipole Δ_2= -1.44 eV, leading to a 1.44 eV downwards shift of the vacuum level (VL) in the doped BCP side, as shown in Fig. 12(ii). Accordingly, the electron injection barrier at the 1:2 Li_2CO_3:PTCDA/1:4 Li_2CO_3:BCP heterojunction is estimated to be 0.16 eV, suggesting that the double n-doping can induce significant realignment of molecular levels of organic-organic heterojunction. Thus, the contact becomes ohmic, resulting in efficient electron current and increasing device performance (the case of device 8).

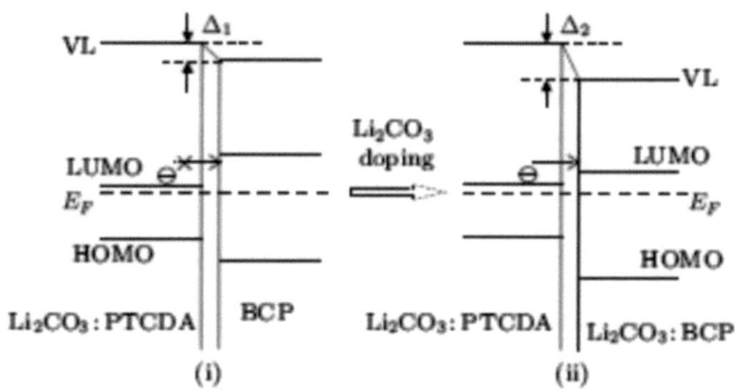

Figure 12. Schematic diagram for depicting the electronic structures of the interfaces of 1:2 Li_2CO_3:PTCDA/BCP (i) and 1:2 Li_2CO_3:PTCDA/1:4 Li_2CO_3:BCP (ii), both deposited onto the ITO substrates. For simplicity, the gap states are not shown. The E_F, HOMO, and horizontal dotted arrow represent the Fermi level, highest occupied molecular orbital, and the electron-injecting process, respectively.

3.2.2. Inverted OLEDs incorporating the combination of MoO_3 and Li_2CO_3:BCP

There has been some controversy about the intrinsic property of MoO_3, which has been intensively investigated in OLEDs for improving the hole injection. It is considered by some groups that MoO_3 should act as a hole conductor with highest occupied molecular orbital (HOMO) and lowest unoccupied molecular orbital (LUMO) levels of ~5.4 eV and ~2.3 eV, respectively [32]. Recently, however, it is suggested [33] that due to being doped by oxygen vacancy defects MoO_3 intrinsically offers n-typed conduction with deep-lying HOMO and LUMO levels of ~9.7 eV and ~6.7 eV, respectively. The enhanced hole injection by the MoO_3 intervention may be attributed to the process that one electron in the HOMO level of organic hole-transporting material can be easily transferred into the LUMO level of MoO_3 because the Fermi level of MoO_3 is necessarily pinned close to the HOMO level of p-typed organic conductor [33]. Here, we investigate MoO_3 as electron injection layer in IBOLEDs and show that MoO_3 in association with an n-doped layer of BCP can enable the efficient electron injection.

3.2.2.1. Conductivity of MoO_3 and Li_2CO_3:BCP

For comparing the conductivity due to the incorporation of MoO_3 and Li_2CO_3:BCP in OLEDs, We have fabricated the following two devices:

Device 11: ITO (cathode)/ MoO_3 10 nm/ 1:4 Li_2CO_3:BCP 5 nm/ Alq3 80 nm/ Al (anode);
Device 12: ITO (cathode)/ MoO_3 5 nm/ 1:4 Li_2CO_3:BCP 10 nm/ Alq3 80 nm/ Al (anode).

The conductivity of an organic thin film can be measured via a vertical sandwiched structure. However, the diffusion of the top metal electrode into organic thin film enables the chemical reaction between metal atoms and organic molecules, leading to the unavoidable alteration for the intrinsic property of organic thin film. Thus, devices 11, 12 are fabricated to qualitatively compare the conductivities of MoO_3 and 1:4 Li_2CO_3:BCP. In the electron-only device with structure of ITO (cathode)/ MoO_3 x nm/ 1:4 Li_2CO_3:BCP 15-x nm/ Alq3 80 nm/ Al (anode), the electrons are firstly injected into MoO_3, then transported through MoO_3, 1:4 Li_2CO_3:BCP, and Alq3 in sequence, and finally transfered into Al. Thus, the study of the I-V characteristics with the varying thickness of MoO_3 can be studied from devices 11 and 12 as shown in Fig. 13. As seen in Fig. 13, the current density of device 11 is greater than that of device 12 in the whole applied voltage range. This clearly indicates that MoO_3 (10 nm thick) is more conducting than 1:4 Li_2CO_3:BCP (10 nm thick) composite. Thus, MoO_3 can be expected to act as a better electron transport layer in the IBOLED

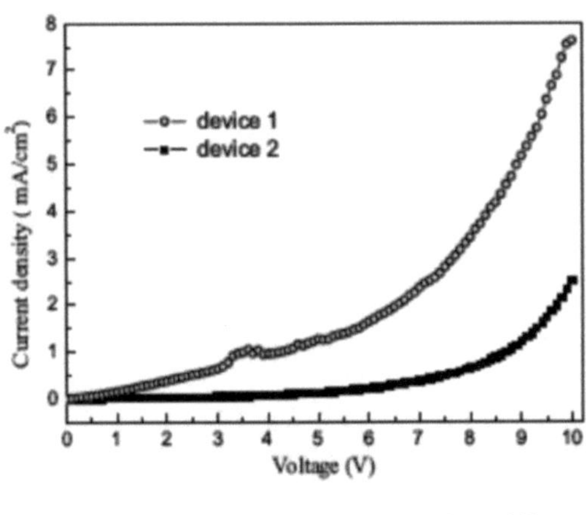

Figure 13. The I-V characteristics of devices 11 (open circles) and 12 (fill squares). No Alq3 electroluminescence is detected in the measurement range for these two devices.

3.2.2.2. Inverted OLEDs using the combination of MoO_3 and Li_2CO_3:BCP

We have fabricated and studied the characteristics of the following IBOLEDs:
Device 13: ITO/ MoO_3 5 nm/ 1:4 Li_2CO_3:BCP 5 nm/ Alq3 40 nm/ NPB 60 nm/ MoO_3 10 nm/Al;
Device 14: ITO/ 1:4 Li_2CO_3:BCP 10 nm/ Alq3 40 nm/ NPB 60 nm/ MoO_3 10 nm/Al;
Device 15: ITO/ MoO_3 5 nm/ BCP 5 nm/ Alq3 40 nm/ NPB 60 nm/ MoO_3 10 nm/Al;

Device 16: ITO/ MoO_3 5 nm/ 1:4 Li_2CO_3:BCP 5 nm/ Alq3 50 nm/ NPB 60 nm/ MoO_3 10 nm/Al;
Device 17: ITO/ MoO_3 5 nm/ 1:4 Li_2CO_3:BCP 5 nm/ Alq3 60 nm/ NPB 60 nm/ MoO_3 10 nm/Al;
Device 18: ITO/ MoO_3 5 nm/ 1:4 Li_2CO_3:BCP 5 nm/ Alq3 70 nm/ NPB 60 nm/ MoO_3 10 nm/Al;

Fig. 14 shows the performances of devices 13 and 14. Device 13 shows nearly the same I-V characteristics as device 14, but the luminance and current efficiency of device 13 are higher than those of device 14. At a luminance of 100 cd/m^2, device 13 achieves a driving voltage of 7.51 V and a current efficiency of 1.39 cd/A, in comparison with 7.72 V and 1.08 cd/A, respectively, reached in device 14. At a luminance of 1000 cd/m^2, device 13 exhibits a driving voltage of 8.86 V and a current efficiency of 2.0 cd/A, in comparison with 9.04 V and 1.62 cd/A obtained in device 14. The improved performance of device 13 over device 14 may be attributed to the reduced dopant diffusion from Li_2CO_3:BCP into Alq3 as a result of the thinner Li_2CO_3:BCP in device 13.

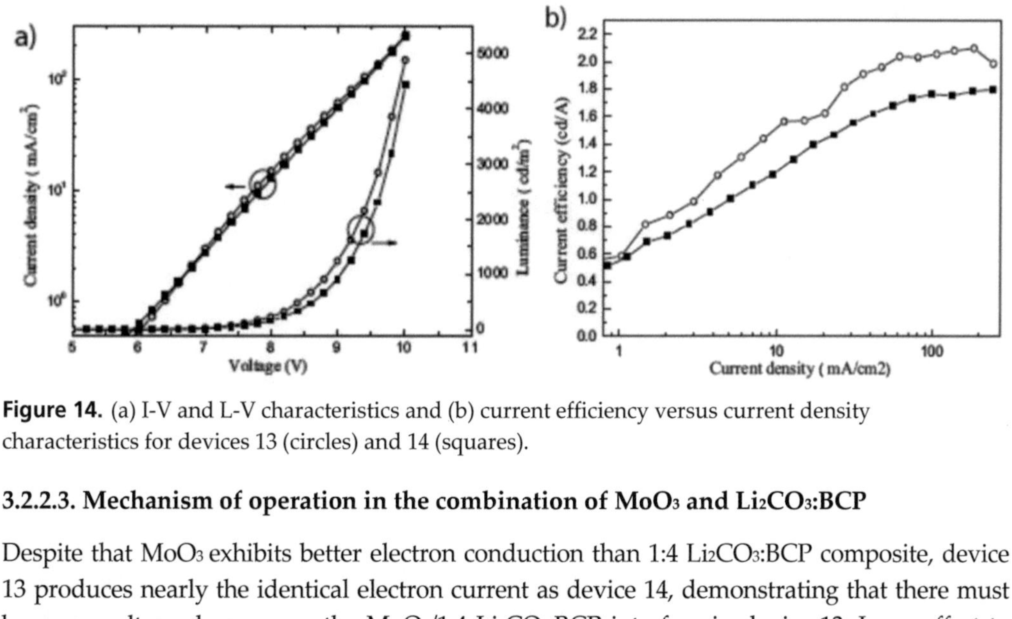

Figure 14. (a) I-V and L-V characteristics and (b) current efficiency versus current density characteristics for devices 13 (circles) and 14 (squares).

3.2.2.3. Mechanism of operation in the combination of MoO_3 and Li_2CO_3:BCP

Despite that MoO_3 exhibits better electron conduction than 1:4 Li_2CO_3:BCP composite, device 13 produces nearly the identical electron current as device 14, demonstrating that there must be some voltage drop across the MoO_3/1:4 Li_2CO_3:BCP interface in device 13. In an effort to investigate the mechanism of the electron transfer from MoO_3 to 1:4 Li_2CO_3:BCP, the characteristics of device 15 is studied as presented in Fig. 15, which shows that device 15 exhibits poor I-V characteristics with no Alq3 electroluminescence observed, suggesting there is a big Schottky barrier for electron injection from MoO_3 to BCP. The marked performance difference between devices 13 and 15 indicates that the LUMO-LUMO offset at the MoO_3/BCP interface can be remarkably lowered due to doping Li_2CO_3 into BCP, which can be explained by the interfacial dipole calculated by Eq. (2). In the case of the MoO_3/BCP heterojunction, taking ε_{MoO3}~20 and ε_{BCP}~1.4 [30], the item of $(1/\varepsilon_1+1/\varepsilon_2)$ is estimated to be ~0.76. Considering that the CNL (work function) of MoO_3 is -6.9 eV [33] and that of BCP~-3.8 eV, Eq. (2) yields an interface dipole Δ_1=-1.92 eV. This results in a 1.92 eV downward shift of the vacuum level on the BCP side. Thus, the Schottky barrier for the electron injection from the LUMO of MoO_3 to that of BCP is roughly 1.78 eV, determining that the electron transport through the MoO_3/BCP interface is very inefficient in the case of device 15. For the MoO_3/1:4 Li_2CO_3:BCP heterojunction, the item of $(1/\varepsilon_1+1/\varepsilon_2)$ is assumed to be ~0.1, because the low-frequency dielectric constant of the doped BCP is larger compared to the undoped one [30]. Provided that the CNL of MoO_3~-6.9 eV and that of n-doped BCP~-3.2 eV [20], Eq. (2) yields an interface dipole Δ_2=-3.52 eV, leading to a 3.52 eV downwards shift of the vacuum level on the doped BCP side. Thus, the electron injection barrier at the MoO_3/1:4 Li_2CO_3:BCP heterojunction is estimated to be roughly 0.18 eV, favoring the efficient electron injection in MoO_3 and hence better device performance (the case of device 13). In addition, the intrinsic Schottky barrier for electron injection from Li_2CO_3:BCP to Alq3 is estimated to be ~0.4 eV according to Eq. (2).

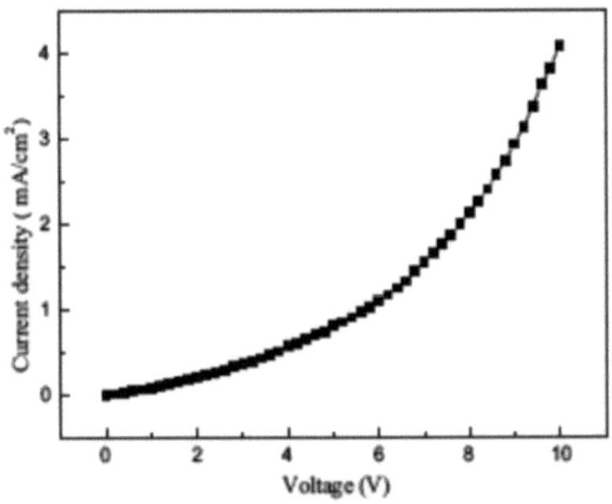

Figure 15. The I-V characteristics of device 15. There was no Aq3 electroluminescence detected in the measurement range, suggesting that the electron current was very inefficient in device 15.

3.2.2.4. The dopant diffusion in the Li_2CO_3:BCP

Fig. 16(a) shows the I-V characteristics of devices 13, 16, 17 and18 and Fig. 16(b) presents the plot of the voltage drop as a function of the thickness of Alq_3 film. The inset of Fig. 16(a) displays the luminance properties of these devices. Fig. 16(b) indicates that the driving voltage across the device varies linearly with the Alq3 thickness when the device current (I) increases from 2 to 100 mA/cm^2. Thus, the average internal electric field in the Alq3 layer ($F_{av, Alq3}$) of device 13 is calculated [34, 35] as shown in Fig. 17. It can be seen in Fig. 17 that $F_{av, Alq3}$ increases monotonously as I increases from 2 to 30 mA/cm^2, and becomes gradually saturated when I > 30 mA/cm^2. The turning point at I = 30 mA/cm^2 of the $F_{av, Alq3}$ versus I characteristics can be due to the efficient diffusion of Li_2CO_3 from Li_2CO_3:BCP into Alq3, which can not only lead to some exciton-quenching effect in the emissive layer, coincident with the observations in Fig. 14(b), but also it can leave the emissive layer n-doped, thereby reducing the interfacial energy barrier [13] and making Alq3 more conductive. Also seen in Fig. 17 is that when I varies between 2 and 30 mA/cm^2, the total device current versus the average electric field in Alq3 appears to follow a Richardson–Schottky (RS) behavior, $I \propto \exp[\beta\cdot(F_{av,Alq3})^{1/2}]$, due to the good linear fit (R^2=0.997) to the log I versus $F^{1/2}_{av, Alq3}$ plot. Accordingly, the coefficient β is calculated to be 28.1 $(MV/cm)^{-1/2}$, approximately one and half times larger than the value of β_{RS}=11.6 $(MV/cm)^{-1/2}$, implied by the RS formula of $\beta_{RS} = (e^3/4\pi\varepsilon_0\varepsilon_r)^{1/2}/k_BT$ (taking $\varepsilon_r \sim 1.6$). This discrepancy is consistent with the other reported observations [34]. It should be stressed here that the electric field at the Li_2CO_3:BCP/Alq3 interface, responsible for electron injection, may be different to the average electric field in the Alq3 layer. Interestingly, when I exceeds 30 mA/cm^2, the total device current versus the average electric field in Alq3 diverts from the RS behavior due to the deteriorated linear fit (R^2=0.989) to the log I versus $F^{1/2}_{av,\ Alq3}$ plot. This diversion is likely to be attributed to the fact that the conclusive electron injection process shifts from the Li_2CO_3:BCP/Alq3 interface to the MoO_3/1:4 Li_2CO_3:BCP heterojunction as a result of the efficient Li_2CO_3 diffusion towards the anode in device 13.

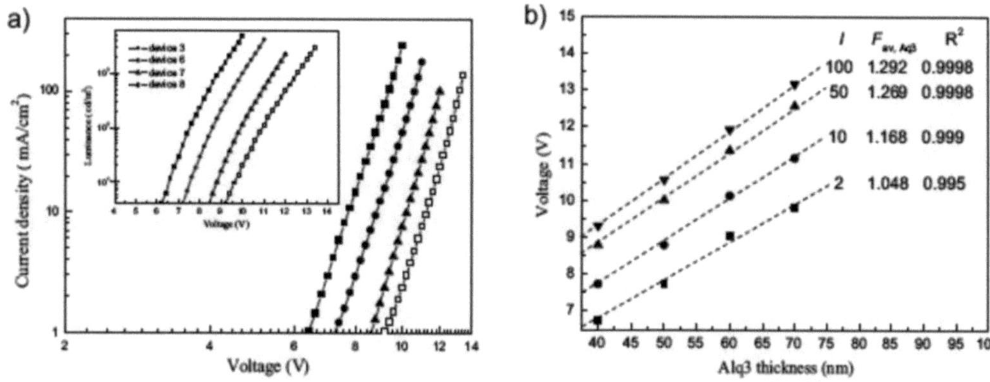

Figure 16. (a) The I-V characteristics of devices 13 (solid squares), 16 (solid circles), 17 (solid triangles), and 18 (open squares). (b) The total voltage drop across device versus the Alq3 thickness at various current densities (I in mA/cm^2). The dashed lines represent the linear fits to the plots, and their slopes denote the the average internal electric fields in the Alq3 layer ($F_{av,\ Alq3}$ in MV/cm). R^2 represents the linear-fit correlation coefficient. The inset (a) shows the L-V characteristics for devices 13, 16, 17 and 18.

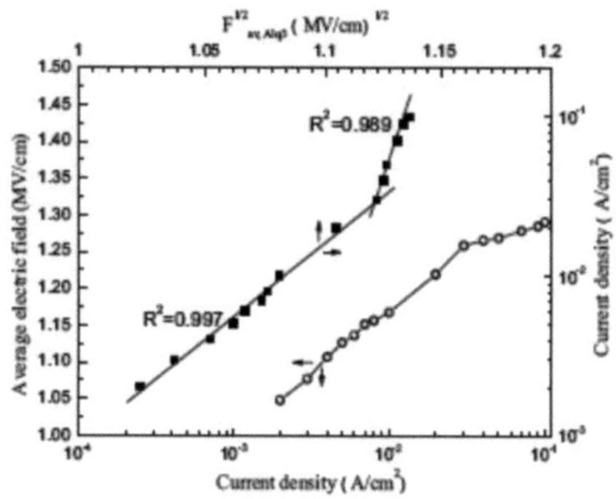

Figure 17. The average internal electric fields in the Alq3 layer (circles, bottom X + left Y) and the dependence of of the total current density on the average internal electric field (squares, top X + right Y) in device 13. The two straight lines with correlation coefficients (R^2) mean the linear fits.

3.2.3. The inverted OLEDs incorporating the combination of n-NTCDA and Li_2CO_3:BCP

To maximize the performance for the combination of two n-ETLs, the n-ETL contacting the cathode needs to meet the two requirements: high conductivity and good transparency over the visible light range. Thus, the n-NTCDA was chosen to function with Li_2CO_3:BCP.

3.2.3.1. The properties of the NTCDA deposited onto the ITO

NTCDA is a class of n-type organic semiconductor with a LUMO level at 4.0 eV [36]. It can have the electronic interactions with the active metals like In or Mg at room temperature, which is attributed to the partial charge transfer from the metal to carbonyl oxygen of NTCDA [37]. It is also reported that NTCDA shows strong electronic interaction with metal atoms in an ITO substrate to form charge transfer (CT) complexes or gap states [38]. The formation of a CT complex can not only lower the energy barrier for hole injection from ITO into N,N'-bis-(1-naphthl)-diphenyl-1,1'-biphenyl-4,4'-diamine, but also can offer a conducting path to assist hole transport [38]. Thus, the introduction of NTCDA onto the ITO anode can significantly improve the hole current and thereby the performance of regular OLED. Nevertheless, it needs to be stressed that the electronic interaction between active metal and NTCDA can leave NTCDA n-doped via the spontaneous charge transfer from metal atom to NTCDA [37].

The four NTCDA thin films have been fabricated and characterized in Fig. 18. Fig. 18(a) shows that the NTCDA thin films both deposited on ITO and glass are crystalline, proven by their same diffraction peaks at $2\theta=11.8^{\circ}$. Fig. 18(b) shows the UV-VIS absorption spectra of the four NTCDA thin films. All the NTCDA thin films gave two peaks at the wavelengths of 368 nm and 390 nm assigned to π-π^*. Compared with the 15 nm NTCDA thin film on glass, the 15 nm NTCDA thin film on ITO shows a broad absorption peak from 420 nm to 600 nm, which becomes weaker with the decreasing thickness of the NTCDA on ITO [39]. The emergence of a new sub-gap absorption indicates the formation of the CT complex between NTCDA and ITO as a result of the upward diffusion of In atoms from ITO into NTCDA and the concomitant interaction between In and anhydride groups of NTCDA. Thus, the NTCDA thin films on the ITO are conclusively crystalline and n-doped. The conductivity of the n-doped NTCDA (n-NTCDA) is reported to be as high as 9.29×10^{-4} S/cm [4], almost two orders of magnitude higher than that of n-BCP [31]. Hence, according to the concept of using two n-ETLs mentioned above, the NTCDA on the ITO may be applied as the first n-ETL to enhance the electron conduction in IBOLEDs.

3.2.3.2. The comparison between the electron injection structures of ITO/ BCP:Li_2CO_3 (4:1) 10 nm and ITO/NTCDA 3 nm/ BCP:Li_2CO_3 (4:1) 7 nm

For this study we fabricated the following IBOLEDs:
Device 19. ITO/ BCP:Li_2CO_3 (4:1) 10 nm/ Alq3 40 nm/ CBP 60 nm/ MoO_3 10 nm/Al;
Device 20. ITO/NTCDA 3 nm/ BCP:Li_2CO_3 (4:1) 7 nm /Alq3 40nm/CBP 60 nm/ MoO_3 10 nm/Al;
Device 21. ITO/NTCDA 3 nm/ BCP 7 nm /Alq3 40nm/ CBP 60 nm/ MoO_3 10 nm/Al;

Device 22. ITO/NTCDA 5 nm/ BCP:Li_2CO_3 (4:1) 5 nm /Alq3 40nm/ CBP 60 nm/ MoO_3 10 nm/Al;

Device 23. ITO/NTCDA 7 nm/ BCP:Li_2CO_3 (4:1) 3 nm /Alq3 40nm/ CBP 60 nm/ MoO_3 10 nm/Al.

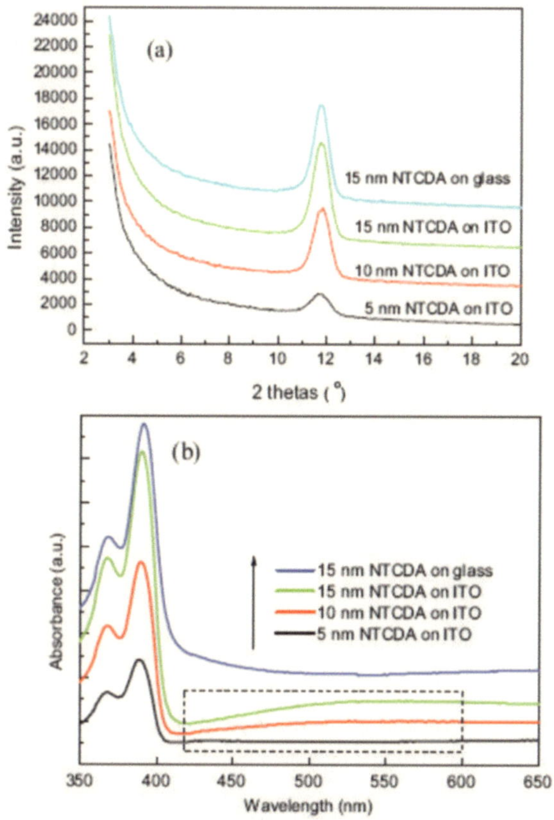

Figure 18. The XRD (a) and UV-Vis absorption spectra (b) of the 5, 10, 15 nm NTCDA thin films on the ITO and the 15 nm NTCDA thin film on the glass. Due to the limitation of the instrument, we were unable to detect the sub-gap absorption in the 5 nm NTCDA on the ITO. Note that, the 3 nm NTCDA thin film on the ITO is also confirmed to be crystalline.

Fig. 19(a) shows the I-V characteristics of devices 19 and 20. As expected, device 20 showed higher current density than device 19 at a given voltage between 3 and 10 V. To generate a current density of 100 mA/cm^2, device 20 needed a driving voltage of 8.0 V, smaller than that for device 19 (9.0 V). Fig. 19(b) shows device 20 also exhibits enhanced luminance than device 19. At a driving voltage of 10 V, the luminance of device 20 is found to be 11706 cd/m^2, in comparison with 4016 cd/m^2 for device 19. Fig. 19(c) shows that the maximum current efficiency in device 20 is 2.29 cd/A, 38% higher than that of 1.66 cd/A in device 19, mostly attributed to the following two factors: firstly, the intervention of the NTCDA improved the hole-electron balance in device 20 relative to device 19; secondly, the NTCDA was nearly transparent in the visible-light range. Hence, it may be concluded that the combination of n-NTCDA and 1:4 Li_2CO_3:BCP excels the single 1:4 Li_2CO_3:BCP in promoting the current conduction and efficiency of the IBOLED.

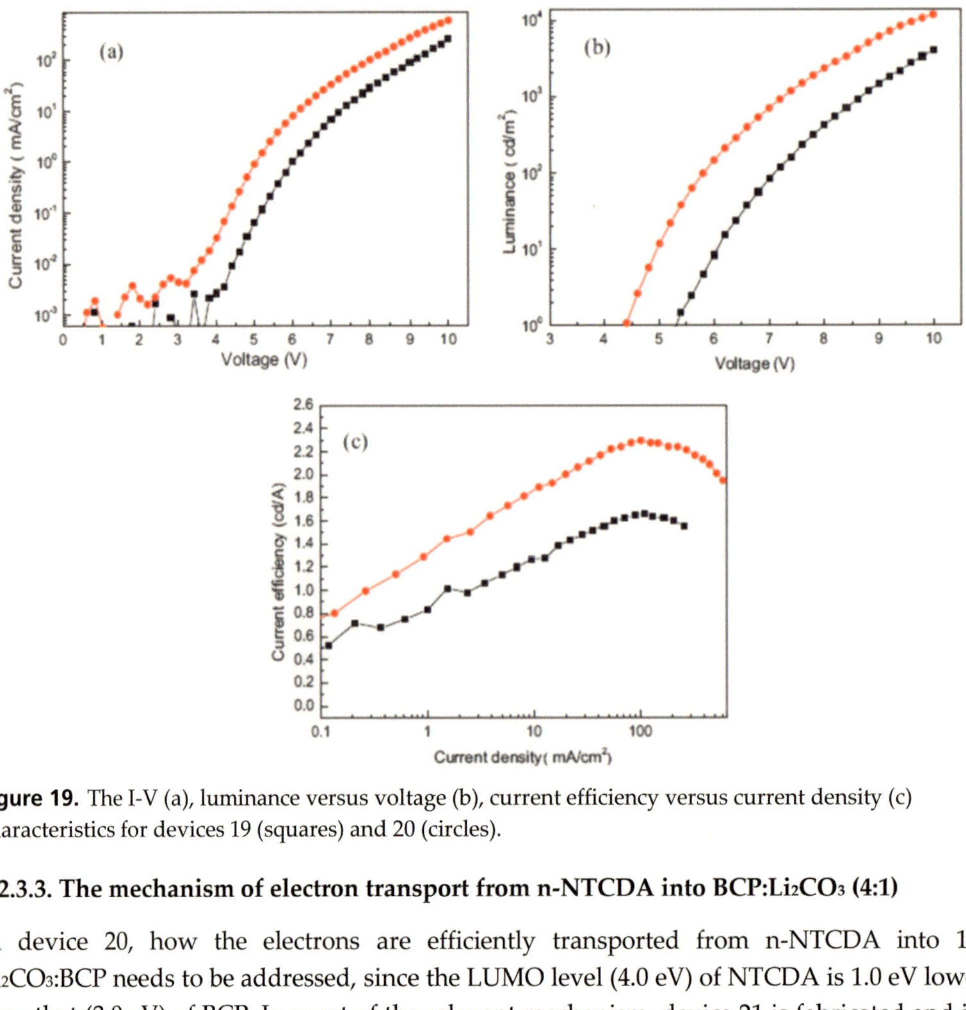

Figure 19. The I-V (a), luminance versus voltage (b), current efficiency versus current density (c) characteristics for devices 19 (squares) and 20 (circles).

3.2.3.3. The mechanism of electron transport from n-NTCDA into BCP:Li_2CO_3 (4:1)

In device 20, how the electrons are efficiently transported from n-NTCDA into 1:4 Li_2CO_3:BCP needs to be addressed, since the LUMO level (4.0 eV) of NTCDA is 1.0 eV lower than that (3.0 eV) of BCP. In quest of the relevant mechanism, device 21 is fabricated and its characteristics are shown in Fig. 20. As seen in Fig. 20, device 21 exhibits poor current--voltage characteristics. At a driving voltage of 10 V, it yields a current of 3.63 mA/cm^2 and no light emission, which is clearly contrary to 601 mA/cm^2 and 11709 cd/m^2 achieved in device 20. The greatly improved performance of device 20 relative to device 21 indicates that the energy barrier for electron injection from n-NTCDA into BCP is significantly reduced by fulfilling the n-doping on the BCP side [13].

The electronic structures of the interfaces of n-NTCDA/BCP and n-NTCDA/1:4 Li_2CO_3:BCP are schematically shown in Fig. 21. The LUMO-LUMO offset at the organic heterojunction can be varied by the interfacial dipole (Δ) expressed as Eq. (2). In the case of n-NTCDA/BCP heterojunction, considering the dielectric constant of n-NTCDA is large

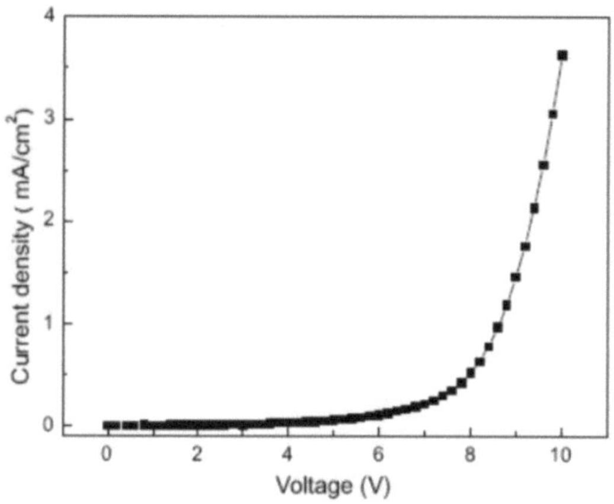

Figure 20. The I-V characteristics of device 21. Note that, there was no Alq3 emission observed in the measurement range.

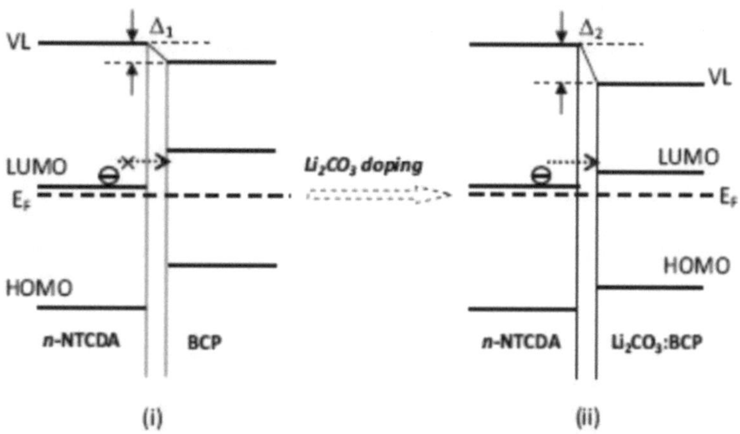

Figure 21. Schematic diagrams for depicting the electronic structures at the interfaces of n-NTCDA/BCP (i) and n-NTCDA/1:4 Li_2CO_3:BCP (ii), both deposited onto the ITO substrates. For simplicity, the gap states are not shown. The E_F, HOMO, and horizontal dotted arrow represent the Fermi level, highest occupied molecular orbital, and the electron-injecting process, respectively.

compared to that (1.6) of the undoped NTCDA and taking ε_{BCP}~1.4, $(1/\varepsilon_1+1/\varepsilon_2)$ is estimated to be ~0.7. Provided that the CNL for n-NTCDA is roughly -4.2 eV due to the Fermi level pinning [22] and taking the CNL of BCP~-3.8 eV, Eq. (2) yields an interface dipole Δ_1=-0.26 eV, which results in a 0.26 eV downward shift of the vacuum level (VL) on the BCP side as seen in Fig. 21(i). Thus, the energy barrier from the LUMO of n-NTCDA to that of BCP is estimated to be about 0.74 eV, determining that the electron transport through the n-NTCDA/BCP interface is very poor in the case of device 22. For the n-NTCDA/1:4 Li_2CO_3:BCP heterojunction, $(1/\varepsilon_1+1/\varepsilon_2)$ is assumed to be ~0.2, because the low-frequency dielectric constants of the doped materials are larger compared to the undoped materials. Provided that the CNL for n-NTCDA ~-4.2 eV and the CNL for n-BCP~-3.2 eV, Eq. (2) yields an interface dipole Δ_2=-0.90 eV, leading to a 0.90 eV downward shift of the VL in the n-BCP side, as as shown in Fig. 21(ii). The electron injection barrier from n-NTCDA to 1:4 Li_2CO_3:BCP is estimated to be roughly 0.10 eV. Thus, the interface becomes ohmic, enabling the efficient electron conduction and hence enhaced device performance (the case of device 20).

3.2.3.4. The effect of the NTCDA thickness on the current conduction of IBOLED

It is found that the device performance is strongly influenced by the thickness of the NTCDA on ITO as a hole injection layer [39]. Therefore, the dependence of the IBOLED performance on the NTCDA on ITO as n-ETL has been investigated and the observed characteristics are shown in Fig. 22 for devices 20, 22, and 23. It can be seen that the current density of the IBOLED using the 3 nm NTCDA is found to be higher than that of the IBOLEDs using the 5 and 7 nm NTCDA at a given voltage between 3 and 10 V. It may be explained as follows. When the thickness of the NTCDA exceeds 3 nm, the formation of the CT complex near the NTCDA top surface starts to vanish [38], that is, the surface density of the CT complex near NTCDA top surface starts to decrease, leading to a reduced generation of free electrons therein. Thus, the electron conduction across the whole NTCDA thin film decreases and gives rise to the decreased IBOLED current.

3.2.4. The tandem OLEDs incorporating the combination of Li_2CO_3:PTCDA and Li_2CO_3:BCP

Tandem OLEDs consisting of two light emitting units stacked vertically in series have been demonstrated [40, 41]. Compared to the single light emitting unit, the two-unit tandem OLED can provide the prolonged working lifetime and higher luminance but nearly the same power efficiency, because the driving voltage of the tandem OLED is twice that of the single light emitting unit at a given current density. Therefore, the tandem OLEDs have been widely recognized as a promising technology for organic flat-panel displays and solid-state lighting sources in the market competition. The future development of the tandem OLEDs is to further reduce their high driving voltage via both the structure optimization of the single light emitting unit and the improvement of the interconnecting structure.

It has been recognized that the voltage drop across the interconnected structure is the key to determining the operation voltage of the tandem OLEDs [40]. In order to minimize the series resistance of the interconnected structure, it is of great necessity to reduce the ohmic loss of the current conduction in this structure provided that the energy losses due to the internal charge generation and injections have already been optimized. Thus, it is meaningful to seek for a higher-conductivity alternative to the conventional n-doped organic electron transporters. Recently, the n-doped PTCDA has been used to reduce the driving voltage of OLEDs due to its higher conductance than the conventional n-doped organic electron conductors (BCP and Alq3) [13, 42], implying that the n-doped PTCDA may act as a potential candidate for the n-type section to reduce the current loss in the tandem structure.

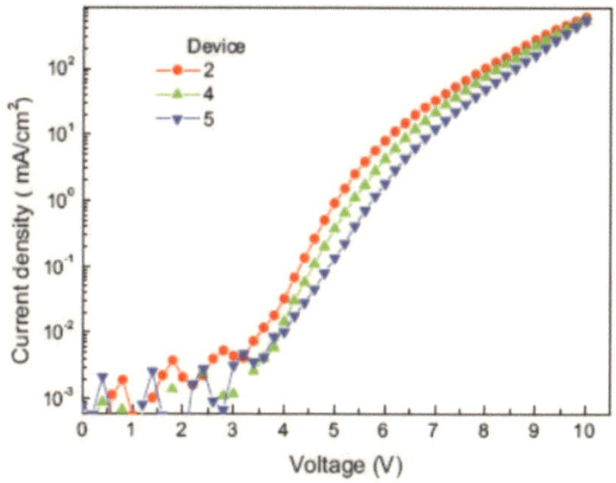

Figure 22. The I-V characteristics of devices 20 (circles), 22 (uptriangles), and 23 (downtriangles).

Here, a 5 nm lithium carbonate doped PTCDA (1:2 Li_2CO_3:PTCDA) has been incorporated in the n-type section of the interconnected structure. Compared to the tandem OLED of 5 nm Li_2CO_3 doped BCP (1:4 Li_2CO_3:BCP)/ MoO_3, the tandem OLED of 5 nm 1:2 Li_2CO_3:PTCDA/ 5 nm MoO_3 showed the enhanced electrical and luminous performance, due to the higher electron conductivity of 1:2 Li_2CO_3:PTCDA than that of 1:4 Li_2CO_3:BCP. The working mechanism of the improved interconnecting structure is also discussed.

3.2.4.1. The optical properties of the IS1 and IS2

Fig. 23 shows the optical transmittance of two interconnected structures. It can be seen that the 5 nm 1:2 Li_2CO_3:PTCDA/ 5 nm MoO3 (IS1) is nearly transparent in the visible-light range, but the 5 nm 1:4 Li_2CO_3:BCP/ 5 nm MoO3 (IS2) shows slight light absorption from 400 to 700 nm, due to the narrow optical band gap of PTCDA (1.7 eV).

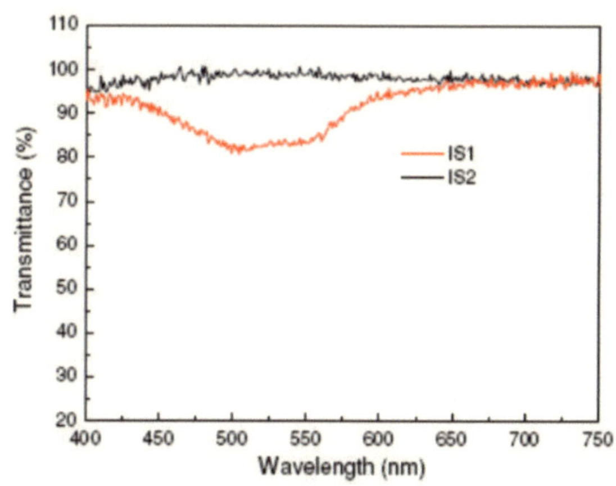

Figure 23. The optical properties of the IS1 and IS2 deposited on the quartz glasses.

3.2.4.2. The electrical properties of the IS1 and IS2

For studying the electrical properties, we have fabricated the following two devices:
Device 24: ITO (anode)/ 1:2 Li_2CO_3:PTCDA 5 nm/ MoO_3 5 nm/ NPB 80 nm/ Al (cathode).
Device 25: ITO (anode)/ 1:4 Li_2CO_3:BCP 5 nm/ MoO_3 5 nm/ NPB 80 nm/ Al (cathode).

The I-V characteristics of devices 24 and 25 are shown in Fig. 24. Because the ITO anode is unable to inject holes into n-doped PTCDA and BCP, and the Al cathode provides very inefficient electron injection into NPB, the current versus voltage plots of devices 24 and 25 mostly represent the internal charge generation and transport in the devices IS1 and IS2. Because devices 24 and 25 have the same interfaces of 5 nm MoO_3 and 80 nm NPB, the internal charge generation and hole conduction for the two devices are identical. Therefore, the difference between the I-V characteristics of devices 24 and 25 exhibit mostly the electron conduction through the IS1 and IS2. The current density of device 24 shown in Fig. 24 is found to be greater than that of device 25, indicating the resistance of IS1 is less than that of the IS2. Hence, the IS1 tandem structure is advantageousthan IS2 because IS1 results in less ohmic loss of electron conduction than IS2.

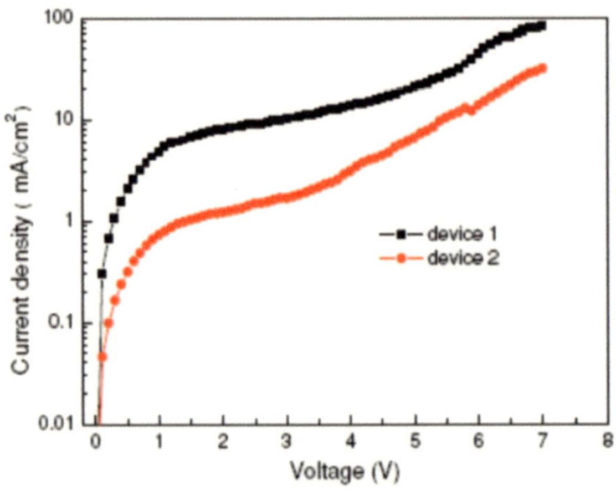

Figure 24. The I-V characteristics of devices 24 (blak curve) and 25 (red curve).

The LUMO-LUMO offset at the 1:2 Li_2CO_3:PTCDA/MoO_3 interface can be estimated via calculating the interfacial dipole through Eq. (2). The term ($1/\varepsilon_1+1/\varepsilon_2$) is assumed to be ~0.1, taking ε_1~20 and ε_2~20 because the low-frequency dielectric constant of the doped PTCDA is larger compared to the undoped one. Provided that the CNL for n-doped PTCDA~-4.8 eV [30] and the CNL for MoO_3~-6.9 eV, Eq. (2) yields an interface dipole Δ=2.0 eV, leading to a 2.0 eV upward shift of the vacuum level on the MoO_3 side. Thus, the electron injection barrier at the 1:2 Li_2CO_3:PTCDA/MoO_3 heterojunction is estimated to be roughly 0.1 eV, favoring the very efficient electron injection from MoO_3 into 1:2 Li_2CO_3:PTCDA. Likewise, the electron transport barrier from MoO3 to 1:4 Li_2CO_3:BCP is estimated to be 0.18 eV [14], meaning that the electron current can flow from MoO3 into 1:4 Li_2CO_3:BCP very efficiently as well. As a result, the increased electron transport in the IS1 over IS2 may be mostly attributed to the higher electron conductivity of 1:2 Li_2CO_3:PTCDA than 1:4 Li_2CO_3:BCP [13].

3.2.4.3. The performance comparison between the S, T1, and T2 devices

A single OLED S with structure of ITO/ NPB 80 nm/ Alq3 55 nm/ 1:4 Li_2CO_3:BCP 5 nm/ Al and two tandem OLEDs, T1 and T2, have been fabricated for this study as shown in Fig. 25.

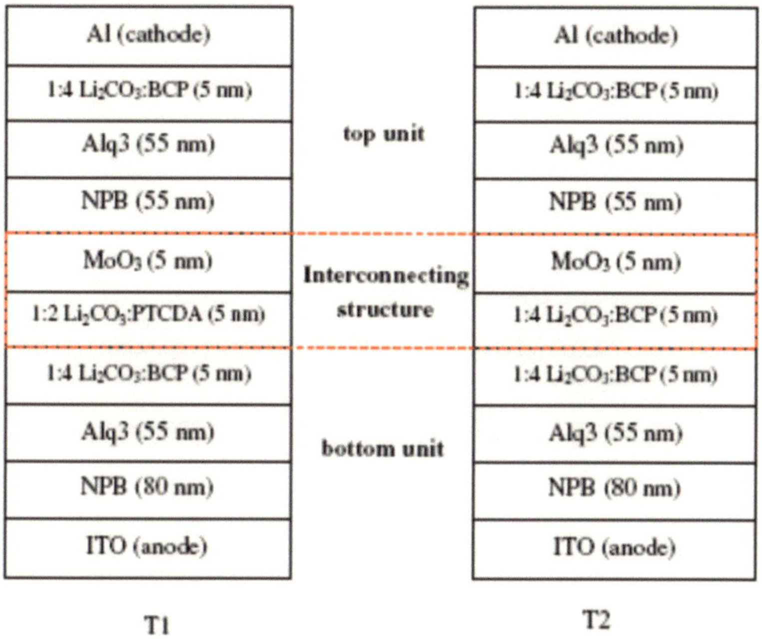

Figure 25. The schematic diagrams for the structures of the T1 and T2 device.

Fig. 26 shows the electroluminescent spectra of S, T1, and T2 devices. It is obvious that these three devices produce nearly the same Alq3 emission peaking at a wavelength of about 520 nm, demonstrating that IS1 and IS2 behave as good optical spacers without bringing any appreciable microcavity effect.

The electrical and luminous properties of the three devices are shown in Fig. 27. As expected, the T1 device gives higher current density than T2 device, which may be mostly attributed to the more efficient current conduction in IS1 than in IS2. Accordingly, the device T1 is also brighter than the device T2. At a driving voltage of 19 V, the luminance of T1 is 3656.8 cd/m^2, brighterr than 2323.9 cd/m^2 of T2 device. Because of some absorption in IS1 over the Alq3 emission, the T1 device shows slightly less current efficiency than T2 device when the current density ranges from 0.1 to 80 mA/cm^2. The maximum current efficiency of T2 device (4.86 cd/A) is found to be almost twice that of S device (2.51 cd/A). However, due to the reduced working voltage required for T1 than for T2, these two tandem OLEDs show comparable power efficiencies when the current density ranges from 0.1 to 80 mA/cm^2. Both of them have the maximum power efficiency of about 0.84 lm/W, slightly greater than that of the single-unit device S (0.80 lm/W). Note that, compared to both T1 and T2 devices, S device gives much lower power efficiency at a current density $\leq$ 10 mA/cm^2 and slightly increased power efficiency at current density $\geq$ 45 mA/cm^2.

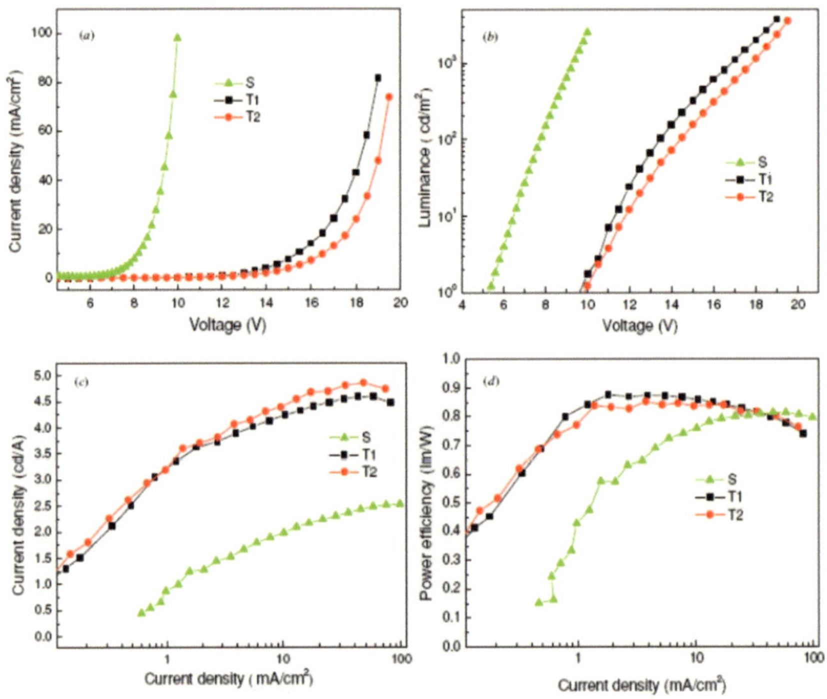

Figure 26. The I-V (a), luminance versus voltage (b), current efficiency versus current density (c), and power efficiency versus current density (d) characteristics of the S, T1, and T2 devices.

In Table I are listed the performance data of S device at a driving voltage of 9 V and those of T1 and T2 devices at a driving voltage of 18 V. The current density of T2 device at 18 V is lower than that of the S device at 9 V, while the luminance and current efficiency of the T2 device are nearly twice of those of the S device, and the power efficiencies of the T2 and S devices were nearly same. This indicates that the IS2 can effectively connect the two units in T2 device but with a marked voltage drop across it. The current density of T1 device at 18 V is 50% higher than that of S device at 9 V, its luminance is about three times that of S device, its current efficiency is twice that of S device, and its power efficiency is equal to that of S device. The current density of T1 device at 18 V is more than that of S device at 9 V. This may be attributed to the following two factors. Firstly, the voltage drop across IS1 is markedly lower compared to IS2 and secondly, the NPB layer in the upper unit is of much less thickness than that in the bottom unit.

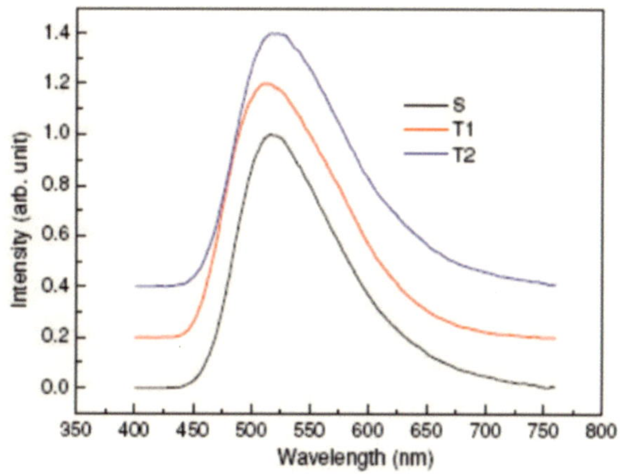

Figure 27. The electroluminescent spectra of the S, T1, and T2 devices.

Device	S	T1	T2
Voltage (V)	9	18	18
I (mA/cm2)	27.7	43.1	23.4
L (cd/m2)	637.6	1974.1	1124.8
CF (cd/A)	2.3	4.59	4.69
PF (lm/W)	0.80	0.80	0.82

Table 1. The performance data of the devices S, T1, and T2 . I, L, CF, and PF represent current density, luminance, current efficiency, and power efficiency, respectively.

It should be pointed out that, due to the poor ability of Li_2CO_3:PTCDA to inject electrons into the traditional electron transport materials (e.g., Alq3, BCP), an n-doped layer with high-lying LUMO level, n-doped BCP must be involved to facilitate the electron injection from n-doped PTCDA into the traditional electron transport materials [13]. Thus, IS1 cannot only reduce the ohmic loss of current conduction, but also can offer comparable electron injection into the bottom unit, relative to IS2.

3.2.5. The design concept of uniting two n-doped layers and its significance in OLEDs

The n-doped organic electron acceptors, e.g., n-NTCDA, n-PTCDA and n-C_{60}, possess markedly higher conductivities but lower capabilities of injecting electrons into electron transport layer (such as BCP, Alq3, etc.), as compared to the frequently used n-doped materials (such as n-BCP, n-Alq3, etc.) in organic light emitting diodes (OLEDs). In this section, we study the combination of the above two classes of n-doped materals, called the structure of uniting double n-doped layers (n-ETL_1/ n-ETL_2/ ETL). The characteristics for the structure of n-ETL_1/ n-ETL_2/ ETL are as follows: n-ETL_1 features the LUMO level greater than 4.0 eV (i.e., the LUMO level lies below 4.0 eV) and possesses higher conductivity of transporting electrons; n-ETL2 features the LUMO level less than 3.4 eV (i.e., the LUMO level lies above 3.4 eV) and possesses better capability of injecting into ETL; at the n-ETL_1/ n-ETL_2 interface, due to the quasi Fermi level alignment. In this caes, the energy barrier for the electron injection from n-ETL_1 to n-ETL_2 is usually less than 0.2 eV. Such a constructed combination cannot only significantly reduce ohmic loss in electron conduction, but also possess strong capability of injecting electrons into ETL. It has been demonstrated that the structure of uniting double n-doped layers enables remarkable enhancement in both electron current and luminous performance for inverted and tandem OLEDs. Therefore, it can be considered as an advanced electron injection technology, which can significantly push forward the commercializations of organic flat-panel displays and solid-state lighting.

4. Conclusions

Currently, the commerical application of organic lighting is being held back by the fact that the power efficiency of OLEDs drops drastically with the increase in active area.This means that the luminous flux of OLEDs does not increase automatically if the emissive area is increased. Thus, it becomes the top priority to develop the charge injection techniques which can increase the power and emittance of OLEDs. The charge injection techniques discussed in this chapter are still in their very early stages, and their further development depends mostly by inventing new materials which can meet all the requirements listed above.

Author details

Dashan Qin
Institute of Polymer Science and Engineering, School of Chemical Engineering,
Hebei University of Technology, Tianjin, People's Republic of China

Jidong Zhang
State Key Laboratory of Polymer Physics and Chemistry, Changchun Institute of Applied Chemistry, Chinese Academy of Sciences, Changchun, Jilin Province, People's Republic of China

Acknowledgement

The authors are grateful for the financial supports from the National Science Foundation of China (Grant No 50803014) and from Open research fund of state key laboratory of polymer physics and chemistry, Changchun Institute of Applied Chemistry, Chinese Academy of Sciences.

5. References

[1] C.W. Tang and S. A. Vanslyke, Organic electroluminescent diodes. Appl. Phys. Lett. 51: 913 (1987).
[2] S. Braun, W. Osikowicz, Y. Wang and W. R. Salaneck W R, Energy level alignment regimes at hybrid organic-organic and inorganic-organic interfaces. Org. Electron. 8: 14 (2007).
[3] S. Braun S, W. R. Salaneck and M. Fahlman, Energy-level alignment at organic/metal and organic/organic interfaces. Adv. Mater. 21: 1450 (2009).
[4] K. Walzer K, B. Maennig, M. Pfeiffer and K. Leo,Highly Efficient Organic Devices Based on. Electrically Doped Transport Layers. Chem.Rev. 107: 1233 (2007).
[5] G. H. Cao, D. S. Qin, J. S. Cao, M. Guan, Y. P. Zeng and J. M. Li, Organic light emitting diodes with an organic acceptor/donor interface involved in hole injection. Chin. Phys. Lett. 24: 1380 (2007).
[6] G. H. Cao, D. S. Qin, J. S. Cao, M. Guan, Y. P. Zeng and J. M. Li, Improved performance in organic light emitting diodes with a mixed electron donor-acceptor film involved in hole injection. J. Appl. Phys. 101: 124507 (2007).
[7] Y. Y. Yuan, S. Han, D. Grozea and Z. H. Lu, Fullerene-organic nanocomposite: a flexible material platform for organic light-emitting diodes. Appl. Phys. Lett. 88: 093503 (2006).
[8] X. Zhou, M. Pfeiffer, J. S. Huang, J. Blochwitz-Nimoth, D. S. Qin, A. Werner, J. Drechesel, B. Maennig and K. Leo, Low-voltage inverted transparent vacuum deposited organic light-emittingdiodes using electrical doping. Appl. Phys. Lett. 81: 922 (2002).
[9] T. Y. Chu, J. F. Chen, S. Y. Chen, C. H. Chen, Comparative study of single and multiemissive layers in inverted white organic light-emitting devices. Appl. Phys. Lett. 89: 113502 (2006).
[10] N. J. Watkins, L. Yan and Y. Gao, Electronic structure symmetry of interfaces between pentacene and metals. Appl. Phys. Lett. 80: 4384 (2002).

[11] W. Song, S. K. So, J. Moulder, Y. Qiu, Y. Zhu and L. Cao, Study on the interaction between Ag and tris(8-hydroxyquinoline) aluminum using x-ray photoelectron spectroscopy. Surf. Interface Anal. 32: 70 (2001).

[12] M. Thomschke, S. Hofmann, S. Olthof, M. Anderson, H. Kleemann, M. Schober, B. Lüssem and K. Leo, Appl. Phys. Lett. 98: 083304 (2011).

[13] C. R. Cheng, Y. H. Chen, D. S. Qin, W. Quan and J. S. Liu, Inverted bottom-emission organic light emitting diode using two n-doped layers for the enhanced performance. Chin. Phys. Lett. 27: 117801 (2010).

[14] D. S. Qin, J. S. Liu, Y. H. Chen, C. R. Cheng and W. Quan, Inverted bottom-emission organic light emitting diodes using MoO_3 for both hole and electron injections. Phys. Solidi Status A 208: 1976 (2011).

[15] D. S. Qin, L. Chen, Y. H. Chen, J. S. Liu, G. F. Li, W. Quan, J. D. Zhang and D. H. Yan, Enhanced performance in inverted organic light emitting diode assisted by an interlayer of crystalline and n-doped 1,4,5,8-naphthalene-tetracarboxylic-dianhydride, Phys. Solidi Status A 209: 790 (2012).

[16] J. S. Liu, Y. H. Chen, D. S. Qin, C. R. Cheng, W. Quan, L. Chen and G. F. Li, Improved interconnecting structure for a tandem organic light emitting diode. Semicond. Sci. Technol. 26: 095011 (2011).

[17] C. W. Tang, Two-layer organic photovoltaic cell. Appl. Phys. Lett. 48: 183 (1986).

[18] C. Rost, S. Karg, W. Riess, M. A. Loi, M. Murgia and M, Muccini, Ambipolar light-emitting organic field-effect transistor. Appl. Phys. Lett. 85: 1613 (2004).

[19] J. Wang, H. B. Wang, J. X. Yan, H. C. Huang and D. H. Yan, Organic heterojunction and its application for double channel field-effect transistors. Appl. Phys. Lett. 87: 093507 (2005).

[20] L. Chkoda, C. Heske, M. Sokolowski and E. Umbach, Improved band alignment for hole injection by an interfacial layer in organic light emitting devices. Appl. Phys. Lett. 77: 1093 (2000).

[21] P. E. Burrows and S. R. Forrest, Electroluminescence from trap-limited current transport in vacuum deposited organic light emitting devices, Appl. Phys. Lett. 64: 2285 (1994).

[22] I. G. Hill, A. Rgjagopal, A. Kahn and Y. Hu, Molecular level alignment at organic semiconductor-metal interfaces. Appl. Phys. Lett. 73: 662 (1998).

[23] S. R. Forrest, W. Y. Yoon, L. Y, Leu and F. F. So, Optical and electrical properties of isotype crystalline molecular organic heterojunctions. J. Appl. Phys. 66: 5908 (1998).

[24] B. P. Rand, J. Xue, S. Uchida and S. R, Forrest, Mixed donor-acceptor molecular heterojunctions for photovoltaic applications. I. Material properties. J. Appl. Phys. 98: 124902 (2005).

[25] P. Liu, Q. Li, M. S. Huang and W. Z. Pan, High open circuit voltage organic photovoltaic cells based on oligothiophene derivatives. Appl. Phys. Lett. 89: 213501 (2006).

[26] V. Bulović, P. E. Burrows, S. R. Forrest, J. A. Cronin and M. E. Thompson, Study of localized and extended excitons in 3,4,9,10-perylenetetracarboxylic dianhydride (PTCDA) I. Spectroscopic properties of thin films and solutions. Chem. Phys. 210: 1 (1996).

[27] Y. H. Chen, J. S. Chen, D. G. Ma, D. H. Yan and L. X. Wang, Effect of organic bulk heterojunction as charge generation layer on the performance of tandem organic light-emitting diodes. J. Appl. Phys. 110: 074504 (2011).

[28] D. R. T. Zahn, G. N. Gavrila and G. Salvan, Electronic and Vibrational Spectroscopies Applied to Organic/Inorganic Interfaces. Chem. Rev. 107: 1161 (2007).

[29] Y. Yuan, D. Grozea, S. Han and Z. H. Lu, Interaction between organic semiconductors and LiF dopant. Appl. Phys. Lett. 85: 4959 (2004).

[30] A. Kahn, W. Zhao, W. Y. Gao, H. Vazquez and F. Flores, Doping-induced realignment of molecular levels at organic–organic Heterojunctions. Chem. Phys. 325: 129 (2006).

[31] G. Parthasarathy, C. Shen, A. Kahn and S. R. Forrest, Lithium doping of semiconducting organic charge transport materials. J. Appl. Phys. 89: 4986 (2001).

[32] C. Tao, S. P. Ruan, X. D. Zhang, G. H. Xie, L. Shen, X. Z. Kong, W. Dong, C. X. Liu and W. Y. Chen, Performance improvement of inverted polymer solar cells with different top electrodes by introducing a MoO_3 buffer layer. Appl. Phys. Lett. 93: 193307 (2008).

[33] M. Kröger, S. Hamwi, J. Meyer, T. Riedl, W. Kowalsky and A, Kahn, Role of the deep-lying electronic states of MoO_3 in the enhancement of hole-injection in organic thin films. Appl. Phys. Lett. 95: 123301 (2009).

[34] E. Tutiš, D. Berner and L. Zuppiroli, Internal electric field and charge distribution in multilayer organic light-emitting diodes. J. Appl. Phys. 93: 4594 (2003).

[35] S. W. Shi and D. G. Ma, Investigation on internal electric field distribution of organic light-emitting diodes (OLEDs) with Eu_2O_3 buffer layer. Phys. Status Solidi A 206: 2641 (2009).

[36] Y. Shirota and H. Kageyama, Charge carrier transporting molecular materials and their applications in devices. Chem. Rev. 107: 953 (2007).

[37] H. Tachikawa, H. Kawabata, R. Miyamoto, K. Nakayama and M. Yokoyama, Experimental and Theoretical Studies on the Organic–Inorganic Hybrid Compound: Aluminum-NTCDA Co-Deposited Film. J. Phys. Chem. B 109: 3139 (2005).

[38] P. Jeon, H. Lee, J. Lee, K. Jeong, J. W. Lee and Y. Yi, Interface state and dipole assisted hole injection improvement with 1,4,5,8,-naphthalene-tetracarboxylic-dianhydride in organic light-emitting devices. Appl. Phys. Lett. 99: 073305 (2011).

[39] Y. M Koo and O. K. Song, Spontaneous charge transfer from indium tin oxide to organic molecules for effective hole injection. Appl. Phys. Lett. 94: 153302 (2009).

[40] L. S. Liao and K. P. Klubek, Power efficiency improvement in a tandem organic light-emitting diode. Appl. Phys. Lett. 92: 223311 (2008).

[41] C. W. Chen, Y. J. Lu, C. C. Wu, E. H. E. Wu, C. W. Chu and Y. Yang, Effective connecting architecture for tandem organic light-emitting devices. Appl. Phys. Lett. 87: 241121 (2005).

[42] C. R. Cheng, Y. H. Chen, D. S. Qin, W. Quan and J. S. Liu, Lithium carbonate doped 3, 4, 9, 10 perylenetetracarboxylic dianhydride for enhanced performance in organic light emitting diode Chin. J. Lumin. 32: 387 (2011).

Chapter 7

Exciplex Electroluminescence of the New Organic Materials for Light-Emitting Diodes

M.G.Kaplunov, S.N. Nikitenko and S.S. Krasnikova

Additional information is available at the end of the chapter

http://dx.doi.org/10.5772/51799

1. Introduction

In typical organic light emitting devices (OLEDs), light originates from radiative recombination of molecular excited states formed by electrons and holes injected from electrodes and localized on individual molecular sites. That is, the results are interpreted as due to Frenkel exciton generation and recombination [1,2]. In particular, this is applied to the bilayer OLEDs composed of metal 8-hydroxyquinolates Mq_3 (M = Al, Ga, In, or Sc) as an electron-transporting and emitting layer and amines like triphenylamine derivative (TPD) as a hole-transporting layer. The electroluminescence (EL) spectra of these devices are close to the photoluminescence (PL) spectra of corresponding Mq_3 molecules [1-3]. The similarity of the EL and PL spectra was also observed for zinc complexes with hydroxy-substituted quinolines, benzothiazoles and related ligands [4-6].

In some bilayer devices, interactions of donor and acceptor molecules at the organic/organic interface can lead to formation of an exciplex state. Exciplex is a kind of excited state complex formed between donor and acceptor, with one in the excited state and the other in the ground state. Exciplex usually leads to the red shifted emission and broadened spectrum relative to the emissions of the individual acceptor or donor [7-10].

Exciplex formation at the solid interface between Alq_3 and the electron-rich multiple triarylamine hole-transporting materials m-MTDATA and t-Bu-TBATA was observed in a study by Itano et al., [11]. Exciplexes can also be observed in the PL spectra of donor-acceptor blends [10,12-15].

Sometimes, another sort of bimolecular excited complex called electroplex can be generated around heterojunction. Unlike exciplex emission which can be observed under both photo-excitation and electric field excitation, electroplex emission can not be typically observed under photo-excitation and can be formed only in the presence of high electric field in some OLEDs. [16-19].

The intrinsic luminescence of the emitting layer is quenched by the formation of exciplexes [20-24]. So, for pure monochromatic OLEDs, exciplexes should be avoided [25-27]. On the other hand, exciplexes were proposed to tune the OLED emission color [28-30] and to design white OLEDs [12,18,31-35]. The use of exciplex emission simplifies the structure of white OLEDs. Efficient white electroluminescence from a double-layer device based on a boron complex was demonstrated by Liu et al. [36]. High-efficiency nondoped white organic light-emitting device based on the triarylamine derivative was demonstrated by Tong et al. [37] and Lai et al. [38]. For some OLEDs, pure exciplex emission was obtained by Wang et al. [39] and Nayak et al. [40].

One of the problems in utilizing the exciplex effects in devices is finding systems with high exciplex EL efficiency, so design of new materials and investigation of the active factors for efficient exciplex emission are a subject of significance.

Recently, spectral properties of the electroluminescent devices based on the novel zinc-chelate complexes of sulphanilamino-substituted quinolines and benzothiazoles were investigated and some exciplex phenomena were found [41-46]. The structures of zinc complexes are shown in Figure 1.

Most presently known metal complexes used for OLEDs contain the chelate cycles including the C-O-M-N chains [2-4,6,22,27,47]. In the amino-substituted complexes, the oxygen atom in the chelate cycles is replaced by a nitrogen atom of the sulphanylamino groups forming the C-N-M-N chains. The presence of a spatially extended, electron-rich amine segment in the zinc complex molecule can enhance its ability of intermolecular interactions with the molecules of the hole-transporting layer and hence magnify the possibility of exciplex forming. This chapter presents a review of electroluminescent properties of sulphanylamino-substituted zinc complexes.

2. EL spectra of OLEDs based on sulphanilamino-substituted zinc-chelate complexes

We have prepared and measured the EL spectra of the following OLED devices based on zinc complexes with sulphanilamino-substituted ligands.

- device D1: ITO/PTA/NPD/Zn(PSA-BTZ)$_2$/Al:Ca
- device D2: ITO/PTA/Zn(PSA-BTZ)$_2$/Al:Ca
- device D3: ITO/PTA/NPD/CBP/Zn(PSA-BTZ)$_2$/Al:Ca
- device D4: ITO/PTA/CBP/Zn(PSA-BTZ)$_2$/Al:Ca
- device D5: ITO/PEDOT:PSS/Zn(PSA-BTZ)$_2$/Al:Ca
- device D6: ITO/PTA/NPD/Zn(TSA-BTZ)$_2$/Al:Ca
- device D7: ITO/PTA/Zn(TSA-BTZ)$_2$/Al:Ca
- device D8: ITO/PTA/NPD/Zn(POPS-BTZ)$_2$/Al:Ca
- device D9: ITO/PTA/NPD/CBP/Zn(POPS-BTZ)$_2$/Al:Ca
- device D10: ITO/PTA/NPD/Zn(DFP-SAMQ)$_2$/Al:Ca
- device D11: ITO/PTA/Zn(DFP-SAMQ)$_2$/Al:Ca
- device D12: ITO/PTA/CBP/Zn(TSA-BTZ)$_2$/Al:Ca

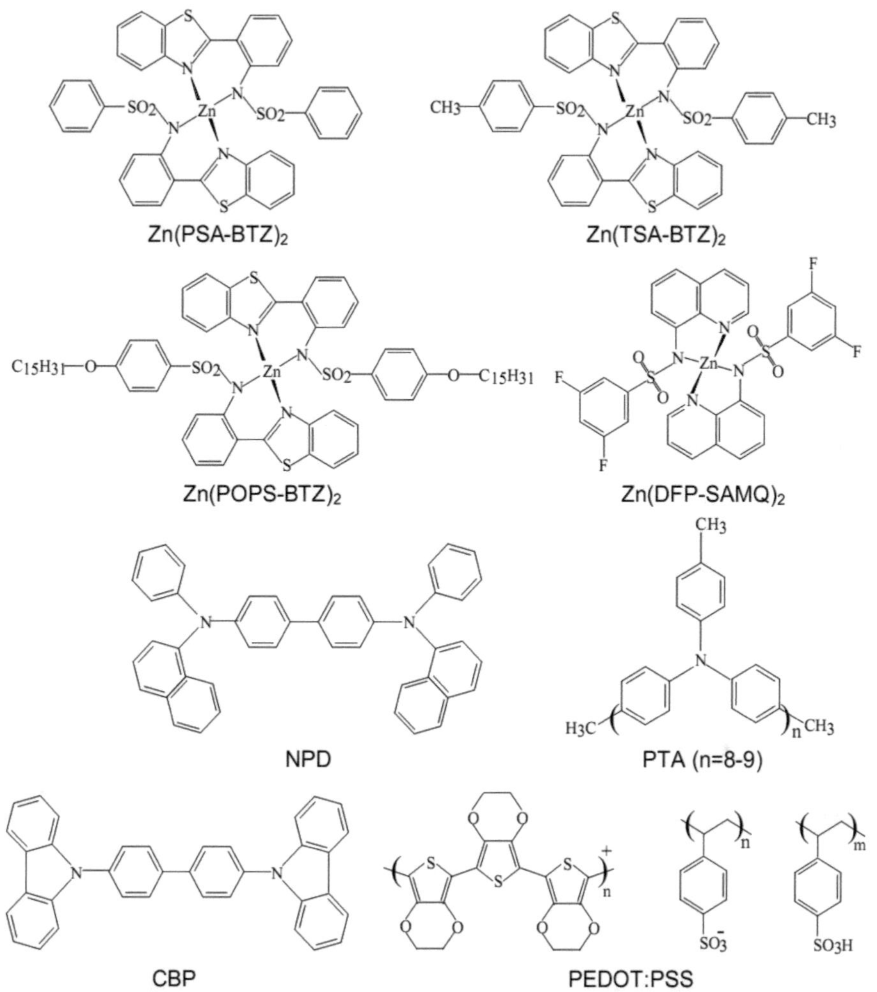

Figure 1. Structures of zinc complexes and of materials for hole-transporting layers. Zn(PSA-BTZ)$_2$: bis{2-[2-(phenylsulphanylamino)phenyl]benzothiazolate}zinc; Zn(TSA-BTZ)$_2$: bis{2-[2-(4-methylphenylsulphanylamino)phenyl]benzothiazolate}zinc; Zn(POPS-BTZ)$_2$: bis{2-[2-(4-penthadecyloxyphenylsulphanylamino)phenyl]-benzothiazolate}zinc; Zn(DFP-SAMQ)$_2$: bis[8-(3,5-difluorophenylsulphanylamino)-quinolato]zinc; NPD: N,N′-bis(1-naphthyl)-(1,1′-biphenyl)-4,4′-diamine; PTA: oligo(4,4′-(4′′-methyl)triphenylamine); CBP: 4,4′-bis(N-carbazolyl)-1,1′-biphenyl; PEDOT:PSS: poly(3,4-ethylenedioxythiophene)-poly(styrenesulfonate).

Methods of preparing the devices and measuring their properties are described elsewhere [42-44]. Materials of hole-transporting layer were triaryl derivatives: PTA, olygomer of triphenylamine with high glass-transition temperature [48] and a well-known N,N′-bis(1-naphthyl)-(1,1′-biphenyl)-4,4′-diamine (NPD). The carbazol derivative 4,4′-bis(N-carbazolyl)-1,1′-biphenyl (CBP) and poly(3,4-ethylenedioxythiophene) poly(styrenesulfonate) (PEDOT:PSS) were also used for forming the hole-transporting layer. The structures of these compounds are also shown in Figure 1. In some devices, both PTA and NPD deposited in succession were used as materials for hole-transporting layers. In any case, the EL spectrum of the device is determined by the hole-transporting material, which is in contact with the zinc complex. The devices are typically characterized by bias voltages of light appearance about 2.5 to 3 V and brightness of 10^3 cd/m^2 at 10 V.

2.1. EL spectra of OLEDs based on Zn(PSA-BTZ)$_2$

Figure 2 shows the EL spectra of Zn(PSA-BTZ)$_2$ in two electroluminescence devices: device D1, ITO/PTA/NPD/Zn(PSA-BTZ)$_2$/Al:Ca, (Figure 2a, curve 1) and device D2, ITO/PTA/Zn(PSA-BTZ)$_2$/AlCa, (Figure 2b, curve 1). For comparison, curve 2 in Figure 2a shows the PL spectrum of Zn(PSA-BTZ)$_2$ powder.

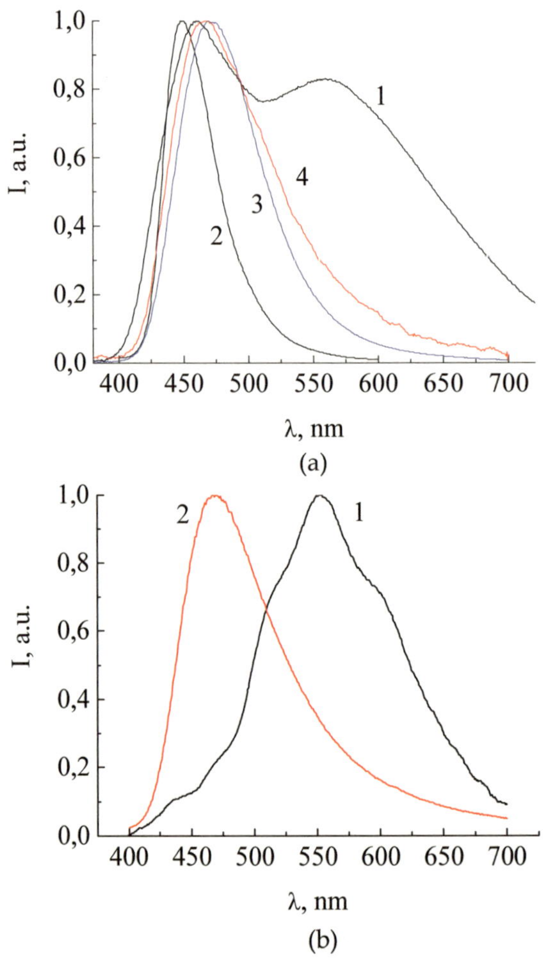

Figure 2. Spectra of Zn(PSA-BTZ)$_2$ in the devices D1 to D5. (a): EL spectrum of the device D1 ITO/PTA/NPD/Zn(PSA-BTZ)$_2$/Al:Ca (1); PL spectrum of Zn(PSA-BTZ)$_2$ powder (2); EL spectrum of the device D3 ITO/PTA/NPD/CBP/Zn(PSA-BTZ)$_2$/Al:Ca (3); EL spectrum of the device D5 ITO/PEDOT:PSS/Zn(PSA-BTZ)$_2$/Al:Ca (4) (b): EL spectrum of the device D2 ITO/PTA/Zn(PSA-BTZ)$_2$/AlCa (1); EL spectrum of the device D4 ITO/PTA/CBP/Zn(PSA-BTZ)$_2$/AlCa (2)

The EL spectrum of device D1 contains two bands with maxima at 460 and 560 nm. Maximum of the first band is close to that of the PL peak of Zn(PSA-BTZ)$_2$ powder at 450 nm and may be attributed to the intrinsic luminescence of Zn(PSA-BTZ)$_2$. The second peak may be probably due to exciplex formation between NPD and Zn(PSA-BTZ)$_2$. For device D2, the EL spectrum exhibits only wide band with a maximum at 553 nm, which may be attributed to exciplex formation between PTA and Zn(PSA-BTZ)$_2$. Exciplex can be formed between the ground state of a donor molecule and the excited state of an acceptor molecule [12]. In our case, the donor molecule is presented by NPD or PTA, and the acceptor molecule by Zn(PSA-BTZ)$_2$ complex. Exciplex band corresponds to the transition from the excited state of the acceptor and the ground state of the donor and has lower transition energy compared to the intrinsic emission band corresponding to the transition between the excited and ground state of the acceptor molecule [7-10,12].

2.2. EL spectra of OLEDs based on Zn(TSA-BTZ)$_2$

Figure 3 shows the EL spectra of Zn(TSA-BTZ)$_2$ in the electroluminescence devices: (a) device D6, ITO/PTA/NPD/Zn(TSA-BTZ)2/Al:Ca, (b) device D7, ITO/PTA/Zn(TSA-BTZ)$_2$/AlCa and device D12, ITO/PTA/CBP/Zn(TSA-BTZ)$_2$/Al:Ca. The PL spectrum of Zn(TSA-BTZ)$_2$ powder is shown for comparison (Figure 3a, curve 2).

For the devices D6 and D7, intensive exciplex EL bands are observed in the yellow region with the maxima around 585 nm. Only a weak shoulder in the region of the intrinsic Zn(TSA-BTZ)$_2$ emission at about 460 nm is observed. For device D7, the EL spectra are shown for different bias voltages from 3.5 to 6.0 V. The spectra are normalized to obtain equal intensities of exciplex bands for all voltages. A small continuous growth of intrinsic emission relative intensity is observed. A small blue shift of exciplex band maximum from 585 nm at 3.5 V to 575 nm at 6.0 V is also observed. This is in contrast with previously reported strong dependence of EL bands positions on bias voltages [11,49,50] where field induced shift of EL band of about 50 nm could be attributed to the overlap of the emission from different excited states. As showed Kalinowski et al. [51], the field dependence of EL spectrum in such systems is a result of electric field mediated interplay among localized (monomolecular) excitons, exciplexes, and electroplexes in conjunction with their specific environment. For the device D12, no exciplex band is observed which is discussed in section 4.

2.3. EL spectra of OLEDs based on Zn(POPS-BTZ)$_2$

Figure 4 shows the EL spectra of Zn(POPS-BTZ)$_2$ in the devices D8 ITO/PTA/NPD/Zn(POPS-BTZ)$_2$/Al:Ca (curves 1) and D9 ITO/PTA/NPD/CBP/Zn(POPS-BTZ)$_2$/Al:Ca (curve 3). The PL spectrum of Zn(POPS-BTZ)$_2$ powder (curve 2) is shown for comparison. Strong exciplex band in the green region with the maximum at about 540 nm and shoulder at about 460 nm due to intrinsic emission of Zn(POPS-BTZ)$_2$ is observed in the

EL spectra of the device D8. The normalized EL spectra are shown for different bias voltages from 4.0 to 6.0 V. A small continuous growth of intrinsic emission relative intensity and a small blue shift of exciplex band maximum from 545 nm at 4.0 V to 535 nm at 6.0 V are also observed. For the device D9, no exciplex band is observed which is discussed in section 4.

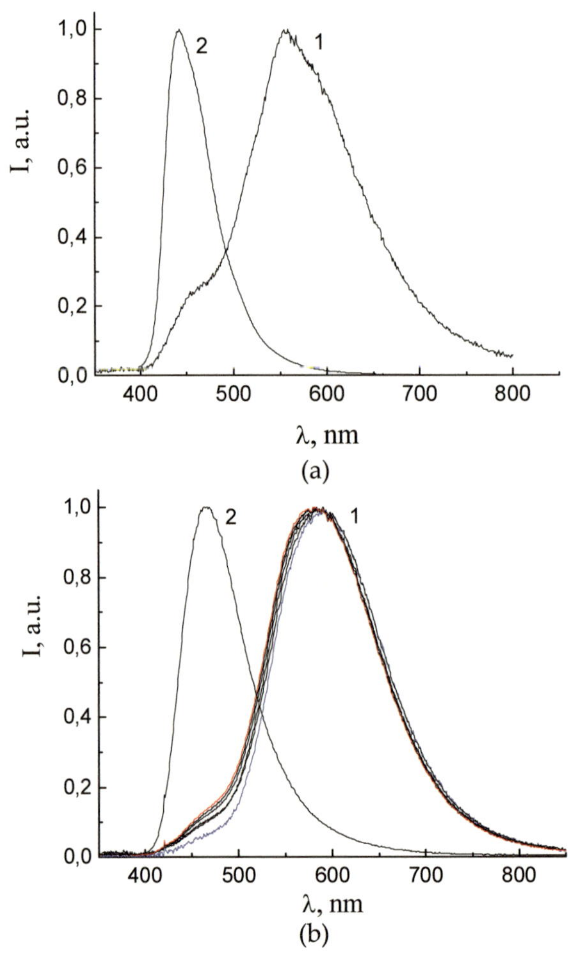

Figure 3. Normalized EL spectra of the devices based on $Zn(TSA-BTZ)_2$ and the PL spectrum of $Zn(TSA-BTZ)_2$ powder. (a): Normalized EL spectrum of the device D6 ITO/PTA/NPD/$Zn(TSA-BTZ)_2$/Al:Ca (1) and the PL spectrum of $Zn(TSA-BTZ)_2$ powder (2); (b): Normalized EL spectra of the device D7 ITO/PTA/$Zn(TSA-BTZ)_2$/Al:Ca for bias voltages 3.5 V (blue curve), 4.0, 4.5, 5.0, 5.5 V (black curves) and 6.0 V (red curve) (1) and normalized EL spectrum of the device D12 ITO/PTA/CBP/$Zn(TSA-BTZ)_2$/Al:Ca (2).

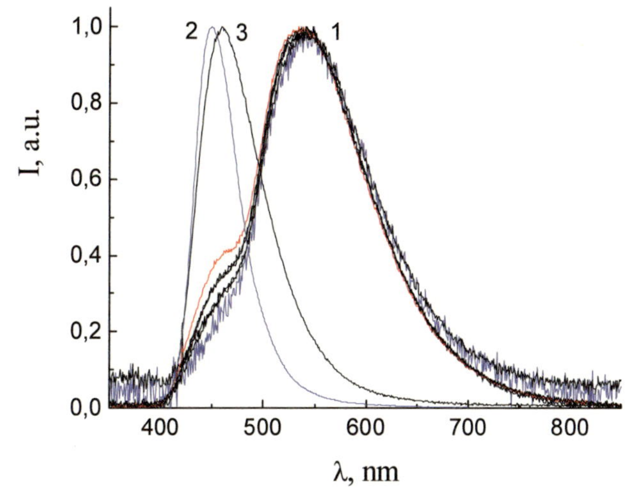

Figure 4. Spectra of $Zn(POPS-BTZ)_2$ based devices: Normalized EL spectra of the device D8 ITO/PTA/NPD/$Zn(POPS-BTZ)_2$/Al:Ca for bias voltages 4.0 V (blue curve), 4.5, 5.0, 5.5 V (black curves) and 6.0 V (red curve) (1), PL of $Zn(POPS-BTZ)_2$ powder (2) and the EL spectrum of the device D9 ITO/PTA/NPD/CBP/$Zn(POPS-BTZ)_2$/Al:Ca (3)

2.4. EL spectra of OLEDs based on $Zn(DFP-SAMQ)_2$

Figure 5 shows the EL spectra of $Zn(DFP-SAMQ)_2$ in the devices D10 ITO/PTA/NPD/Zn(DFP-SAMQ)$_2$/AlCa (curve 1) and D11 ITO/PTA/$Zn(DFP-SAMQ)_2$/AlCa (curve 2). For comparison, the PL spectrum of $Zn(DFP-SAMQ)_2$ powder with maximum at 465 nm is shown (curve 3). Exciplex bands with maxima at about 560 nm and no intrinsic emission are observed in the EL spectra both for devices with both PTA and PTA/NPD hole-transporting layers. Exciplex emission can also be observed in the PL spectra of the films containing blends of zinc complex and hole-transporting material (curve 4) which is discussed in section 3.

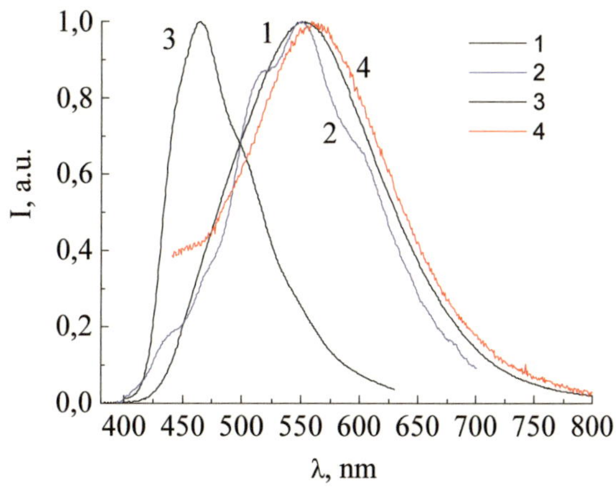

Figure 5. Spectra of $Zn(DFP-SAMQ)_2$ based devices: EL spectra of the devices D10 ITO/PTA/NPD/$Zn(DFP-SAMQ)_2$/AlCa (1) and D11 ITO/PTA/$Zn(DFP-SAMQ)_2$/AlCa (2); PL spectra of $Zn(DFP-SAMQ)_2$ powder (3) and of PTA:$Zn(DFP-SAMQ)_2$ mixed film (4).

3. The PL spectra of the films containing blends of zinc complex and of hole-transporting material

One of the main evidences of the exciplex nature of long-wave bands in the EL spectra is the presence of such bands in the PL spectra of blends of donor and acceptor materials [12-15,17,40,53].

For the zinc-chelate complexes with sulphanilamino-substituted ligands, the exciplex long-wave bands were observed in the PL spectra of their blends with hole-transporting materials by Kaplunov et al. [44-45]. It was shown previously that the PL spectra taken from the layered structure exhibiting the exciplex EL OLEDs do not contain long-wave bands but only the intrinsic bands of components [46]. This is due to the extremely small thickness of the contacting interface of the two layers, which is responsible for EL. To observe the long-wave bands in PL, films containing blends of zinc complex and hole-transporting material were prepared by casting from toluene solutions containing both components in appropriate concentrations. In such films, contacts between the two kinds of molecules take place in the whole volume of the film, unlike the bilayer OLED structure with very thin contact interface.

Kaplunov et al. [44] studied the PL spectra of the films containing blends of $Zn(DFP\text{-}SAMQ)_2$ with PTA. The spectra contain no intrinsic luminescence of zinc complex or PTA and exhibit only the exciplex band with maximum at 555 nm (figure 5, curve 4). Kaplunov et al. [45] studied the PL spectra of the films containing PTA, $Zn(DFP\text{-}SAMQ)_2$ and their blends in different ratios. For the films with a relatively low fraction of PTA where PTA:$Zn(DFP\text{-}SAMQ)_2$ = 0.5:1 and 1:1 (mass), the PL bands are close to that of Zn(DFP-SAMQ)2 with λmax = 490 nm (intrinsic emission). For the films with a higher PTA fraction where PTA:$Zn(DFP\text{-}SAMQ)_2$ = 2.6:1 and 4:1 (mass), the exciplex PL band with λmax in the region of 560 nm is observed. This result shows that the exciplex PL can be observed for donor-acceptor blends with proper relation between components, which guarantees large amount and good quality of intermolecular donor-acceptor contacts.

4. Elimination of exciplex emission for the devices based on zinc complexes with amino-substituted ligands

4.1. Elimination of exciplex emission by introducing an intermediate layer between the hole-transporting and the emitting layers

To prove the exciplex origin of the long-wavelength EL, we have fabricated several control devices in which the long-wave EL bands are eliminated.

One of the methods for preventing exciplex emission is the insertion of an additional layer between the hole-transporting and electron-transporting materials [22,25-27]. CBP is considered as one of the materials appropriate for such layers [27,54,55]. We have fabricated two control devices with $Zn(PSA\text{-}BTZ)_2$ as emitting layer and CBP as the intermediate layer: ITO/PTA/NPD/CBP/$Zn(PSA\text{-}BTZ)_2$/Al:Ca (device D3) and ITO/PTA/CBP/Zn(PSA-

BTZ)$_2$/Al:Ca (device D4). Figure 2 shows the EL spectra of devices D3 and D4 (Figure 2a, curve 3 and Figure 2b, curve 2, respectively). In both cases, the EL spectra contain no wide band around 560 nm and exhibit only one band in the blue region with the maximum at 471 nm (device D3) and 469 nm (device D4), which may be attributed mainly to the intrinsic emission of the Zn(PSA-BTZ)$_2$ complex.

Similar to the devices based on Zn(PSA-BTZ)$_2$, the exciplex band can be eliminated by introducing the intermediate layer of CBP between NPD and Zn(POPS-BTZ)$_2$ or PTA and Zn(TSA-BTZ)$_2$. The EL spectrum of the device ITO/PTA/NPD/CBP/Zn(POPS-BTZ)$_2$/Al:Ca (device D9) is shown in Figure 4 (curve 2). The exciplex band in the region of 540 nm is absent, and only the intrinsic emission of Zn(POPS-BTZ)$_2$ at λmax = 460 nm is observed. The EL spectrum of the device ITO/PTA/CBP/Zn(TSA-BTZ)$_2$/Al:Ca (device D12) is shown in Figure 3b (curve 2). The exciplex band in the region of 580 nm is absent, and only the intrinsic emission of Zn(TSA-BTZ)$_2$ at λmax = 465 nm is observed.

4.2. The role of amino groups in exciplex formation

Exciplex can be formed at the solid interface between a hole-transporting layer and an electron-transporting layer, in case when there is a significant spatial overlap between the lowest unoccupied molecular orbitals (LUMOs) of the constituent species [56].

It should be noted that both NPD and PTA, as well as many other materials usually used to form the hole-transporting layer, are the derivatives of triarylamines. One may suppose that the interaction of the nitrogen atoms in the amino groups of the hole-transporting molecules and the amino groups of the zinc complexes (due to their spatial overlap) determines the exciplex formation in the studied systems. Evidence in favor of this supposition comes from our results on using other materials different from triarylamine derivatives for hole-transporting layers. Figure 2a, curve 4 shows the EL spectrum of device ITO/PEDOT:PSS/Zn(PSA-BTZ)$_2$/Al:Ca (device D5) where the hole-transporting layer is presented by PEDOT:PSS, a hole injecting and transporting material which does not contain nitrogen atoms at all. This spectrum does not contain a wide band around 560 nm and exhibits only one band with a maximum at 466 nm, which is close to the Zn(PSA-BTZ)2 powder PL band (450 nm) and may be attributed mainly to the intrinsic emission of Zn(PSA-BTZ)$_2$ complex. One may suppose that the formation of exciplex in this case is suppressed by the absence of nitrogen atoms in the hole-transporting layers.

Commonly, the reason for preventing the exciplex emission by changing the hole-transporting material is argued to be the relation between the energy levels of the donor and acceptor molecules. Materials like CBP with low highest occupied molecular orbital (HOMO) energy level are considered as appropriate ones [22,25-27]. Really, the HOMO level of CBP is 6.1 to 6.3 eV below vacuum level [47,57,58], which is appreciably lower than that of NPD (5.2 to 5.7 eV) [58-60].

On the other hand, the highest occupied energy level of PEDOT:PSS is 5.2 eV below vacuum level [61], which does not differ from that of NPD. So, the fact that NPD produces exciplexes with the studied complexes and CBP and PEDOT:PSS do not may be explained not only by positions of energy levels but also by other reasons. Good spatial overlap of donor and acceptor molecular orbitals seems to be one of the most important factors promoting the formation of exciplexes.

From this point of view, molecules with amino groups are most appropriate for exciplex formation because of high electron density at nitrogen atoms. Zinc complexes studied in the present work contain amino groups bonded to metal atom and produce exciplexes in pair with triarylamine molecules NPD and PTA. Note that the analogs of our complexes containing oxygen atom bonded to metal such as Mq_3, Znq_2, $Zn(BTZ)_2$ do not exhibit exciplexes in their EL spectra when triarylamine hole-transporting materials like NPD or TPD are used [1,3-6,47]. At the same time, the derivatives of Alq_3 containing amino groups bonded to quinoline species exhibit EL exciplex bands for the devices with NPD [27].

5. The relation between intrinsic and exciplex bands in the devises containing NPD/Zn(TSA-BTZ)$_2$ interface

Some of the EL devices based on amino-substituted zinc complexes demonstrate the EL spectra with only exciplex bands, other demonstrate the intrinsic EL bands either. The relation between intrinsic and exciplex bands can be affected by different factors: materials of contacting layers [22], thicknesses of layers [17,29,30,31,62], applied voltage and current [20,63-64]. The dependence of the relation between intrinsic and exciplex bands in the EL spectra of the devices based on amino-substituted zinc complexes on the thickness of hole-transporting layer and on the applied bias voltage were studied for the devises based on $Zn(TSA\text{-}BTZ)_2$ as an illustrative example.

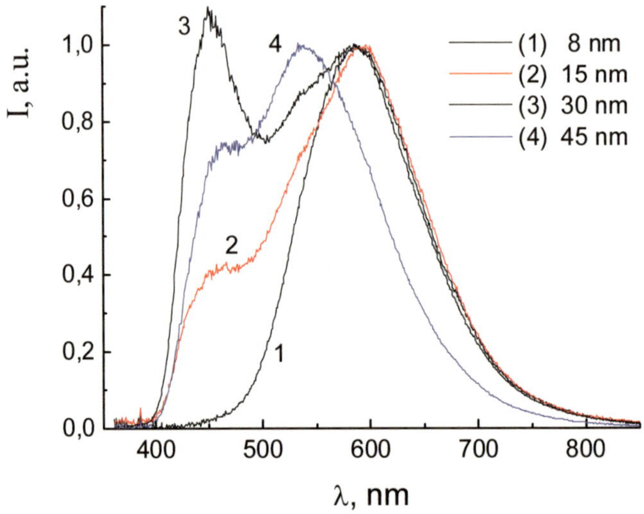

Figure 6. Normalized EL spectra of ITO/PTA/NPD/Zn(TSA-BTZ)$_2$/Al:Ca devices with different thicknesses of NPD layer: 8 nm (1), 15nm (2), 30 nm (3) and 45 nm (4). Spectra are measured at the same current 7.4 mA/cm^2 for all devices.

Figures 6 and 7 show the EL spectra of the devices ITO/PTA/NPD/Zn(TSA-BTZ)$_2$/Al:Ca (device D6) with different thicknesses of NPD layer 0, 8, 15, 30 and 45 nm. Thicknesses of other organic layers are constant: about 100 nm for PTA and about 30 nm for Zn(TSA-BTZ)$_2$. Spectra are measured at different bias voltages from 3.5 V to 10 V.

The shape of the EL band strongly depends on the NPD layer thickness (Fig.6). For thicknesses of 8, 15 and 30 nm, the exciplex band with maximum in the region of 590 nm is observed. Further increase in thickness leads to some shift of the exciplex band maximum position. For the thickness of 45 nm, the exciplex band in the region of 540 nm is observed. Devices with 8 nm NPD layer thickness exhibit no intrinsic band in the EL spectra. Devices with 15, 30 and 45 nm NPD layer thickness exhibit intrinsic EL band in the region of 450-460 nm in addition to exciplex bahds.

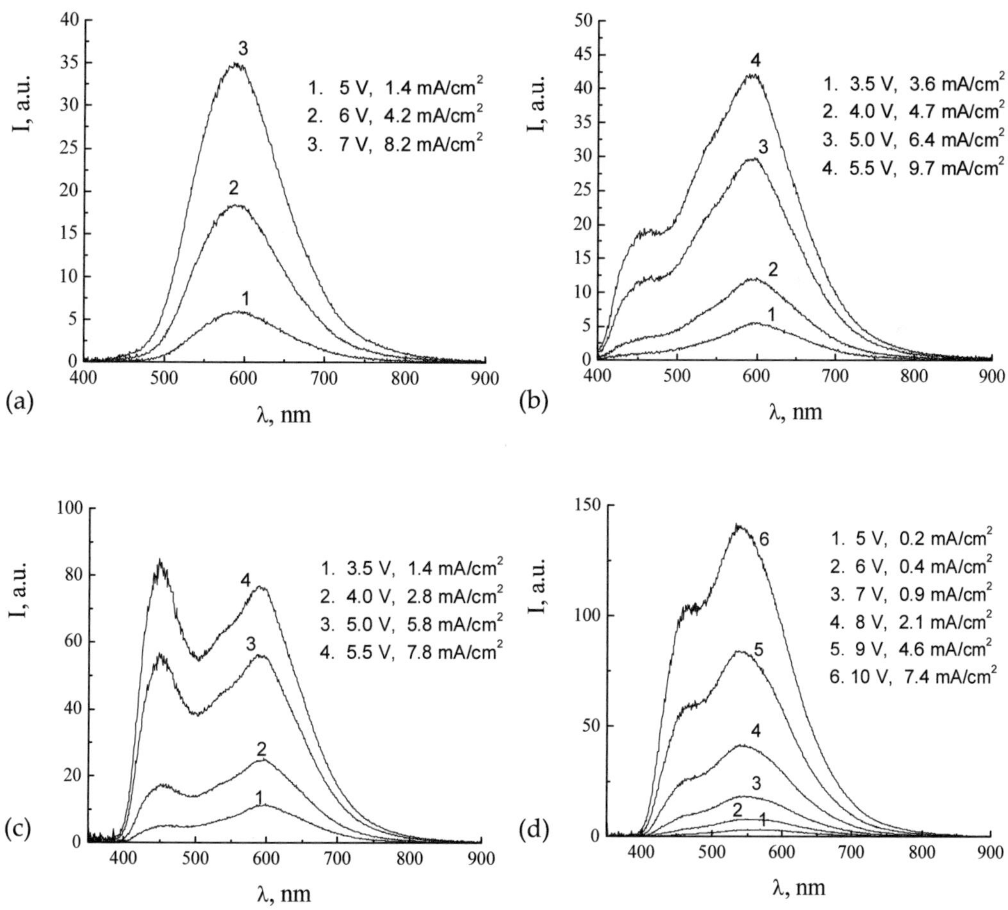

Figure 7. EL spectra of the devices ITO/PTA/NPD/Zn(TSA-BTZ)$_2$/Al:Ca with different thicknesses of NPD layer: 8 nm (a), 15 nm (b), 30 nm (c) and 45 nm (d). The applied bias voltages and the currents through the device are given along with curve numbers.

Figure 7 shows the dependence of the EL spectra of Zn(TSA-BTZ)$_2$ based devices with different thicknesses of hole-transporting NPD layer on the applied bias voltages and corresponding currents through the device. With increasing voltage, maximum of the exciplex band shifts to lower wavelength by 5-8 nm, maximum of the intrinsic band practically does not change. The increasing voltage leads also to appearance of additional exciplex peak at 540 nm in the EL spectra of the devices with the NPD layer thicknesses of 15 and 30 nm (figure 7b,c).

Shift of exciplex bands maxima with change of the layer thicknesses and of the applied voltage may be due to plurality of excited states in the excited donor-acceptor complexes [28,51].

The relation of intensities of intrinsic and exciplex bands depends on applied bias voltage and corresponding currents. With increasing voltage, the intensity of the intrinsic band relative to the exciplex band increases. For the device with 30nm NPD layer thickness, the intrinsic band becomes more intense than the exciplex band at some voltages. Growth of the intrinsic band can be attributed to the shift of the carrier recombination zone from the NPD/Zn(TSA-BTZ)$_2$ interface to the bulk of the emitter layer and increasing number of holes injected into the emitter layer due to increasing electric field [29,63,65].

It should be noted that the intensity of electroluminescence depends on the number of recombinating electrons and holes that is on current through the device. So the dependence of the EL spectra on the applied bias voltage should more properly be considered as the dependence on the current.

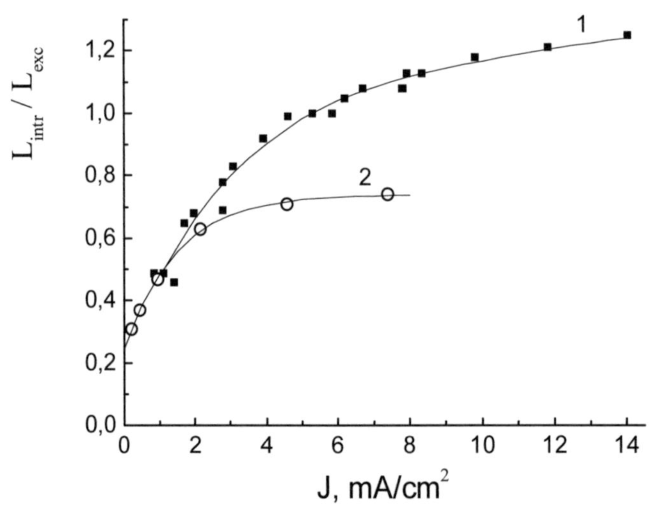

Figure 8. Plot of relative intensities of the intrinsic EL band and that of exciplex band as a function of the current for the devices ITO/PTA/NPD/Zn(TSA-BTZ)$_2$/Al:Ca with NPD layer thicknesses 30 nm (1, squares) and 45 nm (2, circles). Solid curves represent only guide to an eye.

Figure 8 shows the relation of the intensity of the intrinsic EL band (L_{intr}) to that of exciplex band (L_{exc}) for the devices with thicknesses of NPD layer 30 and 45 nm depending on current. The intensity of the intrinsic band relative to that of the exciplex band increases with the increase in current and saturates at large currents. The saturation may be due to extending the carrier recombination zone to the whole $Zn(TSA\text{-}BTZ)_2$ layer.

6. White OLEDs based on zinc complexes with amino-substituted ligands

The combination of narrow intrinsic band and wide exciplex band gives a very wide emission spread over the whole visible spectrum, which is a way to obtain white light emitting diodes [8,12,36,37].

For the novel zinc-chelate complexes of sulphanilamino-substituted ligands, the combination of narrow intrinsic band in blue region and wide exciplex band in yellow region can be observed for some devices with NPD as hole-transporting layer. For example, the device 1 based on $Zn(PSA\text{-}BTZ)_2$, gives the EL emission (Figure 2a, curve1) with the CIE chromaticity coordinates $x = 0.31$ and $y = 0.34$ which is close to that of the white light ($x = 0.33$, $y = 0.33$). Corresponding point on CIE color diagam (Figure 9) is marked by open circle.

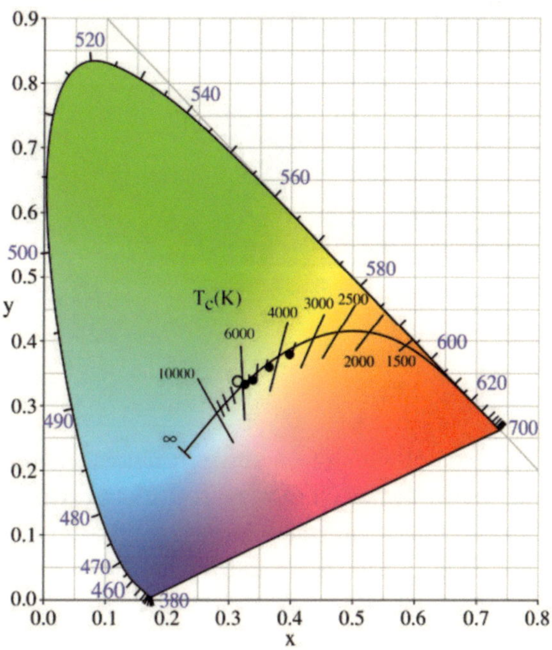

Figure 9. The emission of the devices $ITO/PTA/NPD/Zn(PSA\text{-}BTZ)_2/Al{:}Ca$ (open circle) and $ITO/PTA/NPD/Zn(TSA\text{-}BTZ)_2/Al{:}Ca$ with the NPD layer thickness of 30 nm (filled circles) on the CIE color diagram.

For the devices based on $Zn(TSA\text{-}BTZ)_2$, the presence of both intrinsic and exciplex emission bands in the EL spectra can be observed for the devices with appropriate NPD layer thickness. The EL spectra of the device $ITO/PTA/NPD/Zn(TSA\text{-}BTZ)_2/Al{:}Ca$ with the NPD layer thickness of 30 nm exhibit both intrinsic and exciplex emission bands with the relation depending on bias voltages (figure 7c). The device emits nearly white light. The CIE color coordinates (x, y) corresponding to the emission spectra of the device at bias voltages 3.5, 4.0, 5.0 and 5.5 V are (0.40, 0.38), (0.37, 0.36), (0.34, 0.33) and (0.33, 0.33) respectively. Corresponding points on CIE color diagram (figure 9, filled circles) are close to the black body emission line between color temperatures 3500 and 6000 K.

As an example, Fig. 10 shows a photograph of a light-emitting diode based on the electroluminescent structure ITO/PTA/$Zn(TSA\text{-}BTZ)_2$/Al:Ca. The area of the luminescent surface is 6×6 mm^2 and the operation voltage is 6–8 V. The efficiency of such light-emitting diodes reaches 5–6 lm/W.

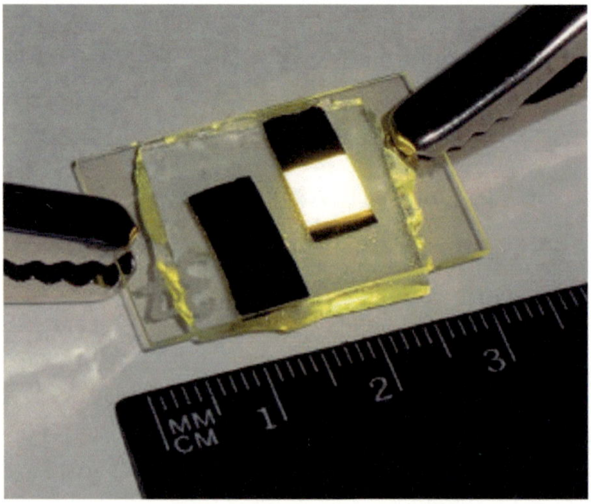

Figure 10. Photograph of an organic light-emitting diode based on an ITO/PTA/$Zn(TSA\text{-}BTZ)_2$/AlCa electroluminescent structure.

7. Conclusions

The novel zinc-chelate complexes of sulphanilamino-substituted quinolines and benzothiazoles are proper materials for OLEDs, with efficient exciplex emission giving rise to white OLEDs and OLEDs of different colors including blue, green, and yellow. Exciplex emission can also be observed in the PL spectra of the films containing blends of zinc complex and hole-transporting material. The main reason of effective exciplex formation for these compounds is probably the presence of a spatially extended, electron-rich amine segment in the zinc complex molecule which can enhance its ability of intermolecular interactions with the molecules of the hole-transporting layer and hence magnify the possibility of exciplex forming. Material of the hole-transporting layer is crucial for the efficiency of exciplex formation. Triarylamine derivatives like NPD or PTA seem to be the most proper materials for exciplex formation. Exciplexes can be eliminated with only intrinsic bands remaining in the EL spectra when the hole-transporting layer is not a triarylamine drivative. This may be due not only to positions of energy levels but also to good spatial overlap and high electron density on amino groups of both zinc complex and triarylamine derivatives. The relation between intrinsic and exciplex EL bands for the EL devices depends not only on the material of hole-transporting layer but also on the applied voltage. This may be due to to the shift of the carrier recombination zone from the interface of hole-transporting and emitting layers to the bulk of the emitter layer.

Author details

M.G.Kaplunov, S.N. Nikitenko and S.S. Krasnikova
Institute of Problems of Chemical Physics RAS, Chernogolovka, Russia

8. References

[1] Burrows PE, Sapochak LS, McCatty DM, Forrest SR, Thompson ME. Metal ion dependent luminescence effects in metal tris-quinolate organic heterojunction light emitting devices. Appl Phys Lett 1994; 64(20) 2718-2720.

[2] Burrows PE, Shen Z, Bulovic V, McCarty DM, Forrest SR, Cronin JA, Thompson ME. Relationship between electroluminescence and current transport in organic heterojunction light-emitting devices. J Appl Phys 1996; 79(10) 7991-8006.

[3] Tang CW, VanSlyke SA. Organic electroluminescent diodes. Appl. Phys. Lett. 1987; 51(12) 913-915.

[4] Hamada Y, Sano T, Fujita M, Fujii T, Nishio Y, Shibata K. Organic electroluminescent devices with 8-hydroxyquinoline derivative-metal complexes as an emitter. Jpn. J. Appl. Phys. 1993; 32, L514-L515.

[5] Hamada Y, Sano T, Fujii H, Nishio Y, Takahashi H, Shihata K. White light emitting material for organic electroluminescent devices. Jpn J Appl Phys 1996; 35(10b), L1339-L1341.

[6] Tanaka H, Tokito S, Taga Y, Okada A. Novel metal–chelate emitting materials based on polycyclic aromatic ligands for electroluminescent devices. J Mater Chem 1998; 8(9) 1999-2003.

[7] Cocchi M, Virgili D, Giro G, Fattori V, Marco PD, Kalinowski J, Shirtota Y. Efficient exciplex emitting organic electroluminescent devices. Appl Phys Lett 2002; 80(13) 2401-2403.

[8] Chao C-I, Chen S-A. White light emission from exciplex in a bilayer device with two blue light-emitting polymers. Appl Phys Lett 1998; 73(4) 426-428.

[9] Bolognesi A, Botta C, Cecchinato L, Fattori V, Cocchi M Poly(3-pentylmethoxythiophene)/Alq3 heterostructure light emitting diodes. Synthetic Metals 1999; 106(3) 183-186

[10] Tian WJ, Wu F, Zhang LQ, Zhang BW, Cao Y. Light emission from exciplex of organic electroluminescent device. Synthetic Metals 2001; 121(1-3) 1725-1726.

[11] Itano K, Ogawa H, Shirota Y. Exciplex formation at the organic solid-state interface: yellow emission in organic light-emitting diodes using green-fluorescent tris(8-

quinolinolato)aluminum and hole-transporting molecular materials with low ionization potentials. Appl. Phys. Lett. 1998; 72(6) 636-638.

[12] Thompson J, Blyth RIR, Mazzeo M, Anni M, Gigli G, Clinigolani R. White light emission from blends of blue-emitting organic molecules: a general route to the white organic light-emitting diode?. Appl. Phys. Lett. 2001; 79(5) 560-562.

[13] Salerno M, Blyth RIR, Thompson J, Cingolani R, Gigli G. Surface morphology and optical properties of thin films of thiophene-based binary blends Journal of Applied Physics 2005; 98(1) 013512-5.

[14] Mazzeo M, Pisignano D, Della Sala F, Thompson J, Blyth RIR, Gigli G, Cingolani R, Sotgiu G., Barbarella G. Organic single-layer white light-emitting diodes by exciplex emission from spin-coated blends of blue-emitting molecules Appl. Phys. Lett. 2003; 82(3) 334-336

[15] Mazzeo M, Pisignano D, Favaretto L, Sotgiu G, Barbarella G, Cingolani R, Gigli G. White emission from organic light emitting diodes based on energy down-convertion mechanisms. Synthetic Metals 2003; 139 (3) 675-677.

[16] Granlund T, Pettersson LAA, Anderson MR, Inganäs O. Interference phenomenon determines the color in an organic light emitting diode. J. Appl. Phys. 1997; 81(12) 8097-8104.

[17] Zhang W, Yu J, Yuan K, Jiang Y, Zhang Q, Cao K. Acceptor thickness effect of exciplex and electroplex emission at heterojunction interface in organic light-emitting diodes Proceedings of SPIE 2010; 7658, 76583V-6.

[18] Zhu H, Xu Z, Zhang F, Zhao S, Wang Z, Song D. White organic light-emitting diodes via mixing exciplex and electroplex emissions Synthetic Metals 2009; 159(23-24), 2458-2461.

[19] Zhao D-W, Xu Z, Zhang F-J, Song S-F, Zhao S-L, Wang Y, Yuan G-C, Zhang Y-F, Xu HH. The effect of electric field strength on electroplex emission at the interface of NPB/PBD organic light-emitting diodes. Applied Surface Science 2007; 253(8) 4025-4028

[20] Matsumoto N, Adachi C. Exciplex formations at the HTL/Alq3 interface in an organic light-emitting diode: Influence of the electron-hole recombination zone and electric field. Journal of Physical Chemistry C 2010; 114(10) 4652-4658.

[21] Seo JH, Kim HM, Choi EY, Choi DH, Park JH, Yoo HS, Kang HJ, Lee KH, Yoon SS, Kim YK. Phosphorescent organic light-emitting diodes with simplified device architecture Japanese Journal of Applied Physics 2010; 49(8 part 2) 08JG04-4.

[22] Su WM, Li WL, Xin Q, Su ZS, Chu B, Bi DF, He H, Niu JH. Effect of acceptor on efficiencies of exciplex-type organic light emitting diodes. Appl. Phys. Lett. 2007; 91(4) 043508-3.

[23] Wang H, Klubek KP, Tang CW. Current efficiency in organic light-emitting diodes with a hole-injection layer Applied Physics Letters 2008; 93 (9), 093306-3.

[24] Lai SL, Tong QX, Chan MY, Ng TW, Lo MF, Ko CC, Lee ST. Lee CS. Carbazole-pyrene derivatives for undoped organic light-emitting devices Organic Electronics 2011; 12(3) 541-546

[25] Noda T, Ogawa H, Shirota Y. A blue-emitting organic electroluminescent device using a novel emitting amorphous molecular material, 5,5′-bis(dimesitylboryl)-2,2′-bithiophene. Advanced Materials 1999; 11(4) 283-285.

[26] Li G, Kim CH, Zhou Z, Shinar J, Okumoto K, Shirota Y. Combinatorial study of exciplex formation at the interface between two wide band gap organic semiconductors. Appl. Phys. Lett. 2006; 88(25) 253505-3.

[27] Liao S-H, Shiu J-R, Liu S-W, Yeh S-J, Chen Y-H, Chen C-T, Chow TJ, Wu C-I: Hydroxynaphthyridine-derived group III metal chelates: wide band gap and deep blue analogues of green Alq3 (Tris(8-hydroxyquinolate)aluminum) and their versatile applications for organic light-emitting diodes. J Am Chem Soc 2009, 131:763.

[28] Li M, Li W, Chen L, Kong Z, Chu B, Li B, Hu Z, Zhang Z. Tuning emission color of electroluminescence from two organic interfacial exciplexes by modulating the thickness of middle gadolinium complex layer. Appl. Phys. Lett. 2006; 88(9) 091108-3.

[29] Yu J, Lou S, Wen W, Jiang Y, Zhang Q. Color-tunable organic light-emitting diodes based on exciplex emission. Proceedings of SPIE 2009; 7282, 728234-7.

[30] Kulkarni, A.P., Jenekhe, S.A. Blue-green, orange, and white organic light-emitting diodes based on exciplex electroluminescence of an oligoquinoline acceptor and different hole-transport materials. Journal of Physical Chemistry C 2008; 112(13) 5174-5184.

[31] Feng J, Li F, Gao W, Liu S. White light emission from exciplex using tris-8-hydroxyquinoline aluminum as chromaticity-tuning layer. Appl Phys Lett 2001; 78(25) 3947–3949

[32] Yang S, Jiang M. White light generation combining emissions from exciplex, excimer and electromer in TAPC-based organic light-emitting diodes. Chemical Physics Letters 2009; 484(1-3) 54-58.

[33] Zhu J, Li W, Han L, Chu B, Zhang G, Yang D, Chen Y, Su Z, Wang J, Wu S, Tsuboi T. Very broad white-emission spectrum based organic light-emitting diodes by four exciplex emission bands. Optics Letters 2009; 34(19) 2946-2948.

[34] Zhou G, Wong W-Y, Suo S. Recent progress and current challenges in phosphorescent white organic light-emitting diodes (WOLEDs). Journal of Photochemistry and Photobiology C: Photochemistry Reviews 2010; 11(4)133-156.

[35] Kumar A, Srivastava R, Bawa SS, Singh D, Singh K, Chauhan G, Singh I, Kamalasanan MN. White organic light emitting diodes based on DCM dye sandwiched in 2-methyl-8-hydroxyquinolinolatolithium Journal of Luminescence 2010; 130(8) 1516-1520

[36] Liu Y, Guo J, Zhang H, Wang Y. Highly efficient white organic electroluminescence from a double-lay device based on a boron hydroxyphenylpyridine complex. Angew Chem Int Ed 2002; 41(1) 182-184.

[37] Tong QX, Lai SL, Chan MY, Tang JX, Kwong HL, Lee CS, Lee ST. High-efficiency nondoped white organic light-emitting devices. Appl. Phys. Lett 2007; 91(2) 023503-3.

[38] Lai SL, Chan MY, Tong QX, Fung MK, Wang PF, Lee CS, Lee ST. Approaches for achieving highly efficient exciplex-based organic light-emitting devices Appl. Phys. Lett. 2008; 93(14) 143301-3.

[39] Wang D, Li W, Chu B, Su Z, Bi D, Zhang D, Zhu J, Yan F, Chen Y, Tsuboi T. Highly efficient green organic light-emitting diodes from single exciplex emission Applied Physics Letters 2008; 92(5) 053304-3

[40] Nayak PK, Agarwal N, Periasamy N, Patankar MP, Narasimhan KL. Pure exciplex electroluminescence in blended film of small organic molecules. Synthetic Metals 2010; 160(7-8) 722-727.

[41] Yakushchenko IK, Kaplunov MG, Krasnikova SS, Roshchupkina OS, Pivovarov AP. A New class of electroluminescent metal complexes based on quinoline ligands containing the sulfanylamino group. Russ. J. Coord. Chem. 2009; 35(4) 312-316.

[42] Kaplunov MG, Yakushchenko IK, Krasnikova SS, Pivovarov AP. Electroluminescent devices based on novel zinc complexes of sulphonylamino substituted heterocycles. Mol Cryst Liq Cryst 2008; 497(1) 211-217.

[43] Kaplunov MG, Yakushchenko IK, Krasnikova SS, Pivovarov AP, Balashova IO. Electroluminescent materials based on new metal complexes for organic light-emitting diodes. High Energy Chemistry 2008; 42(7) 563-565.

[44] Kaplunov MG, Krasnikova SS, Balashova IO, Yakushchenko IK. Exciplex Electroluminescence Spectra of the New Organic Materials Based on Zinc Complexes of Sulphanylamino-Substituted Ligands, Molecular Crystals and Liquid Crystals 2011; 535(1) 212-219.

[45] Kaplunov MG, Krasnikova SS, Nikitenko SL, Sermakasheva NL, Yakushchenko IK. Exciplex electroluminescence and photoluminescence spectra of the new organic materials based on zinc complexes of sulphanylamino-substituted ligands. Nanoscale Research Letters 2012; 7(1), 206-8.

[46] Krasnikova SS, Kaplunov MG, Yakushchenko IK: Exciplex electroluminescence spectra of organic light emitting diodes based on zinc complexes with sulfonylamino substituted ligands. High Energ Chem 2009; 43(7) 536-539.

[47] Adamovich V, Brooks J, Tamayo A, Alexander AM, Djurovich PI, D'Andrade BW, Adachi C, Forrest SR, Thompson ME. High efficiency single dopant white electrophosphorescent light emitting diodes. New J Chem 2002; 26,1171-1178.

[48] Yakushchenko IK, Kaplunov MG, Efimov ON, Belov MY, Shamaev SN. Polytriphenylamine derivatives as materials for hole transporting layers in electroluminescent devices. Phys. Chem. Chem. Phys. 1999; 1, 1783-1785.

[49] Yap CC, Yahaya M, Salleh MM. The effect of driving voltage on the electroluminescent property of a blend of poly(9-vinylcarbazole) and 2-(4-biphenylyl)-5-phenyl-1,3,4-oxadiazole. Current Applied Physics 2009; 9(5) 1038-1041.

[50] Giro G, Cocchi M, Kalinowski J, Di Marco P, Fattori V. Multicomponent emission from organic light emitting diodes based on polymer dispersion of an aromatic diamine and an oxadiazole derivative. Chem. Phys. Lett. 2000; 318(1-3) 137-141

[51] Kalinowski J, Cocchi M, Di Marco P, Stampor W, Giro G, Fattori V. Impact of high electric fields on the charge recombination process in organic light-emitting diodes. J. Phys. D. Appl. Phys. 2000; 33(19), 2379-2387.

[52] Zhao D, Zhang F, Xu C, Sun J, Song S, Xu Z, Sun X. Exciplex emission in the blend of two blue luminescent materials Applied Surface Science 2008; 254(11) 3548-3552.

[53] Qian J, Yu J, Lou S, Jiang Y, Zhang Q. Influence of hole transporter doping on electroluminescent property of novel fluorene molecular material. Proceedings of SPIE 2009; 7282, 72823F-5

[54] Xu X, Chen S, Yu G, Di C, You H, Ma D, Liu Y. High-efficiency blue light-emitting diodes based on a polyphenylphenyl compound with strong electron-accepting groups Advanced Materials 2007; 19(9) 1281-1285.

[55] Zhang W, Yu J-S, Huang J, Jiang Y-D, Zhang Q, Cao K.-L. Exciplex elimination in an organic light-emitting diode based on a fluorene derivative by inserting 4,4'-N,N'-dicarbazole-biphenylinto donor/acceptor interface Chinese Physics 2010; B19(4) 047802-7.

[56] Horvath A, Stevenson KL. Transition metal complex exciplexes. Coord. Chem. Rev. 1996; 153, 57-82.

[57] Hill IG, Rajagopai A, Kahn A. Energy-level alignment at interfaces between metals and the organic semiconductor 4,48-N, N8-dicarbazolyl-biphenyl. J Appl Phys 1998; 84(6) 3236-3241.

[58] Baldo MA, Lamansky S, Burrows PE, Thompson ME, Forrest SR. Very high-efficiency green organic light-emitting devices based on electrophosphorescence. Appl Phys Lett 1999; 75(1) 4-6.

[59] Brutting W, Berleb S, Muckl AG. Device physics of organic light-emitting diodes based on molecular materials. Organic Electronics 2001; 2(1) 1–36.

[60] Lee M-T, Yen C-K, Yang W-P, Chen H-H, Liao C-H, Tsai C-H, Chen CH. Efficient green coumarin dopants for organic light-emitting devices. Organic Letters 2004; 6(8) 1241-1244.

[61] Kim JY, Kim M, Kim HM, Joo J, Choi J-H. Electrical and optical studies of organic light emitting devices using SWCNTs-polymer nanocomposites. Optical Materials 2002; 21, 147-151.

[62] Divayana Y, Sun XW, Chen BJ, Sarma KR. Improved organic light-emitting device with tris-(8-hydroxyquinoline) aluminium inserted between hole-injection layer and hole-transporting layer. Journal of Physics D: Applied Physics 2007; 40(1) 183-186.

[63] Lee KS, Choo DC, Kim TW. White organic light-emitting devices with tunable color emission fabricated utilizing exciplex formation at heterointerfaces including m-MDATA Thin Solid Films 2011; 519(15) 5257-5259.

[64] Yang S, Zhang X, Lou Z, Hou Y. Influence of heterojunction interface on exciplex emission from organic light-emitting diodes under electric fields Applied Physics A: Materials Science and Processing 2008; 90(3) 475-478.

[65] Chen, T-R. Luminescence and electroluminescence of bis (2-(benzimidazol-2-yl) quinolinato) zinc. Exciplex formation and energy transfer in mixed film of bis (2-(benzimidazol-2-yl) quinolinato) zinc and N,N'-bis-(1-naphthyl)-N, N'-diphenyl-1,1'-biphenyl-4,4'-diamine. Journal of Molecular Structure 2005; 737(1) 35-41.

Chapter 8

Synthesis and Physical Properties of Red Luminescent Glass Forming Pyranylidene and Isophorene Fragment Containing Derivatives

Elmars Zarins, Aivars Vembris, Valdis Kokars and Inta Muzikante

Additional information is available at the end of the chapter

http://dx.doi.org/10.5772/53462

1. Introduction

Low molecular mass organic compounds with internal charge transfer properties are widely adopted for organic photonics such as materials for the creation of molecular electronics elements, organic magnets, solar cells and organic light emitting diodes (OLEDs) for full display panels [1-3]. One of the most widely used red light-emitting materials contains pyranylidene (4*H*-pyran-4-ylidene) or isophorene (5,5-dimethylcyclohex-2-enylidene) fragments as backbone of the molecule (see Fig.1), which are conjugated in a system with electron acceptor and electron donor fragments [1,4-24]. In many cases the light-emitting layer from such commercially available compounds is prepared by thermal evaporation in vacuum [1-2, 25-27]. Some of them are used as dopants in a polymer matrix and spin-coated onto a hole transport layer from solution [1,12]. However the doping amount of luminescent compound is limited by self crystallization and photoluminescence quenching at higher concentrations which reduce the quantum efficiency of fabricated devices significantly [11-12]. Therefore it is important to synthesize low molecular mass light-emitting organic compounds which do not crystallize and form thin amorphous solid films from volatile organic solvents. Such compounds, which can make a solid-state glassy structure prepared from solutions, could facilitate technological processes in the production of many devices in optoelectronics, for example, light emitting devices by low-cost deposition such as wet casting methods and easier light-emitting material synthesis. Some of these red light-emitting compounds have been introduced by us [28-32].

In this chapter we present complete synthesis, thermal, optical, photoelectrical and glass forming properties of new organic glass-forming pyranylidene and isophorene fragment containing derivatives with bulky trityloxy groups in their molecules. The optical properties, both in solution and solid state, are compared. The dependance of photoelectrical properties and energy structure of glassy films on molecular structure will be discussed. The most popular derivatives of pyranylidene and isophorene used in OLEDs are shown in Fig.1.

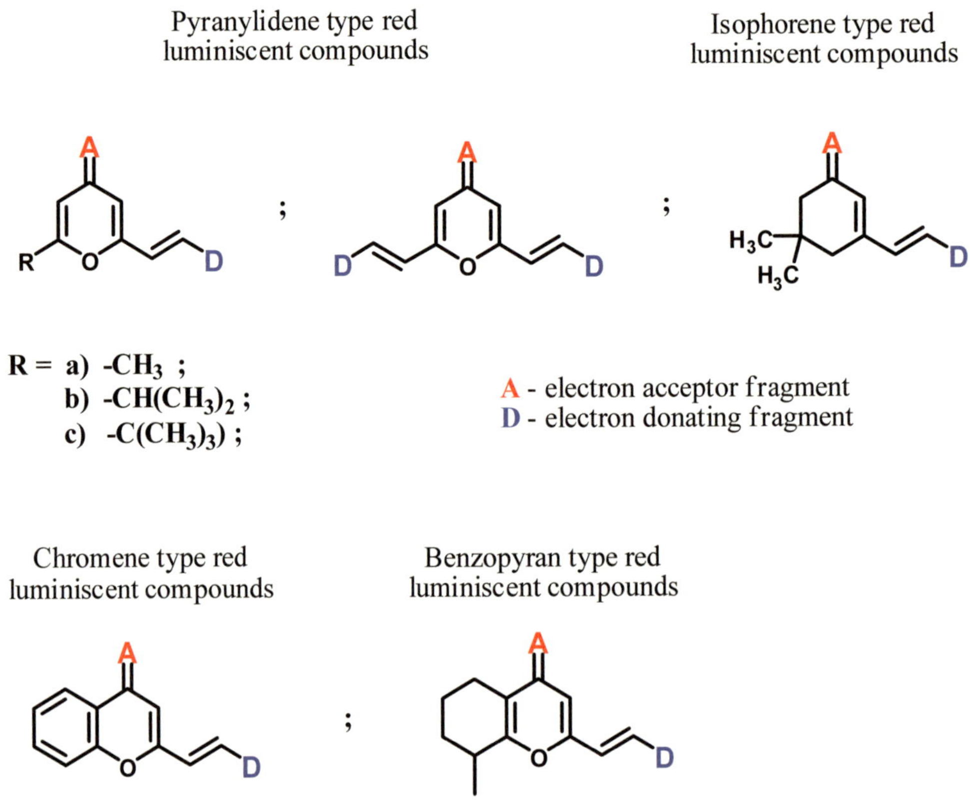

Figure 1. Most widely used pyranylidene and isophorene type red-emitters used as OLED emission layer materials

2. Synthesis

The synthesis procedure of pyranylidene and isophorene D-π-A type luminophores (see Fig.1) with either one or two electron donor fragments can be divided into three main parts:

1. Synthesis of a backbone fragment: Synthesis of derivatives of 4*H*-pyran-4-one, which in their molecules contain not only a carbonyl group, but also at least one methyl group and are able to react further with aromatic aldehydes.
2. Addition of an electron acceptor fragment to the backbone: Condensation reaction of 4*H*-pyran-4-one derivatives synthesized in 1) with active methylene group containing compounds.
3. Synthesis of pyranylidene and isophrene D-π-A type red emitters: Final addition of electron donor group containing aromatic aldehydes to compounds obtained in 2).

2.1. Synthesis of the backbone fragment: 2,6-disubstituted-4*H*-pyran-4-ones

The simplest of 2,6-disubstituted-4*H*-pyran-4-ones is 2,6-dimethyl-4*H*-pyran-4-one (compound 2 in Fig.2), which is obtained in 86% yield from dehydroacetic acid (compound 1 in Fig.2) by acidic rearrangement with following decarboxylation (see Fig.2) [32-33].

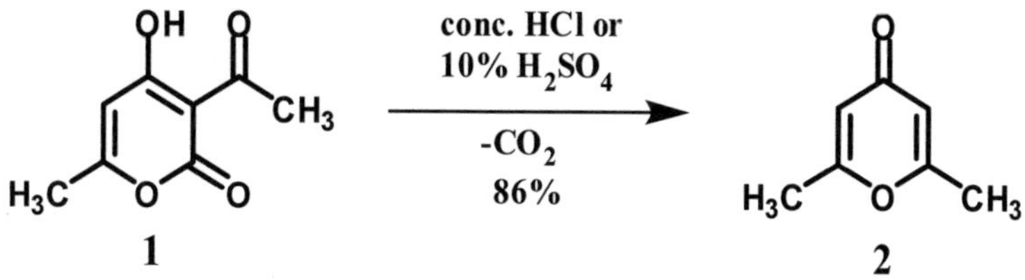

Figure 2. Synthesis of 2,6-dimethyl-4*H*-pyran-4-one. Dehydroacetic acid (compound **1**) is suspended either in concentrated hidrochloric acid (conc. HCl) or 10% aqueous sulfuric acid (10% H_2SO_4) and heated. During the heating carbon dioxide (CO_2) is liberated and 2,6-dimethyl-4*H*-pyran-4-one (compound **2**) is formed.

2,6-Dimethyl-4*H*-pyran-4-one (compound **2** in Fig.2) has one carbonyl group which can further react with active methylene group containing compounds in *Knoevenagel* condensation reactions. It also has two activated methyl groups, which can react in the same type of condensation reactions with one or two molecules of aromatic aldehydes.

Another method for the syntheis of 2,6-disubstituted-4*H*-pyran-4-ones, which contain at least one active methyl group, is using 4-hydroxy-6-methyl-2*H*-pyran-2-one (compound **3** in Fig.3) as starting material [11,34]. Its further reaction with isobutyryl chloride (compound **4** in Fig.3) in trifluoroacetic acid (TFA) gives 6-methyl-2-oxo-2*H*-pyran-4-yl isobutyrate (compound **5** in Fig.3). Without separating the compound **5** from the reaction mixture it was subjected to *Fries rearrangement* resulting in 4-hydroxy-3-isobutyryl-6-methyl-2*H*-pyran-2-one (compound **6** in Fig.3). In its decarboxylation and further acidic cyclization reactions 2-isopropyl-6-methyl-4*H*-pyran-4-one (compound **8** in Fig.3) is obtained with 80% yield. Compound **8** also contains a carbonyl group, just as the previously synthesized 2,6-dimethyl-4*H*-pyran-4-one (compound **2** in Fig.2). Since it now contains just one activated methyl group, only one aromatic aldehyde containing fragment can be added to the **backbone** of pyranylidene derivative **8** (shown in Fig.3).

One of the most preferred 2,6-disubstituted-4*H*-pyran-4-ones is 2-*tert*-butyl-6-methyl-4*H*-pyran-4-one (compound **13** in Fig.4) [7,11,35]. The first synthesis method starts from 3,3-dimethylbutan-2-one (compound **9** in Fig.4). Treating it with acetic anhydryde (Ac_2O) and boron trifluoride (BF_3) a boron enolate (compound **10** in Fig.4) is obtained. Its further condensation reaction with 1,1-dimethoxy-N,N-dimethylethanamine (compound **11** in Fig.4) produces N,N-dimethylamino-vinyl group containing boron enolate (compound **12** in Fig.4). Then following an acidic treatment gives 2-*tert*-butyl-6-methyl-4*H*-pyran-4-one (compound **13** in Fig.4). However this method has a drawback because two synthetic reactions towards our target compound had low yields (30-40%), which results in a very low overall yield for synthesis of 2-*tert*-butyl-6-methyl-4*H*-pyran-4-one (compound **13** in Fig.4).

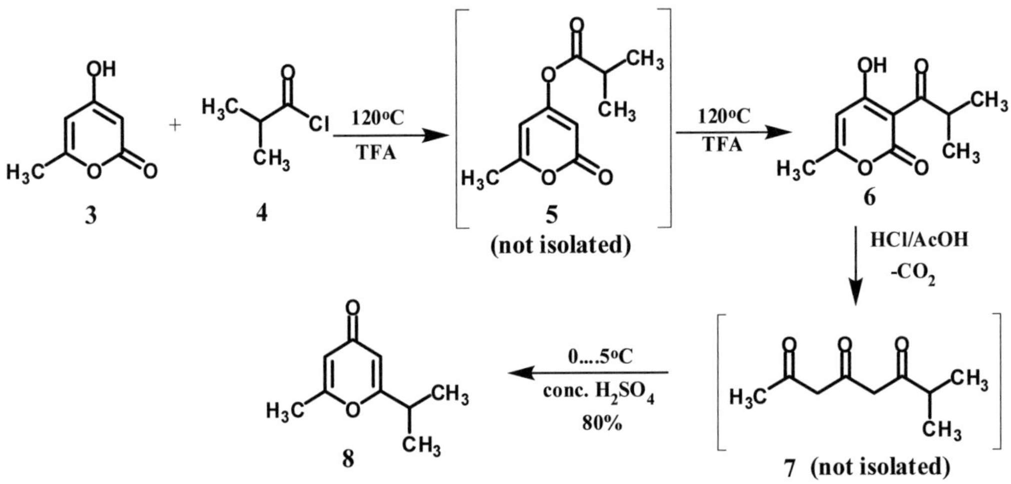

Figure 3. Synthesis of 2-isopropyl-6-methyl-4*H*-pyran-4-one (compound **8**). TFA - trifluoroacetic acid, HCl - hidrochloric acid, AcOH - acetic acid, CO_2 - carbon dioxide, conc. H_2SO_4 - concentrated sulfuric acid.

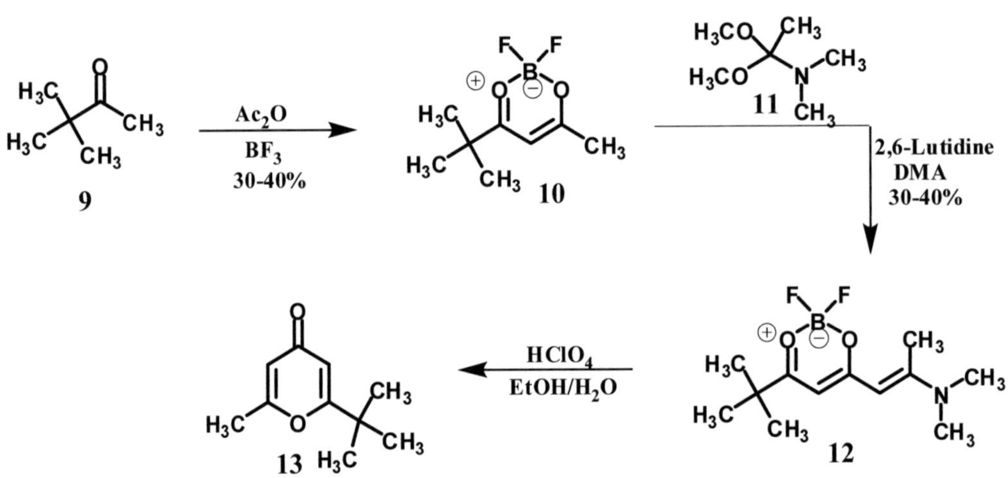

Figure 4. Conventional synthesis of 2-*tert*-butyl-6-methyl-4*H*-pyran-4-one (compound **13**). Ac_2O - acetic anydryde, BF_3 - bornon trifluoride, DMA - dimethylamine, $HClO_4$ - perchloric acid, EtOH - ethanol.

Fortunately, there is another method for synthesizing 2-*tert*-butyl-6-methyl-4*H*-pyran-4-one (compound **13** in Fig.4) with good yields [7] using pentane-2,4-dione (compound **14** in Fig.5) as starting reactant.

In its *Aldol reaction* with methyl pivalate (compound **15** in Fig.5) a 7,7-dimethyloctane-2,4,6-trione (compound **16** in Fig.5) was formed. Without separating the compound **16** from reaction mixture it was subjected to acidic cyclization producing 2-*tert*-butyl-6-methyl-4*H*-

pyran-4-one (compound **13** in Fig.5) with a good overall yield (60%). As with 2-isopropyl-6-methyl-4*H*-pyran-4-one (compound **8** in Fig.3), the resulting 2-*tert*-butyl-6-methyl-4*H*-pyran-4-one (compound **13** in Fig.5) also contains one carbonyl group and one activated methyl group with the possibility of also adding **only one** aromatic aldehyde containing fragment.

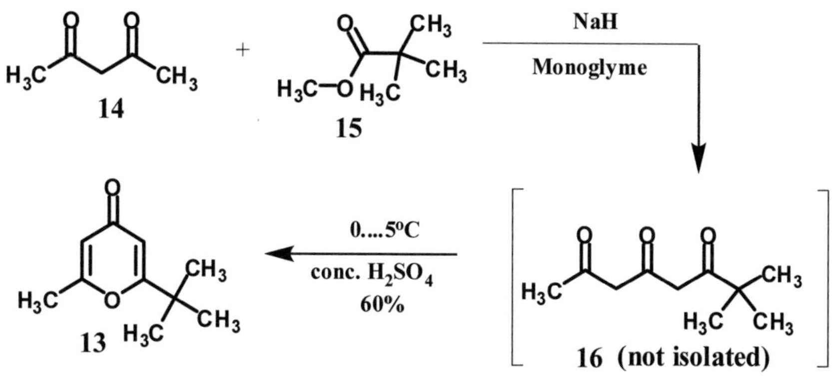

Figure 5. Improved synthesis of 2-*tert*-butyl-6-methyl-4*H*-pyran-4-one (compound **13**). NaH - sodium hydryde, conc. H_2SO_4 - concentrated sulfuric acid.

One of oldest, but no less important methods known for the synthesis of 2,6-disubstituted-4*H*-pyran-4-ones is to obtain them from 3-substituted-vinylcarbonyl-4-hydroxy-6-methyl-2*H*-pyran-2-ones (compounds **17** in Fig.6) [33]. Compounds **17** are obtained from dehydroacetic acid (compound **1** in Fig.6), in which the methyl group in the acetyl fragment is activated to react preferentially with aromatic aldehydes (see Fig.6) giving 3-substituted-vinylcarbonyl-4-hydroxy-6-methyl-2*H*-pyran-2-ones (compounds **17** in Fig.6) [33, 36]. Details on the obtained compounds **17** and their dependence on substituents (R) in their molecules are given in Table 1. They serve as precursors for further synthesis of pyranylidene type compounds.

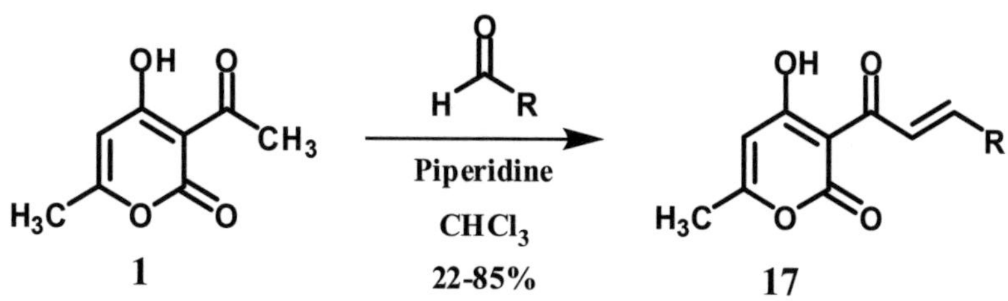

Figure 6. Synthesis of 3-substituted-vinylcarbonyl-4-hydroxy-6-methyl-2*H*-pyran-2-ones (Compounds **17**). Above the arrow are different aromatic aldehydes with different substituents R (see Table 1), which all react with dehydroacetic acid (compound **1**) the same way. $CHCl_3$ - chloroform.

Using this approach it is possible to obtain many different mono-styryl-substituted 4*H*-pyran-4-ones (compounds **17** in Fig.7). However only a few of previously synthesized compounds **17** give 2-styryl-substituted-6-methyl-4*H*-pyran-4-ones (compounds **18** in Fig.7) by acidic decarboxylation under the reaction conditions reported in [30, 33] (see Fig.7) as summarised in Table 2.

R (of compounds 17)	Yield, %	M.p., °C	Recrystallized from
Phenyl	55	130-132	methanol
o-Nitrophenyl	65	161-163	acetic acid/water
m-Nitrophenyl	60	192	chloroform
p-Nitrophenyl	22	165-167	chloroform/ethyl acetate
p-Nitrophenyl	47	246-247	dioxane
p-Dimethylaminophenyl	71	198-200	Chloroform, ethyl acetate, benzene
p-Diethylaminophenyl	58	150	Chloroform, ethyl acetate
o-Hydroxyphenyl	67	186-188	methanol
m-Hydroxyphenyl	61	181-183	ethanol
p-Hydroxyphenyl	69	260-262	dioxane
p-Methoxyphenyl	73	153-154	ethanol
2,3-Dimethoxyphenyl	47	147	ethyl acetate
3,4-Dimetoxyphenyl	46	185	benzene/ethyl acetate
3,4-Diethoxyphenyl	43	163	ethyl acetate
o-Chlorophenyl	36	116-117	ethanol
p-Chlorophenyl	54	155-156	ethanol
3,4-Dichlorophenyl	46	185	ethyl acetate, benzene/chloroform
p-Isopropylphenyl	65	139-141	methanol
1-Naphtyl	62	190	ethyl acetate
β-styryl	57	185	chloroform/ethyl acetate
2-Furyl	85	144	benzene/ethyl acetate

R - substituents of aromatic aldehydes which also remain in the structure of compounds **17** after reactions.
Yield - the practical production of the particular compound **17** in the reaction of dehydroacetic acid (compound **1** in Fig.6) with the corresponding aromatic aldehyde. M.p. - melting point of the particular compound **17**. Recrystallized from - organic solvent, which is used for the particular compound **17** final purification.

Table 1. Synthetic information on pyranylidene compounds **17** (see Fig.6).

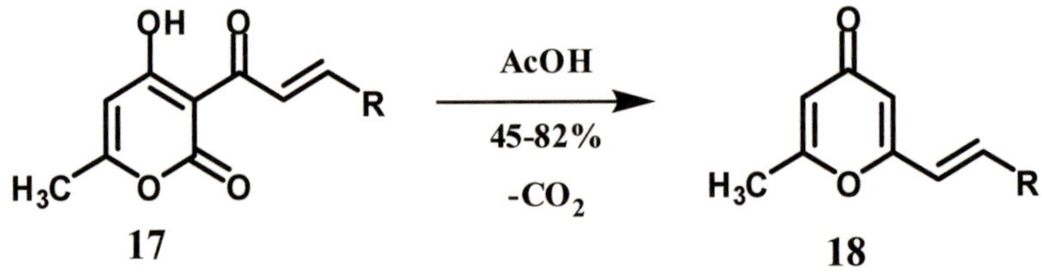

Figure 7. Synthesis of 2-styryl-substituted-6-methyl-4*H*-pyran-4-ones (compounds **18**).

R (of compounds 18)	Yield, %	M.p., °C	Recrystallized from
p-Dimethylaminophenyl	82	156	ethyl acetale/petroleum ether
p-Diethylaminophenyl	68	128-130	methanol/water
o-Nitrophenyl	53	187-189	methanol/water
p-Isopropylphenyl	45	110-112	ethanol/water

R - substituents of aromatic aldehydes which also remain in the structure of compounds **17** after reactions.
Yield - the practical production of the particular compound **17** in the reaction of dehydroacetic acid (compound **1** in Fig.6) with the corresponding aromatic aldehyde. M.p. - melting point of the particular compound **17**. Recrystallized from - organic solvent, which is used for the particular compound **17** final purification.

Table 2. Fries rearrangement possibility [33] of pyranylidene precursors **17** (Fig.7).

Some works can be found on red luminescent compounds where the pyranylidene fragment is hidden as a substructure in the molecule [8, 23-24]. For example, chromene type derivatives of pyranylidene are synthesized from 1-(2-hydroxyphenyl)ethanone (compound **19** in Fig.8) [23-24]. In the *Claisen condensation* reaction (see Fig.8) with ethyl-acetate in the presence of a strong base, 1-(2-hydroxyphenyl)butane-1,3-dione (compound **20** in Fig.8) is obtained. After separation it was subjected to acidic dehydrocyclization giving 2-methyl-4*H*-chromen-4-one (compound **21** in Fig.8) with an overall 45% yield.

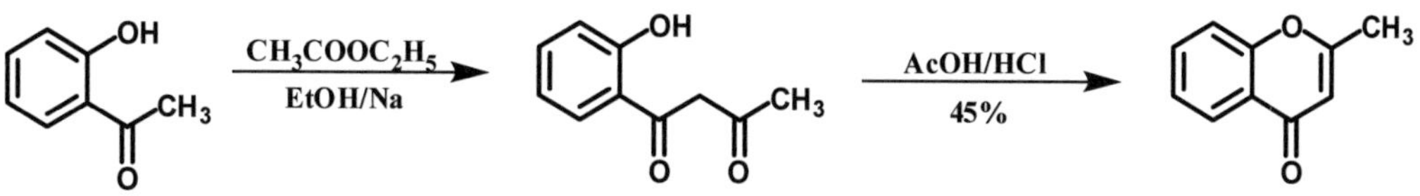

19 **20** **21**

Figure 8. Synthesis of chromene fragment containing derivative of pyranylidene (compound **21**).

For obtaining the benzopyran derivative of pyranylidene [8, 24], a two-stage synthesis procedure is started from 2-methylcyclohexanone (compound **22** in Fig.9).

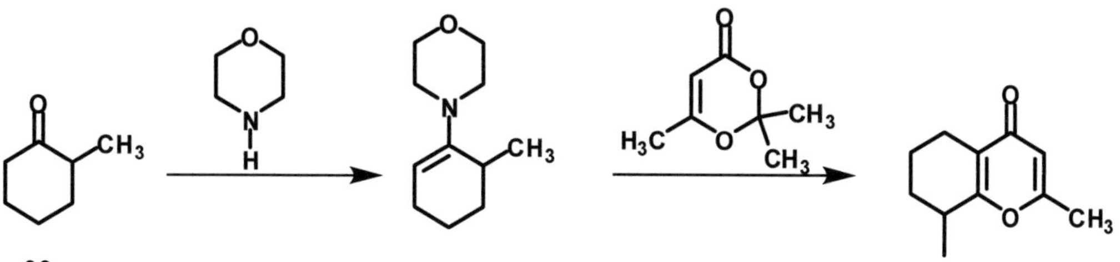

22 **23** **24**

Figure 9. Synthesis of the benzopyran fragment containing derivative of pyranylidene (compounds **24**).

In the first stage of synthesis, treatment with morpholine gives us enamine **23** (4-(6-methylcyclohex-1-enyl)morpholine). In the second stage of synthesis in reaction with 2,2,6-trimethyl-4*H*-1,3-dioxin-4-one, a 2,8-dimethyl-5,6,7,8-tetrahydro-4*H*-chromen-4-one (compound **24** in Fig.9) is sucessfully obtained. Once the desired pyranylidene compound is obtained, the addition of electron acceptor and electron donor fragments becomes a more simplified process, which will be described in detail below in this chapter.

2.2. Addition of electron acceptor fragments to derivatives of 4*H*-pyran-4-ones and 3,5,5-trimethylcyclohex-2-enone

The next step towards synthesizing fully functional pyranilydene and isophorene type red luminescent organic compounds is the addition of electron acceptor fragments to the previously obtained 2,6-disubstituted-4*H*-pyran-4-ones (see Fig.10) and 3,5,5-trimethylcyclohex-2-enone (see Fig.11).

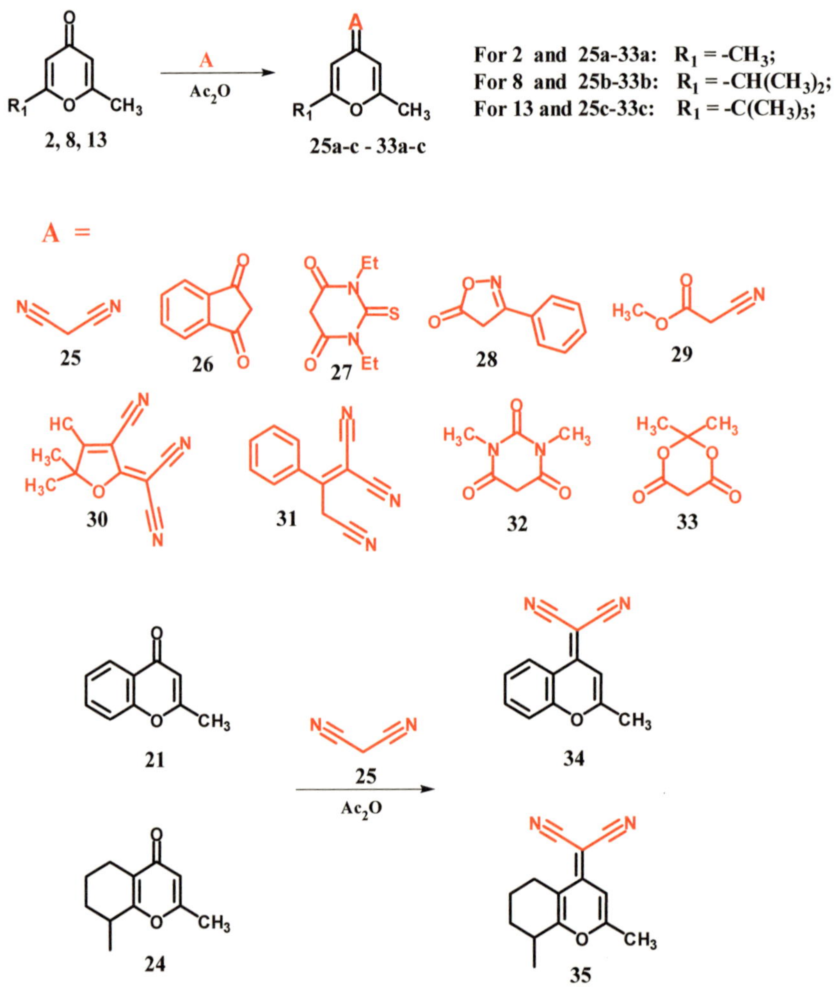

Figure 10. Synthesis of electron acceptor fragment containing derivatives of pyranylidene. Electron acceptors are marked in red while structure backbone, which serves as π-conjugated system remain in black.

Many different electron acceptor fragments (compounds **25-35** in Fig.10) can be introduced in 2,6-disubstituted-4*H*-pyran-4-ones [1,4-18, 28-30,32] using acetic anhydride (Ac_2O) as solvent and catalyst. From these, malononitrile (compounds **25** in Fig.10) is the most commonly used. Since isophorene (3,5,5-trimethylcyclohex-2-enone) (compound **36** in Fig.11) is an inexpensive reagent, which can be purchased from chemical suppliers - such as ACROS and ALDRICH, all that remains is to add electron acceptor and electron donor fragments. As with 2,6-disubstituted-4*H*-pyran-4-ones, the electron acceptor fragments are added in *Knoevenagel* condensation reactions [18-21, 31, 37] with active methylene group containing compounds **37-39** (see Fig.11).

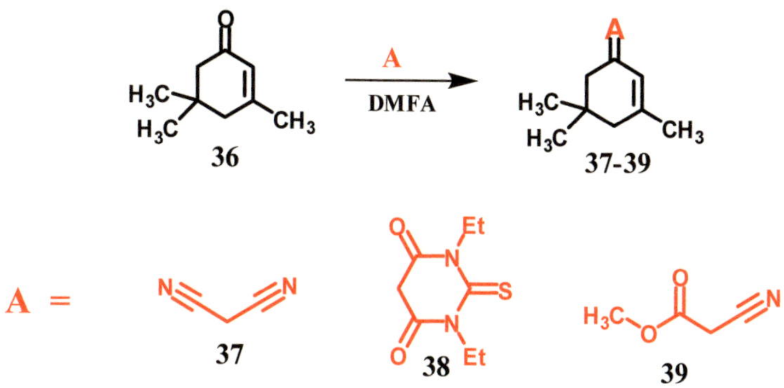

Figure 11. Synthesis of electron acceptor fragment containing derivatives of isophorene (compounds 37-39). As in Figure 10, the electron acceptors are marked in red while the structure backbone remains in black.

The electron acceptor fragment containing derivatives of isophorene (3,5,5-trimethylcyclohex-2-enone) (compounds **37-39** in Fig.11) thus obtained are not always isolated from the reaction mixture [31, 37]. Once they are formed, the electron donor fragment containing aromatic aldehyde is added in the mixture for further reaction with the aldehyde.

2.3. Synthesis of pyranilydene and isophorene type red luminescent compounds by final addition of electron donor fragments

Once the electron acceptor fragment is introduced, the last step for obtaining a fully functional pyranylidene and isophorene red luminescent compounds is to add one or two electron donor fragment containing aldehydes. They are added in *Knoevenagel* condensation reactions with electron acceptor fragment containing derivatives of isophorene as shown in Fig.12 and pyranylidene shown in Fig.13, which contain one or two activated methyl groups.

For isophorene type compounds one electron donor fragment (**40-44**) is always introduced after an electron acceptor fragment is already in the molecule (see Fig.12) [18-21, 31, 37]. Many different structures of electron donor fragments are introduced (compounds **45-57** in Fig.13) in the pyranylidene backbone after introducing the electron acceptor fragment [1,4-18,27-29,31]. In cases where only one methyl group reacts with the aldehyde, a mono-styryl derivative of pyranylidene is obtained (see Fig.13). However, as all possible combinations shown in Fig.13 have not yet been synthesized, it presents a working opportunity for many organic chemists to contribute. If a pyranylidene type compound has two active methyl groups, like compound **25a**, (see Fig.10) it will react with one or two aromatic aldehyde molecules producing chromophores **58-66** (see Fig.14). The reaction product will most likely be a mixture of mono- and bis- condensation products, which are difficult to separate and purify [32]. In reaction with two methyl groups bis-styryl derivatives of pyranylidene are obtained (see Fig.14).

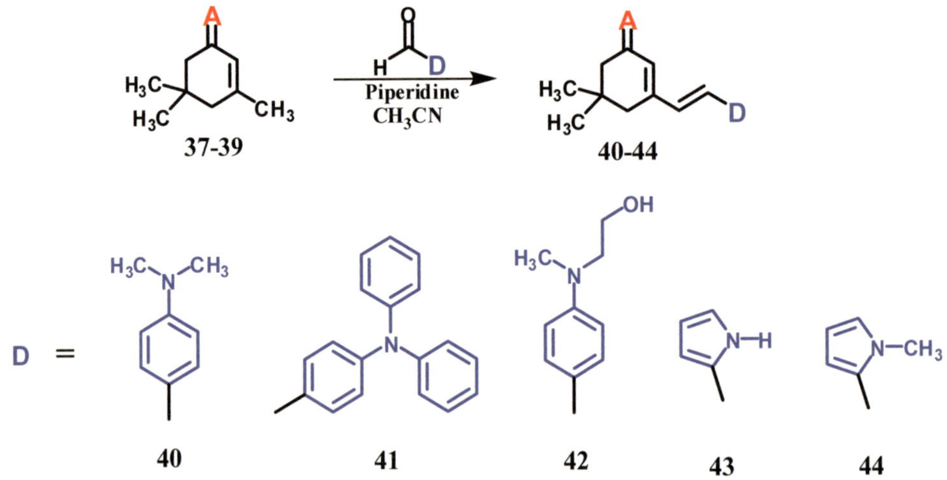

Figure 12. Synthesis of fully functional derivatives of isophorene (compounds **40-44**). CH_3CN - acetonitrile. Electron acceptor fragments are marked in red and electron donor fragments are marked in blue, while the backbone structure fragments remain in black and serve as a π-conjugated system.

A good summary on dicyanomethylene-pyranylidene type red-emitters has been made by Guo et al. [24], according to which the mono electron donor fragment containing pyranylidene-type materials (**45a-c** to **57a-c** in Fig.13) usually have high luminescence quantum yield but their chromaticity is not sufficiently good. At the same time two electron donor fragment derivatives of pyranylidene (compounds **58-66** in Fig.14) have better chromaticity, but their luminance efficiency is relatively low, particulary those with larger conjugations leading to a broad light-emission peak above 650 nm extending to the NIR region, which decreases the efficiency of red electroluminescent materials.

Both chromene (compounds **47,49-50** in Fig.13) and benzopyran (compounds **47,49,51** in Fig.13) type derivatives of pyranylidene have only one electron donor fragment in their molecules, but their optical properties are different. Since chromene type derivatives of pyranylidene have an additional conjugated aromatic ring in its molecule, its optical properties are similar to those with two electron donor fragment derivatives of pyranylidene (compounds **58-66** in Fig.14). At the same time benzopyran pyranylidene compounds **45,46,49** have a simple cyclohexene ring without additional conjugation, so their optical properties are more similar to pyranylidene-type red-emitters, compounds **45a-c** to **57a-c**.

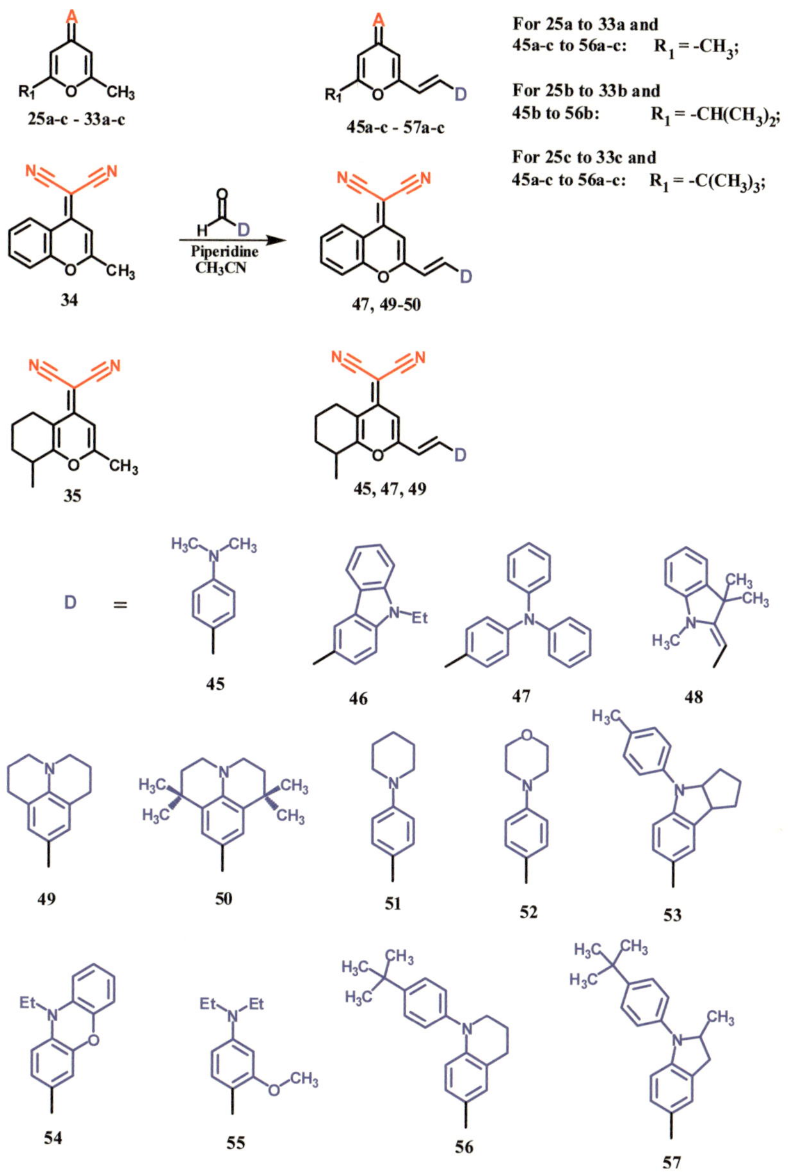

Figure 13. Synthesis of fully functional mono-styryl substituted derivatives of pyranylidene. Electron acceptors are marked in red, electron donors is blue and structure backbone remains in black.

If a pyranylidene backbone with different electron acceptor fragments contains two active methyl groups, then in reaction with a two aldehyde group containing compounds a polymer is formed during the reaction (see Fig.15) [38]. The resulting polymers **70-72** are also reported to be red light-emitting materials.

All derivatives of pyranylidene and isophorene reported so far in this chapter are deposited on the OLED hole transport layer either by thermal evaporation in vacuum or used as dopants in a polymer matrix in limited concentrations.

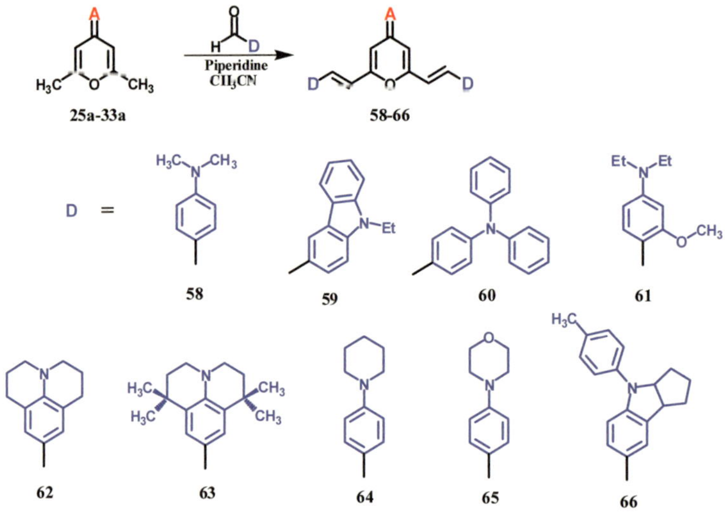

Figure 14. Synthesis of fully functional di-styryl substituted derivatives of pyranylidene. Color significance is the same as for previous figures.

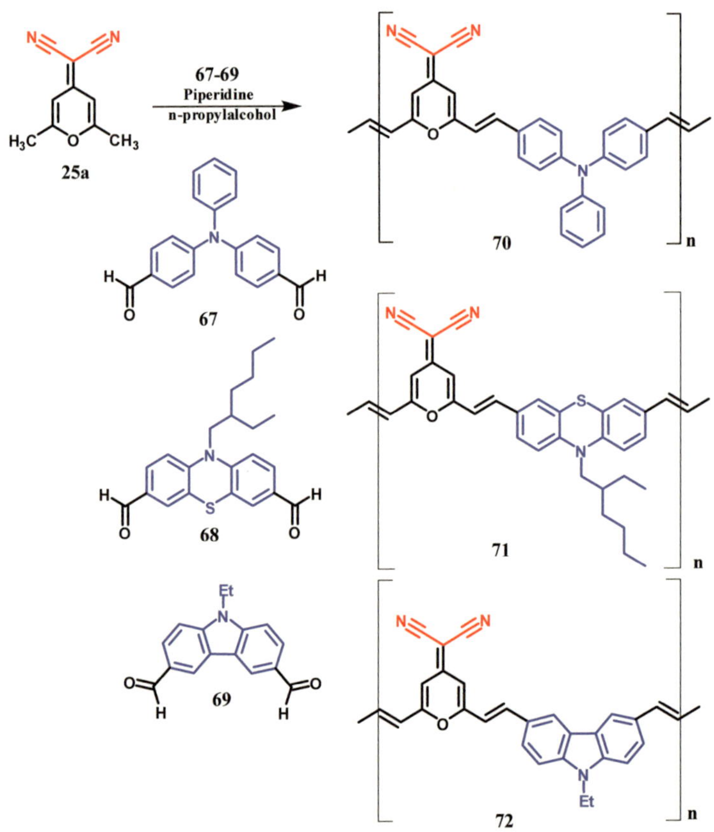

Figure 15. Synthesis of polymeric derivatives of pyranylidene. Color significance is the same as for previous figures.

3. Synthesis and properties of trityloxy group containing glassy derivatives of pyranylidene and isophorene

Our key for obtaining glass forming materials is the synthesis of such electron donor substituent containing aldehyde which would ensure the formation of an amorphous structure of our newly synthesized derivatives of pyranylidene and isophorene. We have synthesized such a compound - 4-(bis(2-(trityloxy)ethyl)amino) benzaldehyde [31-32] **75**, in Fig.16.

3.1. Preparation of molecular glasses

For obtaining a red luminescent glass forming derivative of isophorene, we start with (3,5,5-trimethylcyclohex-2-enone) (compound **29** in Fig.16) as already described in Fig.9. It is subjected to the *Knoevenagel* condensation reaction with malononitrile (**28**). However, 2-(3,5,5-trimethylcyclohex-2-enylidene)malononitrile (**61**) which is formed during the reaction is not isolated because 4-(bis(2-(trityloxy)ethyl)amino) benzaldehyde (**75**) is added to the reaction mixture after 2 hours [31, 37] for further reaction. 2-(3-(4-(Bis(2-(trityloxy)ethyl)amino)styryl)-5,5-dimethylcyclohex-2-enylidene)malononitrile (**IWK**) was obtained in good yield after its separation and purification by liyquid column chromatography as described in [31].

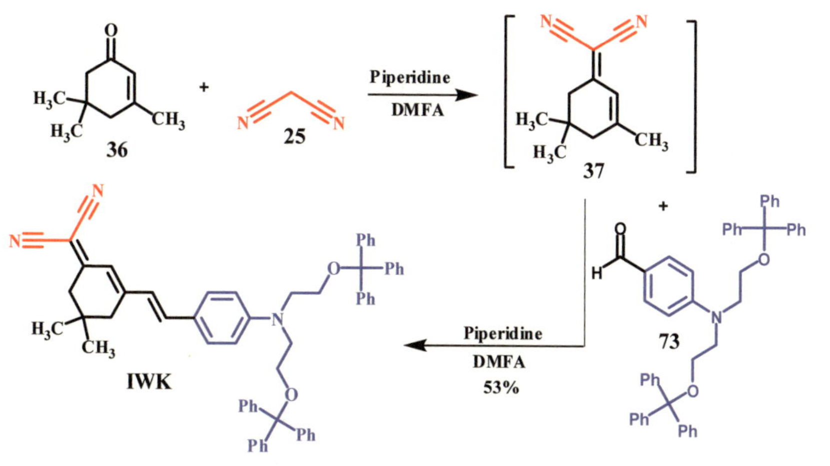

Figure 16. "One pot" synthesis of **IWK.** (See previous figures for explanation of color significance).

For obtaining red luminescent glass forming derivatives of pyranylidene, we use three different electron acceptor fragment containing derivatives of pyranylidene (compounds **25a** in Fig.17). Malononitrile (in compounds **74a and 75a**), indene-1,3-dione (in compounds **74b and 75b**) and barbituric acid (in compounds **74c and 75c**) are used as electron acceptor fragment carrying compounds [32].

In the *Knoevenagel* condensation reaction with compound **25a** and 4-(bis(2-(trityloxy)ethyl)amino) benzaldehyde (**73**) a mixture of mono- (**ZWK-1, DWK-1, JWK-1**) and bis- (**ZWK-2, DWK-2, JWK-2**) condensation products is obtained. Their separation is complicated but nevertheless a large part of each product was separated by liquid column chromatography (silicagel and dichloromethane for **ZWK-1** and **ZWK-2**, dichloromethane: hexane = 4:1 for **DWK-1** and **DWK-2**, dichloromethane: ethyl acetate = 4:1 for **JWK-1** and **JWK-2**). The physical properties of compounds **WK-1**, **WK-2** and **IWK** are described in detail further in this chapter.

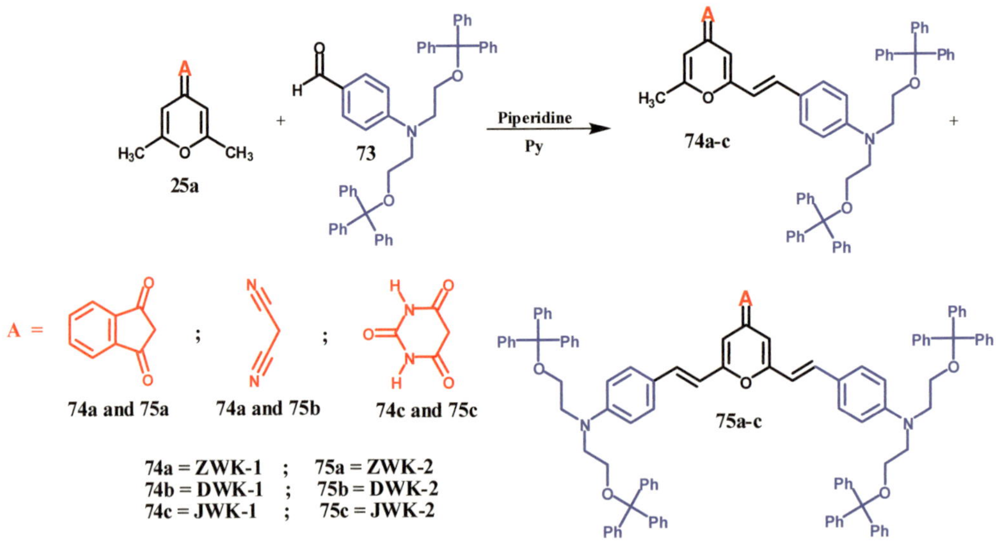

Figure 17. Synthesis of glass forming derivatives of pyranylidene. Py - pyridine. (See previous figures for explanation of color significance). Since compounds **74a-c** and **75a-c** are our obtained red light-emitting materials, we have assigned specific names for each (**ZWK-1, ZWK-2, DWK-1, DWK-2, JWK-1** and **JWK-2**) [28-30, 32, 46].

3.2. Thermal properties

The thermogravimetric analysis (TGA) of trityl group containing pyranylidene type compounds is used to measure their thermal decomposition temperatures (T_d). T_d of compounds **WK-1** and **WK-2** are determined in the temperature range from +30°C to +510°C at a heating rate of 10°C/min [32] at the level of 10% weight loss (see Fig.18).

Pyranylidene type compounds with two N,N-ditrityloxyethylamino electron donor fragments (**ZWK-2**, **DWK-2**, **JWK-2**) are slightly more thermally stable than compounds containing only one such fragment, i.e. **ZWK-1**, **DWK-1** or **JWK-1**. The increase in thermal stability of pyranylidene type compounds by adding another electron donor fragment is as high as 10°C from **ZWK-1** to **ZWK-2**, 19°C from **JWK-1** to **JWK-2** and 29°C from **DWK-1** to **DWK-2**. The most thermally stable compound is a two electron donor fragment containing derivative of pyranylidene with malononitrile as electron acceptor in it (**DWK-2**).

Differential scanning calorimetry (DSC) measurements are used to measure the glass transition temperatures (T_g) of the compounds **WK-1** and **WK-2.** Three thermo cycles are performed for the determination of T_g. The first scan was done within the temperature range

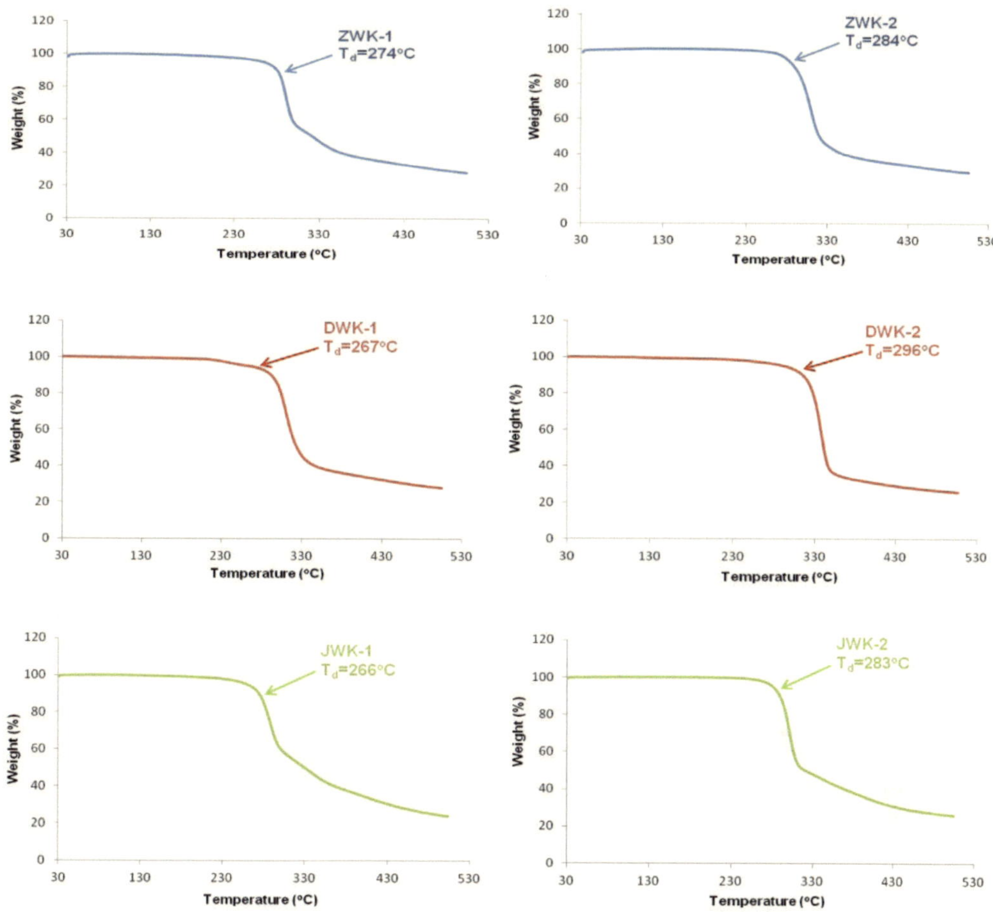

Figure 18. Thermogravimetric analysis of compounds **WK-1** and **WK-2.** A sample of each compound is constantly weighed during heating. At some temperature (T_d) the mass of the sample starts to decrease rapidly - this indicates when the respective compound starts to decompose and is no longer thermally stable.

from +25°C to +250°C at a heating rate of 10°C/min [32]. After the first heating scan samples of the compounds were cooled to 25°C at a rate of 50°C/min and heated for a second time from +25°C to +250°C at a rate of 10°C/min. The T_g value is obtained from the second heating scan (see Fig.19) and for almost all compounds is higher than 100°C. We could not obtain usable DSC curves for **DWK-1**. The compounds with two N,N-ditrityloxyethylamino electron donor fragments have higher T_g compared to those with only one electron donor fragment, which may be attributed to the different numbers of bulky trityloxyethyl groups attached to the two electron donor fragment. In a larger number of bulky groups T_g increases by 8°C from **ZWK-1** to **ZWK-2** and 7°C from **JWK-1** to **JWK-2**. Pyranylidene type compounds with barbituric acid as electron acceptor, e.g. **JWK-1** and **JWK-2** have the highest T_g values compared to **ZWK-1, ZWK-2** and **DWK-2**, which may be due to the additional formation of intermolecular hydrogen bonds by N-H groups of barbituric acid fragments in the molecules.

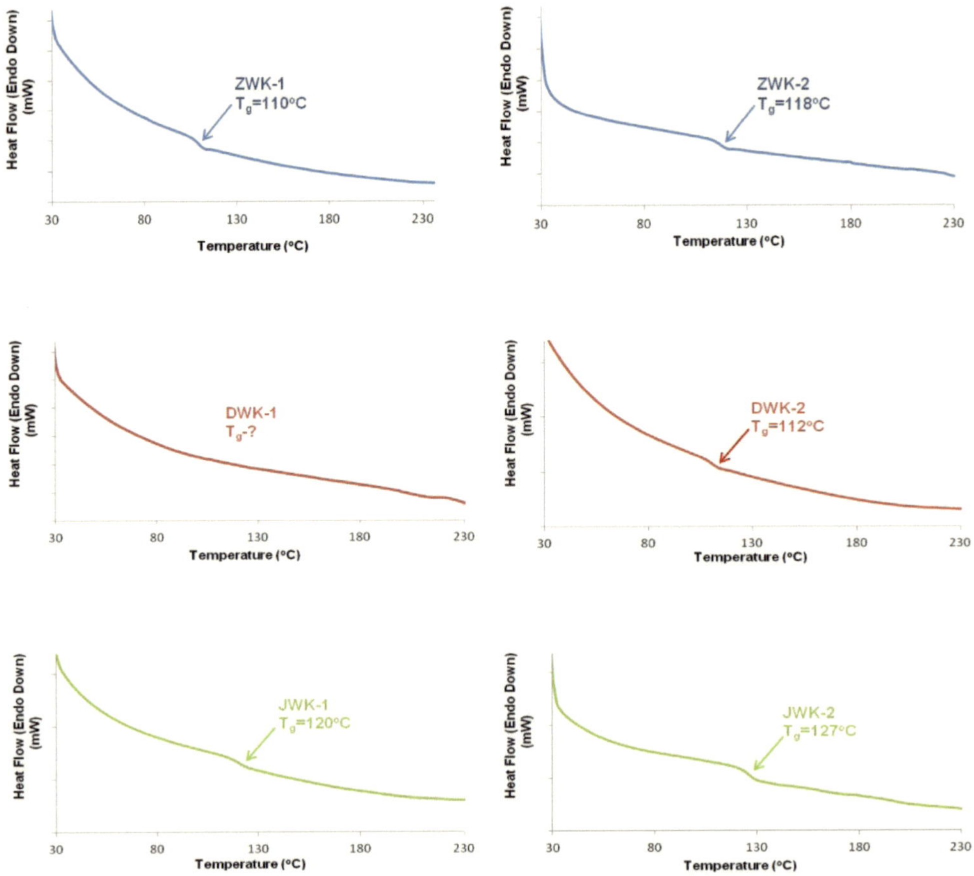

Figure 19. DSC thermogramms of compounds **WK-1** and **WK-2.** Since amorphous compounds have several solid state phase modifications, the glass transition temperature (T_g) indicates when compound solid structure transitions from a more kinetically stable phase (with more free volume) to a more thermodynamically stable phase (with less free volume). During such phase transitions some ammount of heat is absorbed (endothermic process) which appears as a small drop on the DSC curves.

The TGA analysis of **IWK** is conducted as previously described [32]. The thermal decomposition temperature (T_d) of **IWK** is found to be even higher than that of pyranylidene type compounds **WK-1** and **WK-2** (see Fig.20). However its glass transition temperature (T_g) is lower by 18°C to 35°C degrees compared to that of pyranylidene type glasses. Despite the lower thermal stability, the pyranylidene type compounds **WK-1** and **WK-2** have better glass forming properties than the isophorene type compound **IWK**.

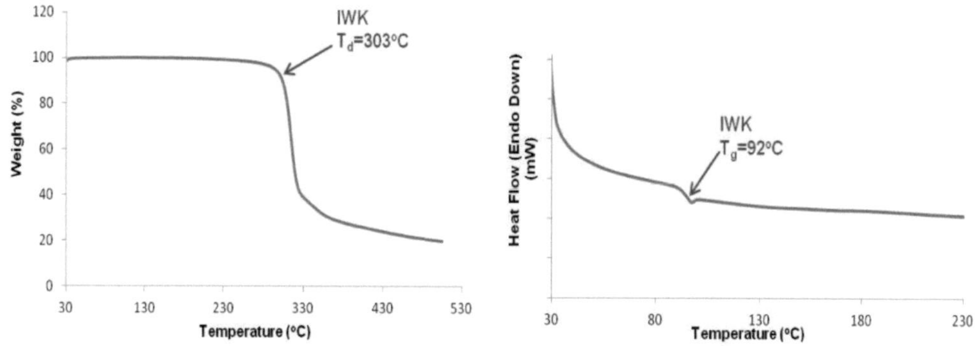

Figure 20. TGA and DSC analysis of **IWK.** (Please see Fig.18 and Fig.19 for a more detailed explanation).

3.3. Glass forming properties

Thin films are deposited on quartz glass by the spin-coating technique. Before the deposition of the layers, the quartz glass substrates are cleaned in dichloromethane. The solutions are spin-coated onto the substrates for 40 s at 400 rpm and acceleration 200 rpm/s.

In all cases, pure films obtained from two electron donor fragment containing pyranylidene compounds (**ZWK-2**, **DWK-2** and **JWK-2**) have an almost pure smooth and amorphous surface, but pyranylidene compounds with one electron donor fragment (**ZWK-1**, **DWK-1** and **JWK-1**) show several crystalline state areas (see Fig.21). Both glasses containing barbituric acid as an electron acceptor fragment (**JWK-1** and **JWK-2**) show the least amount of small crystal formations on their pure film surface. The higher stability of their amorphous state could be explained by an enchancement of N-H group hydrogen bonds in the molecules. Pure films obtained from malononitrile electron acceptor fragment containing compounds (**DWK-1** and **DWK-2**) contain small crystal dots, especially **DWK-1.** This could be due to small steric dimensions of malononitrile group, which allows more **DWK-1** molecules to be concentrated in the same volume to allow closer interaction with other molecules enabling higher possibility to form agreggates and crystallites.

Information obtained from the surfaces of the pure films is consistent with the measured glass transition temperatures (T_g). Glasses having higher T_g values are found to have less crystalline dots on their pure film surface. As we were unable to determinate T_g for **DWK-1**, according to above mentioned trend its glass transition temperature is expected to be below 110°C.

Thin film containing only pyranylidene type compound **WK-1** and **WK-2** are amourphous despite of small crystalline dots in it. Till now only way to prepare amourphse films which contain pyranyliden derivatives was doping them in glass forming compound. In that case maximum doping concentration was considered to be 2wt% due to self crystallization [11-12]. However, incorporation of bulky trityloxy groups in their molecules or using glasses **WK-1** and **WK-2** could increase this concentration limit more then 10 times.

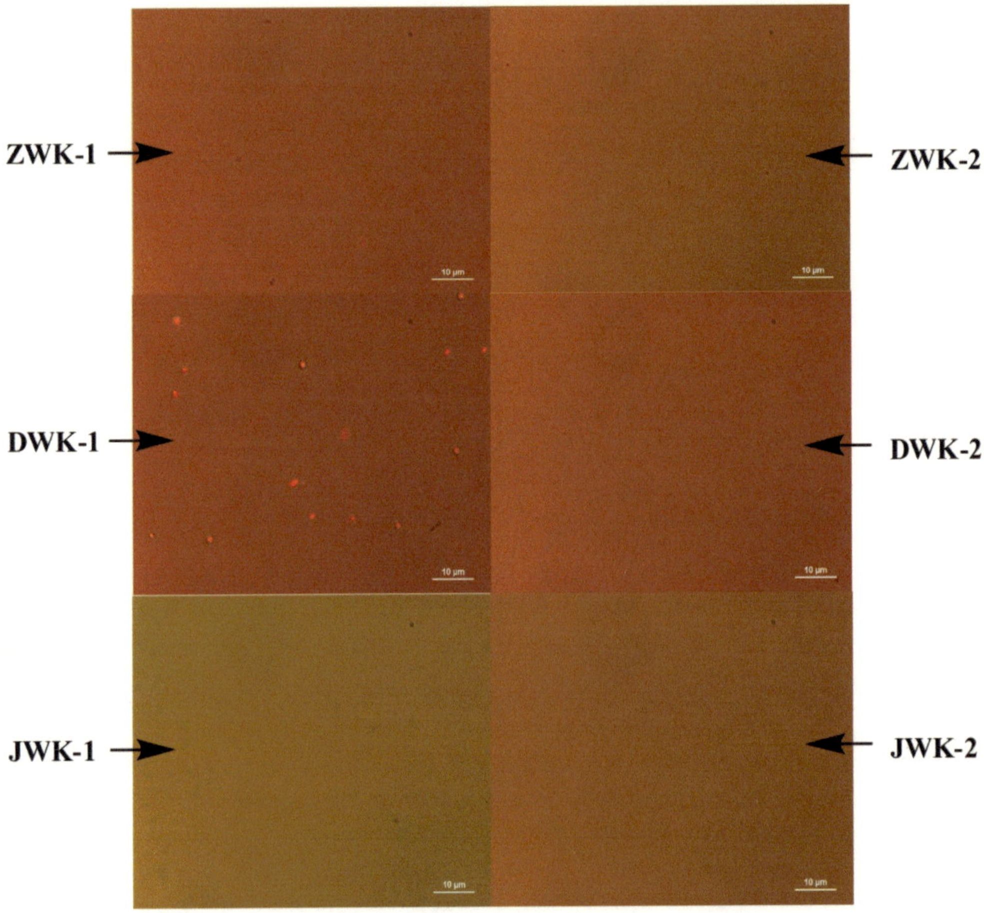

Figure 21. Optical mircroscope images of the pure films of the compound **WK-1** and **WK-2**. Dots on the pure film surface represent compound crystalline state while the remaining smooth area shows amorphous solid state.

3.4. Absorption and luminescence properties

The absorption and fluorescence spectra of the synthesized compounds in diluted dichloromethane solution and pure films are shown in Figs. 22 and 23.

A **DWK-1** molecule, whose backbone consists of the laser dye 4-(dicyanomethylene)-2-methyl-6-[p-(dimethylamino)styryl]-4H-pyran (**DCM**), in dichloromethane solution has its absorption maximum at 472 nm, which is 8 nm red shifted with respect to the pure **DCM** molecule in the same solution [9]. It shows that the bulky trityloxyethyl group has only a small influence on the energy structure of the molecule. The peaks of the absorption spectra in solution of the molecules with indene-1,3-dione (**ZWK-1**) and barbituric acid (**JWK-1**) electron acceptor substituents in the backbone are red-shifted by approximately 40 nm compared to **DWK-1**. A stronger electron acceptor group gives larger red shifts.

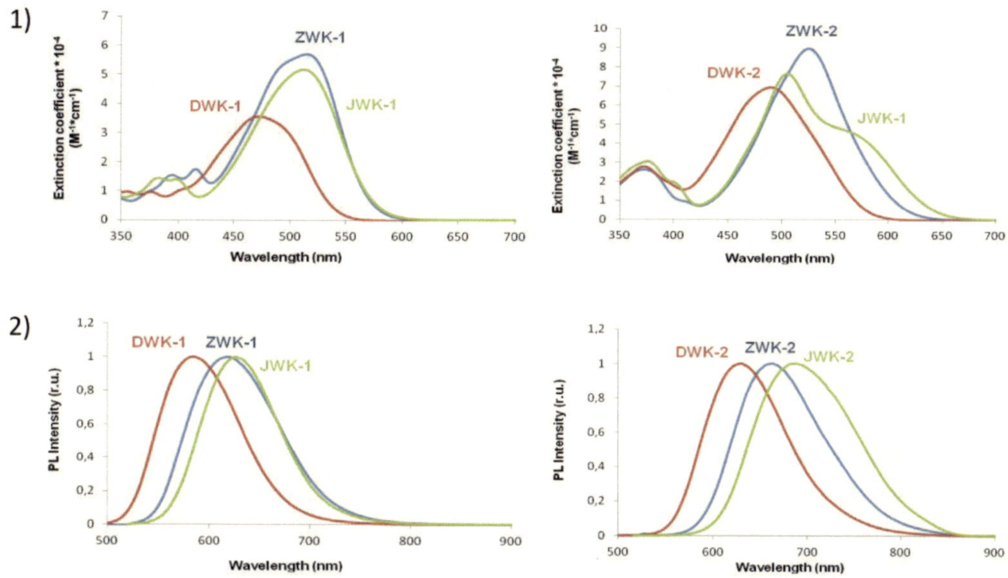

Figure 22. 1) Absorption and 2) Photoluminescence spectra of compounds **WK-1** and **WK-2** in dichloromethane solution

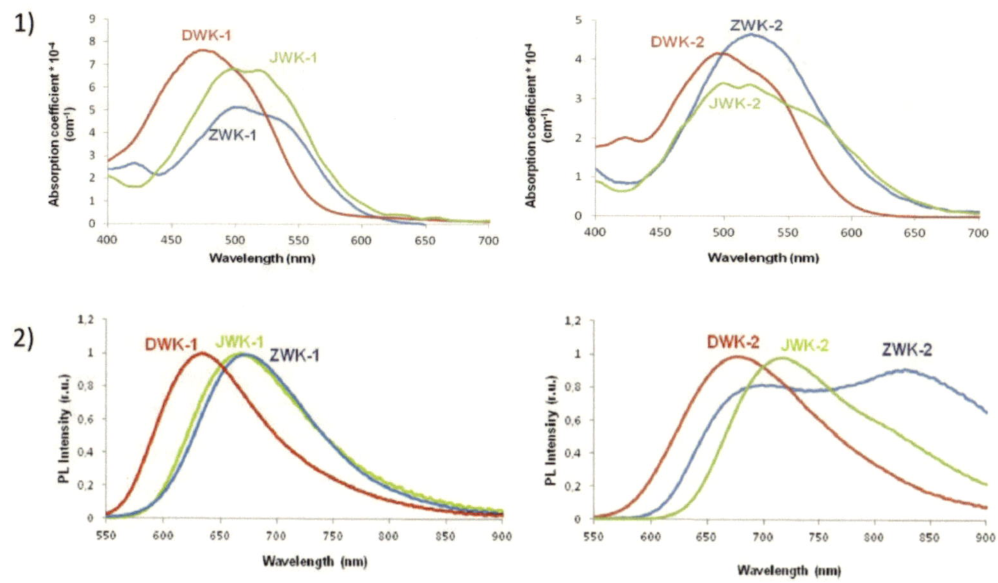

Figure 23. 1) Absorption and 2) Photoluminescence spectra of compounds **WK-1** and **WK-2** in thin solid films

The photoluminescence (PL) spectrum of the **DWK-1** solution was found to be Stokes shifted by about 115 nm (peak position at 587 nm) with respect to the absorption spectra (see Fig.22). The PL spectra of **JWK-1** and **ZWK-1** molecules exhibited similar shapes, with their maxima red-shifted to 635 and 627 nm, respectively. The photoluminescence spectra are unstructured and strongly Stokes shifted in accordance with intramolecular charge-transfer nature of the excited states [39]. For compounds containing two 4-((N,N-ditrityloxyethyl) amino)styryl electron donor fragments the absorption and luminescence spectra of the solution are observed to be red-shifted and have larger extinction coefficients, which is due to the larger absorption cross section of these molecules. The peaks of the absorption spectra of **DWK-2** and **ZWK-2** are red shifted by 17 and 11 nm, respectively, compared to molecules with a single electron donor fragment. A similar red shift has been reported for the molecule with two electron donor fragments **bis-DCM** compared with **DCM** molecules with a single electron donor fragment [40]. It is observed that molecules with two electron donor fragments have a larger conjugation length. A second reason could be simultaneously functioning two donor groups which give stronger electron donor properties. The shape of the absorption spectrum of **JWK-2** is found to be different from that of **JWK-1** and the oscillator strength of the absorption band of **JWK-2** at about 502 nm becomes more intense (see Fig.22(1)).

The fluorescence spectra of molecules with two electron donor fragments are broader and further Stokes shifted than molecules with only one electron donor fragment. This may be attributed to the different conjugation lengths as indicated by the absorption spectra. The peak positions of **DWK-2**, **ZWK-2** and **JWK-2** are observed at 640, 678 and 701 nm, respectively. The red shift of the absorption spectra of solutions increases corresponding sequentially to **ZWK**, **JWK** and **DWK**, as stronger electron acceptor fragments induce larger red shifts. This could be explained by their electron withdrawing properties, which differ among our investigated electron acceptor fragments. The shift of luminescence spectra did not maintain the same sequence due to the larger Stokes shift for the **JWK** molecules.

The absorption spectra of thin solid films of the molecules with one electron donor fragments are practically unchanged with respect to the solutions spectra. They are slightly broader with small red-shift indicating a weak excitonic interaction in the solid state, which is typical for glass-forming amorphous materials. For the molecules with two electron donor fragments **ZWK** and **DWK** the absorption spectra are found to shift by 21 nm and 22 nm, respectively. The peak positions of the absorption spectra for **JWK** molecules remain unchanged by the incorporation of a second electron donor fragment. However, the fluorescence spectra of all films are red-shifted in comparison with those of solution.

For molecules with one electron donor fragment, the shape of the fluorescence spectra of thin films is very similar to that in solution, which confirms that for these compounds the excited states in the aggregates in the solid state are not very different from those in molecules. However, the derivatives with two electron donor fragments exhibit an additional band at longer wavelengths in thin films, which becomes more intense going from weaker to stronger electron acceptor fragments in the studied molecules. In the case of **ZWK-2** in thin films the additional band even becomes dominant.

In the case of **IWK**, the absorption and luminescence spectra of thin solid films are also found to be practically unchanged compared to its solutions spectra as shown in Fig.24.

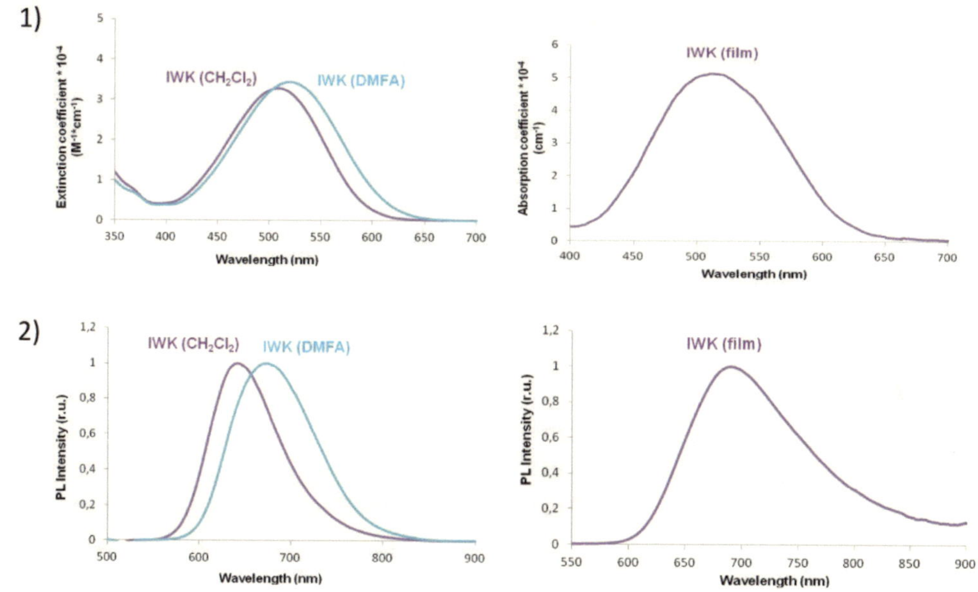

Figure 24. 1) Absorption and 2) Emission of **IWK** in solutions and thin solid film

The same relation is observed for **IWK** emission properties in solution as well as in thin films. However, in solid state its emission is very weak compared to pyranylidene type compounds, which may limit the usefulness of **IWK** in OLED applications.

3.5. Photoluminescence quantum yields

Photoluminescence quantum yield (PLQY) of the investigated compounds in solution and in thin films is measured by using an integrating sphere (Sphere Optics) coupled to a CCD spectrometer [41]. PLQY thus measured for all compounds are summarised in Table 3. Compounds with more polar groups attached exhibit PLQY up to 0.54 in dilute solutions, which is slightly higher than for **DCM** dye in similar surroundings [42, 43]. PLQY depends slightly on the acceptor group as can be seen from Table 3. That means that compounds with a stronger electron acceptor group have higher PLQY. **JWK** and **ZWK** molecules with two electron donor groups have lower PLQY in comparison with one electron donor group. However, the opposite is observed with **DWK** compounds, as molecules with two electron donor groups exhibit larger PLQY. This may be due to the shielding of the acceptor group by bulky trityloxyethyl groups. PLQY of pure films is found to be more than one order of magnitude lower than that in solution. This reduction is particularly strong in the case of molecules with two donor groups. PLQY values of these compounds correlate with the intensity of the long wavelength fluorescence band, as PLQY is lower in materials with a stronger low energy fluorescence band. Molecular distortions taking place during formation of solid films are probably responsible for both of these effects. Compound molecules with two bulky acceptor groups are probably strongly distorted in solid films, so that molecular chains connecting acceptor and donor moieties are twisted. Such twisting usually leads to a red-shift in the molecular fluorescence and to fast non-radiative relaxation [44]. The twisted molecules form energy traps in solid films, which may be populated during the excitation diffusion. Therefore, even a small fraction of distorted molecules may significantly affect the fluorescence spectrum and PLQY. We were unable to measure PLQY in **IWK** pure thin solid films. Moreover, it also shows the lowest value in solution and therefore cannot be used as a light-emitting material.

	Solution	Thin film
DWK-1	0.32	0.026
DWK-2	0.43	0.009
JWK-1	0.47	0.011
JWK-2	0.32	0.007
ZWK-1	0.54	0.01
ZWK-2	0.4	0.003
IWK	0.098	-

Table 3. Photoluminescence quantum yield of investigated molecules in dichlormethane solutions and pure thin films.

It is worth mentioning that **DCM** molecules do not show any photoluminescence from pure films due to the small distance between molecules which results in high molecular interaction. Therefore, host-guest films of transparent polymethylmethacrylat (**PMMA**) polymer with varying dye doping were prepared in order to observe the impact of concentration on photoluminescence quenching. The dependence of PLQY on concentration of **DWK-1** and **DWK-2** molecules is shown in Fig.25.

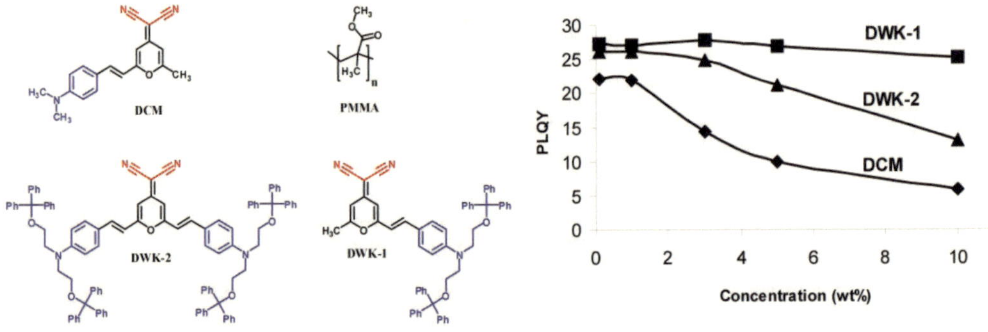

Figure 25. The dependence of PLQY on concentration of **DWK-1**, **DWK-2** and **DCM** dyes in **PMMA** matrix.

For comparison the PLQY of **DCM** in **PMMA** are also included in Fig.25. **PMMA** films doped with **DWK-1** and **DWK-2** at low concentration (<1 wt%) exhibit somewhat lower PLQY as compared to that obtained in solution (See Fig.24 and Table 3). This discrepancy may be attributed to the sensitivity of molecules to the polarity of the surrounding media.

At higher concentrations (>3 wt%) the **DWK-1** molecule shows negligible photoluminescence quenching dependence on concentration. On the other hand molecules with two donor groups exhibit pronounced quenching. Fluorescence efficiency of the polymer film doped with 10 wt% of **DWK-2** molecules decreases 2-times compared to that of films doped with 10 wt% **DWK-1** molecules. The reason for the lower PLQY could be the same as for different PLQY of the pure films. The laser dye **DCM** dispersed in the polymer matrix at high concentration shows a remarkable fluorescence quenching. For example, at a 10 wt% concentration of **DCM** molecules, up to a 4-time decrease of quantum yield is observed in comparison with the same concentration of **DWK-1** molecules. Thus, incorporation of bulky trityloxyethyl groups prevents the formation of aggregates of the dye molecules and remarkably reduces the fluorescence quenching dependence on concentration, enabling the use of higher doping levels in emissive layers.

3.6. Amplified spontaneous emission properties

DCM molecule is a well known laser dye. In a previous work light amplification was demonstrated in **DCM:Alq$_3$** (see Fig.25 and Fig.26) thin films [45].

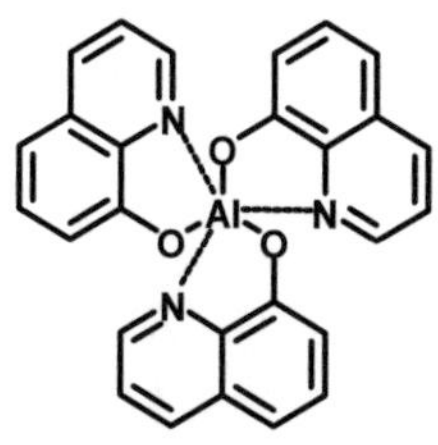

Alq3

Figure 26. Tris(8-hydroxyquinolinato)aluminium (**Alq3**) is a well known light-emitting material.

In order to test the light amplification prospects of our synthesized compounds, we prepared pure thin films of all the compounds on a quartz substrate and measured their amplified spontaneous emission (ASE). Such emission was observed only for four of six compounds, **DWK-1, DWK-2, JWK-1** and **ZWK-1,** as shown in Fig.27 [46].

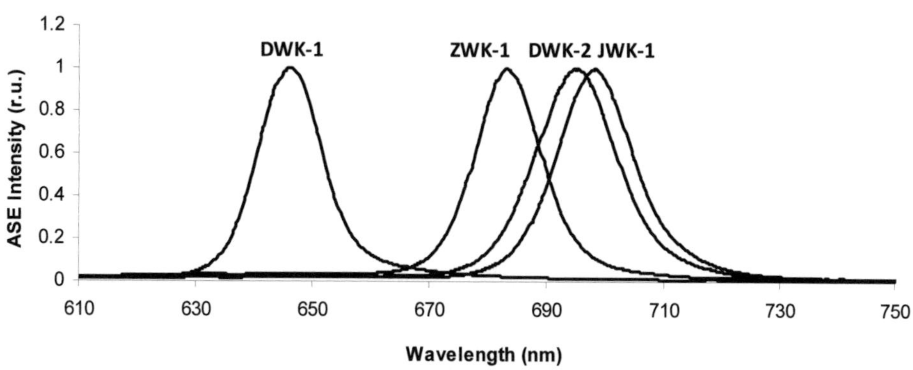

Figure 27. ASE spectrum in pure films of compounds **ZWK-1, DWK-1, DWK-2** and **JWK-1**

From the other two samples of **JWK-2** and **ZWK-2** no ASE signal has been observed. The peak positions of ASE are red shifted as compared to the fluorescence band maxima (see Fig.27 and Fig.22). The red shift values were found to be 14, 18, 10 and 31 nm for **DWK-1**, **DWK-2**, **JWK-1** and **ZWK-1**, respectively. Variations in the peak intensity of ASE spectra as a function of the pump beam pulse energy are shown in Fig.28, from which ASE threshold values are estimated to be 90±10, 330±20, 95±10, 225±20 μJ/cm² for **DWK-1**, **DWK-2**, **JWK-1** and **ZWK-1**, respectively. These values are larger in comparison with the threshold values (of the order of micro joules per square centimeter) reported for some other materials [46, 47].

However, a direct comparison is difficult because the ASE threshold, in addition to material properties, depends also on the sample and excitation geometries, film thickness, optical quality and excitation pulse duration.

Nevertheless it has not been observe ASE in pure **DCM** films, but we have measured it in **DWK-1** which is the same **DCM** with additional trityloxyethyl group. It should also be noted that some sample degradation has been observed at the highest excitation intensities; however no noticeable degradation is observed when excitation intensity is 1.5 - 2 times exceeding the ASE threshold.

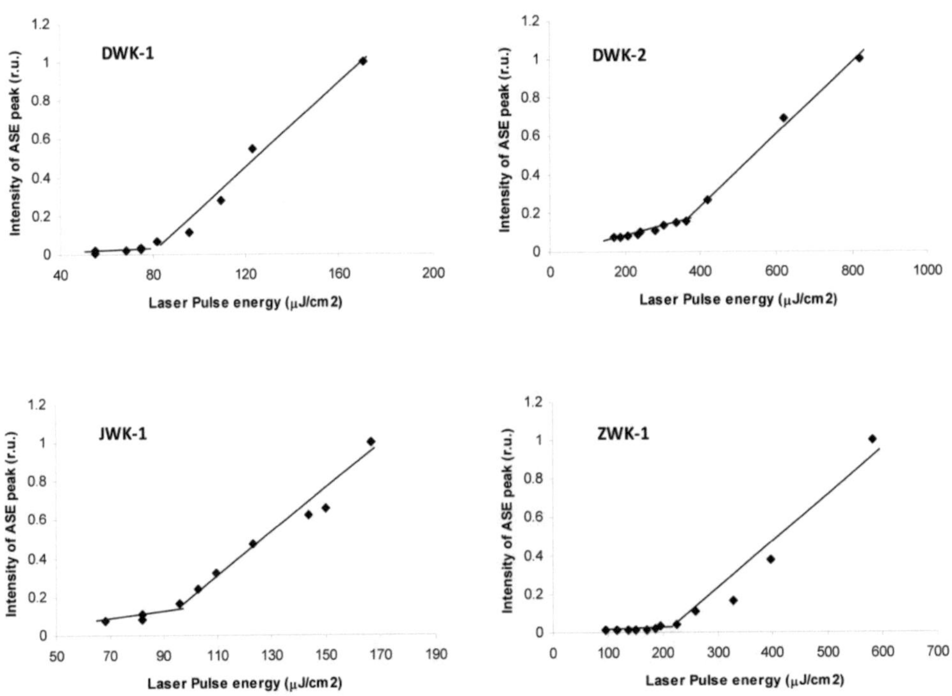

Figure 28. ASE intensity as a function of irradiation pulse energy in **DWK-1, DWK-2, JWK-1, ZWK-1** compounds in thin solid film. Lines are guides for the eye.

ASE develops in the spectral position where the light amplification coefficient has the maximal value. The amplification coefficient may be described as:

$$a(\lambda) = n^* [(\sigma_{em}(\lambda) - \sigma^*(\lambda)] - (N - n^*)\sigma_0(\lambda) \quad (1)$$

where n^* is the density of excited molecules, N is the total density of molecules, $\sigma_0(\lambda)$,$\sigma_{em}(\lambda)$and $\sigma^*(\lambda)$ are cross-sections of the ground state absorption, stimulated emission and excited state absorption, respectively. As it can be seen from Eq. (1) even weak ground state absorption may strongly reduce the amplification coefficient or make it negative. This is because only a small fraction of molecules is usually excited even under high intensity excitation conditions, i.e., N>>n*. Thus, the absorption band tails, which overlap with fluorescence band, are evidently responsible for the red shifts in ASE spectra in comparison with the maxima of the fluorescence. Note, that the light propagation length is limited by the film thickness in the absorption measurements, while ASE emission can propagate a much longer way along the film.

3.7. Photoelectrical properties and energy structure of glassy thin films

Information about the location of energy levels enables one to determine the best sample structure for electroluminescence measurements. To characterize the energy gap in organic solids several methods are applied. In organic crystals as well as amorphous solids charge carriers do not emerge as "bare" quasi-free electrons and holes but as a polaron type quasi-particle, dressed "in electronic and vibronic polarization clouds" [48, 49]. Electronically relaxed charges may be formed far enough from each other which give rise to a wider optical band gap E_G^{Opt} [49, 50]. The optical energy gap E_G^{Opt} may be obtained from the low energy threshold of the absorption spectra of organic thin films. The vibrationally and electronically relaxed charge carrier states contribute to the adiabatic energy gap E_G^{Ad}. It could be attributed to the threshold energy of photoconductivity E_{th} which can be estimated from the spectrum of the quantum efficiency of photoconductivity $\beta(h\upsilon)$ [49]:

$$\beta(h\nu, U) = \frac{j_{ph}(h\nu, U)}{k(h\nu)I(h\nu)g(h\nu)} \quad (2)$$

where j_{ph} is the density of photocurrent at a given photon energy $h\nu$ and applied voltage U, $I(h\nu)$ is the intensity of light (photons/cm^{2}s), $k(h\nu)$ is the transmittance of the semitransparent electrode and $g(h\nu)$ is the coefficient which characterizes the absorbed light in the organic layer.

E_{th} can be determined from a sample where the organic compound is sandwiched between two semitransparent electrodes, which in our case are ITO and thermally evaporated aluminum. The sample is irradiated through the electrodes and current changes are measured as shown in Fig.29(1). Efficiency of photoconductivity at different light energy is calculated using Eq. (2) and is plotted as a function of the photon energy in Fig.29(2). The sample is illuminated from both aluminum and ITO side when positive and negative voltage is applied to them. E_{th} is determined by plotting $\beta^{2/5}$ as a function of the photon energy. The intersections of tangents at low photon energy on the curve of $\beta^{2/5}$ plotted as a function of the photon energy and photon energy axis gives E_{th} as shown in Fig.29(3).

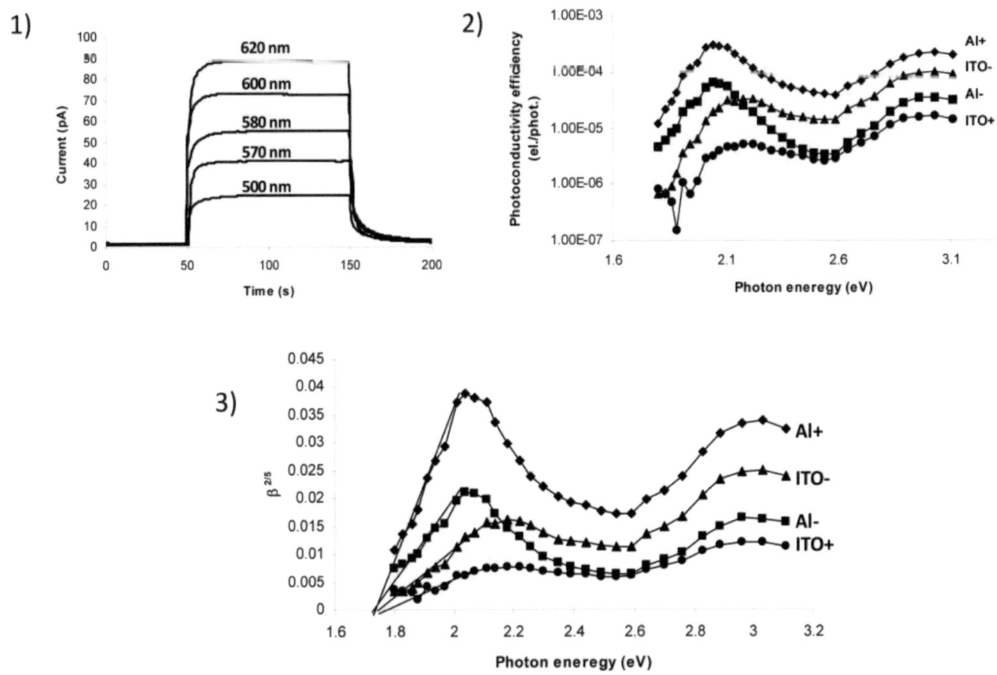

Figure 29. 1) Photocurrent at different wavelength for **JWK-2** compound ,2) Dependency of photoconductivity efficiency on photon energy for **JWK-2** compound,3) Determination of Eth from photoconductivity efficiency spectral dependence.

Optical band gap E_G^{Opt}, photoconductivity threshold value E_{th} and reduction-oxidation potential U_{redox}, determined from cyclic voltamperogramme, for investigated compounds are presented in Table 4.

	E_G^{Opt} (eV)	E_{th} (eV)	U_{redox} (V)
DWK-1	2.20	1.92	2.35
DWK-2	2.10	-	1.99
JWK-1	2.08	1.78	2.01
JWK-2	1.88	1.62	1.90
ZWK-1	2.08	1.78	2.04
ZWK-2	1.96	1.68	2.00

Table 4. Optical band gap E_G^{Opt}, photoconductivity threshold value E_{th} and red-ox potential U_{redox} for the compunds **DWK-1, DWK-2, JWK-1, JWK-2, ZWK-1, ZWK-2.**

According to Table 4, the redox potential of **DWK**, **JWK** and **ZWK** is higher for compounds with one electron donor group compared to compounds with two electron donor groups (see Fig.17). The same relation is found for optical band gap as well.

The photoconductivity threshold value cannot be obtained for **DWK-2** thin films due to the low value of photocurrent. For other compounds we obtain an excellent linear correlation between optical band gaps and photoconductivity threshold values with correlation coefficient 0.993. The slope of this linear relation is found to be 1 and intercept 0.28 as shown in Fig.31

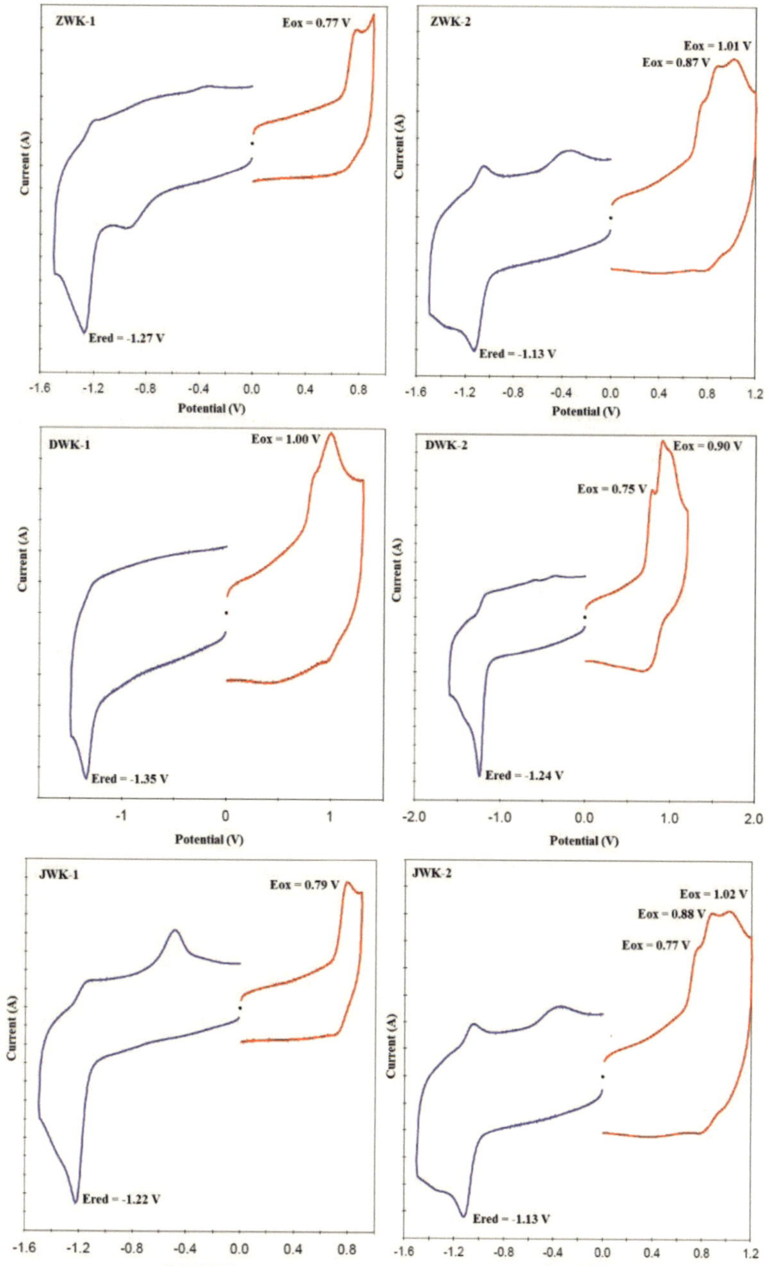

Figure 30. Cyclic voltamperogramme curves of compounds **WK-1** and **WK-2**. Posistive values are oxidation potential and negative values reduction potential.

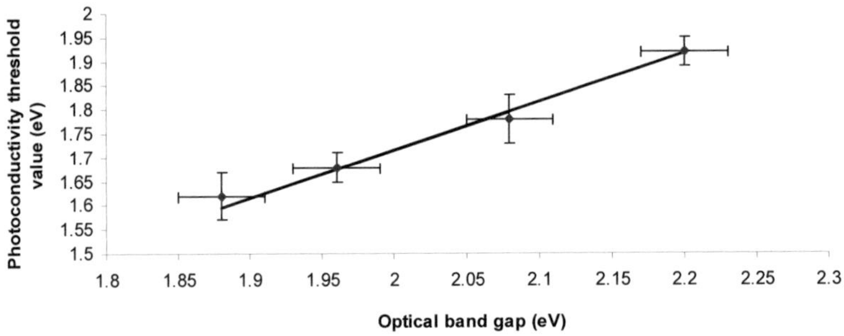

Figure 31. Linear correlation between optical band gap and photoconductivity threshold value. Line is the best fit with slope coefficient one.

The energy of the photoconductivity threshold is defined as the difference between the conduction levels of holes and electrons [51]. The value of the intercept implies that the optical band gap is 0.28 eV larger than the difference between the conduction levels of holes and electrons. It shows a constant energy difference between optical band gap and adiabatic gap despite the various molecule structures.

3.8. Electrical properties

Electrical properties of **WK-1** and **WK-2** compounds are investigated in the regime of space charge limited current (SCLC) [52-54]. Similar sandwich type samples as used for photoelectrical measurements are prepared for this study as well. The thickness of the organic thin film is at least 500 nm.

The current-voltage characteristics of compounds **ZWK-1**, **ZWK-2**, **JWK-1**, **JWK-2** and **DWK-2** in thin solid films are shown in Fig.32.

The current-voltage characteristic of **DWK-1** films could not be measured due to unstable current. This may be due to formation of small crystallites (see Fig.21) around 1 μm in size. Such aggregates are found throughout the sample and induce instability in the current. In all other cases the current-voltage characteristics have similar shapes with three regions. In the first region, 0-2 volts, the current is found to depend linearly on voltage. In the second range, 2 to 50 volts the current increases superlinearly with voltage, following Child's law. In the third region, > 50 V, the current depends on voltage to the power of at least ten, which may be attributable to charge trapping in the local trap states. More details of this aspect will be discussed further below.

Usually the work function of ITO should be near the ionisation energy level of the organic compound while that of aluminium (Al) should be around the middle of the energy gap. This provides efficient hole injection from ITO and electron injection from aluminium when a positive voltage is applied to ITO. Holes may also be injected from the aluminium when positive voltage applied to it. Electron injection may be more difficult in the second case due to the large difference between the ITO work function and electron affinity potential of the organic compound. This is confirmed by the current voltage characteristics shown in Fig.32. A similar current is observed at the lower voltage where only holes are injected either from ITO or aluminium when biased with a positive voltage. At higher voltage current is higher when ITO is positive in comparison with positive aluminium.

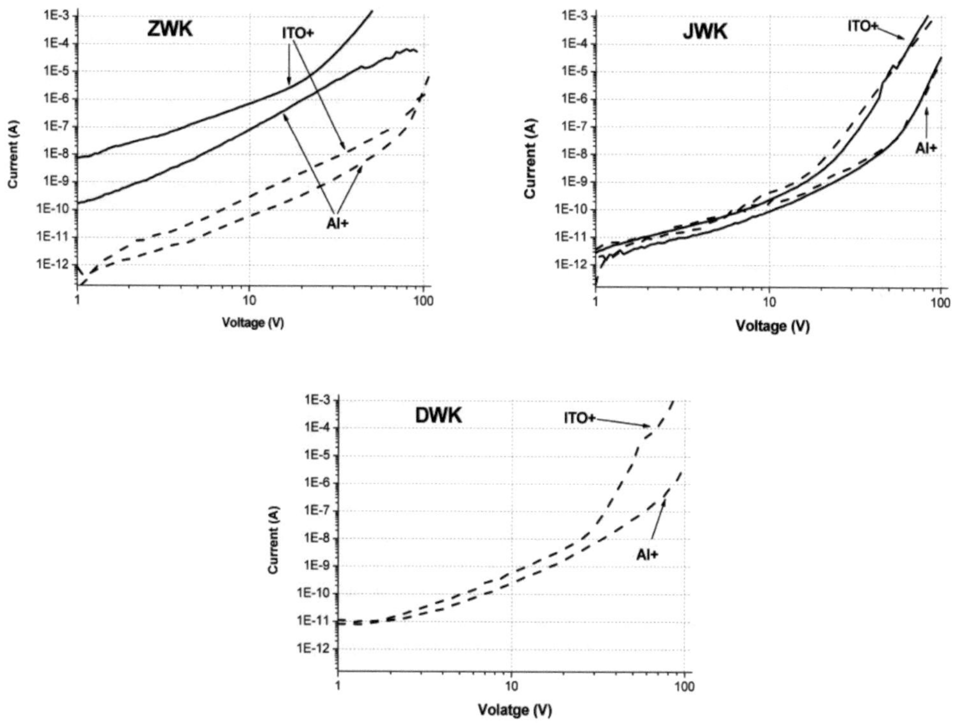

Figure 32. Current-voltage characteristics of pure thin films of **ZWK**, **JWK**, **DWK** compounds. Solid line – compounds with one electron donor group, dashed line - compounds with two electron donor group.

The temperature modulated space charge limit current (TM SCLC) method is used to analyse the charge carrier local trapping states in solid films [55]. The condition for using this method (TM SCLC) is monopolar injection, which is achieved in our case when a positive voltage is applied to the aluminium electrode. The measured activation energy is plotted as a function of the applied voltage for the investigated compounds as shown in Fig.33.

No charge carrier local trap states are found in films of compounds with one electron donor group due to only one plateau which reaches zero. All compounds with two electron donor groups are found to have charge carrier trap states. The additional plateau of activation energy, which can be clearly seen from Fig.33 means that the thin films contain local trap states. The hole shallow trap depths are found to be 0.1, 0.24 and 0.3 eV in **ZWK-2 JWK-2** and **DWK-2,** respectively. Such trap states decrease the efficiency of electroluminescence and should be avoided in fabricating high efficiency light emitting diodes. The activation energy increases at lower voltage for compounds **JWK-1**, **JWK-2** and **DWK-2**. This is indicative of a contact problem where the electrode – organic interface also works as additional charge carrier traps.

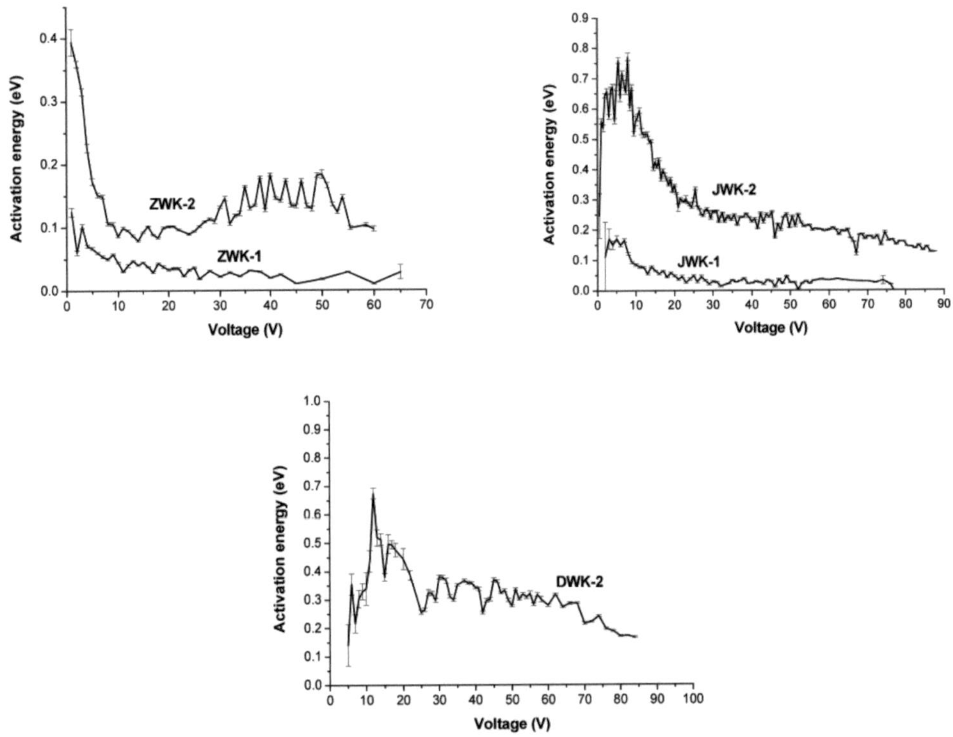

Figure 33. Activation energy dependence on applied voltage of the investigated compounds in solid films. Positive voltage was applied to aluminium electrode.

3.9. Electroluminescence of ZWK-1 and ZWK-2

A multilayer structure is used for electroluminescence (EL) measurements. Polyethylenedioxythiophenne:polystyrenesulfonate (PEDOT:PSS) (from H.C. Starck) is used as the hole injection layer and LiF as electron injection layer. PEDOT:PSS and organic compounds are sequentially spin coated on ITO glass. Then LiF and Al are thermally evaporated in vacuum. The final structure of the device has a structure of ITO/PEDOT:PSS(40nm)/ZWK1 or ZWK-2(~90nm)/LiF(1nm)/Al(100nm) and is not encapsulated.

The EL spectrum of the device is estimated in International Commission on Illumination (CIE) coordinates: x=0.65 and y=0.34 for **ZWK-1** and x=0.64 and y=0.36 for **ZWK-2**. The spectral maximum peak is observed at 667 nm and 705 nm in **ZWK-1** and **ZWK-2**, respectively, as shown in Fig.34. These peaks are slightly red shifted compared with those of PL spectrum of **ZWK-1** and **ZWK-2** thin films. This red shift may be attributed to the interaction of molecules and injected charges.

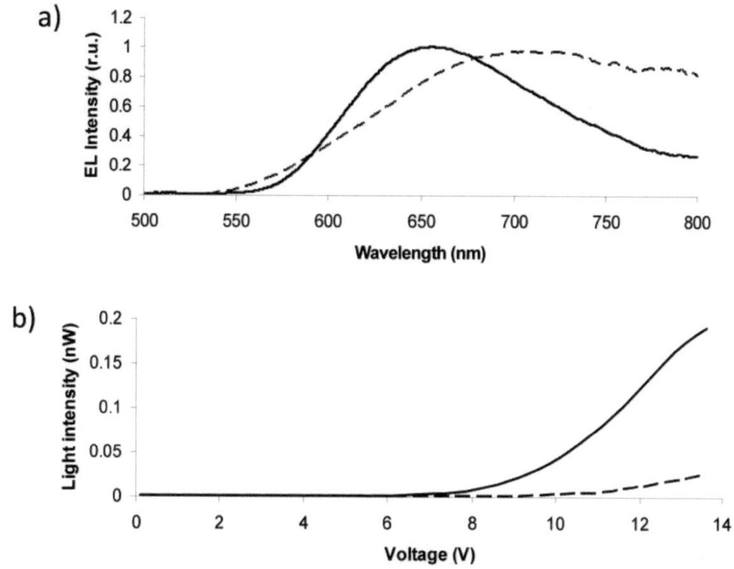

Figure 34. a) Electroluminescence spectrum and b) light intensity dependence on voltage of **ZWK-1** (line) and **ZWK-2** compounds (doted line)

The light emission is observed at 6 V in the electroluminescent device with **ZWK-1** molecules and 9 V in with **ZWK-2** molecules. The light intensity is one order less in **ZWK-2** molecules compared to that in **ZWK-1**. This may be due to the lower PLQY and shallow charge carrier trap states in **ZWK-2**.

4. Conclusions

The absorption and emission bands of the synthesized pyranylidene type compounds **ZWK-1, DWK-1, JWK-1** are comparable with those of other already known one electron donor fragment **DCM** and benzopyran type derivatives of pyranylidene within the spectral region studied here. Similar conclusions can be drawn about **ZWK-2, DWK-2, JWK-2,** which have similar properties to **IWK** and two other already known electron donnor group containing derivatives of pyranylidene. These properties are also similar to those of one electron donor fragment chromene red-emitters. However, incorporation of bulky trityloxy groups in such molecules not only enchances glass transition temperatures by 5° to 20°C compared to previously published pyranylidene type compounds containing one and two electron donor groups, but also enables the formation of a glassy structure in the solid state from volatile organic solvents. In addition, no glass transition values have been observed so far for low molecular mass isophorene type compounds. The photoluminescence quantum yield of investigated molecules in solution is up to 0.54 and is also comparable with the quantum yield of pyranylidene and isophorene derivatives already reported. Most of the thin solid films obtained from **WK-1, WK-2** have almost no crystals in the sample. Newertheless the photoluminescence quantum yield is reduced by one order of magnitude due to the closer intermolecular distance between molecules, resulting in strong excitonic interaction.

Emission from the **IWK** film is too weak to detect, which may be attributed to the higher photoluminescence quenching in **IWK** than in glassy pyranylidene films. However, using the doping approach, the compounds we have introduced enable up to 3 times higher doping concentration without losing optical properties compared to other already known red-emitters.

Four investigated compounds - **ZWK-1**, **JWK-2**, **DWK-1** and **DWK-2** show amplified spontaneous emission from pure solid films. Obtained threshold values are larger than those previously reported, but it should be mentioned that for pyranylidene type compounds, amplified spontaneous emission has been observed only in the doped systems until now.

Electrical properties are found to be better in compounds with one electron donor group due to absence of local trap states in their thin films. In the case of molecules with two electron donor groups shallow hole trap states have been observed, which may decrease efficiency of electroluminescence and should therefore be avoided in fabricating high efficiency light emitting diodes.

Even though we are able to prevent pyranylidene and isophorene type red-emitters from self crystallization in the solid state, their concentration in the emission layer would still be limited due to photoluminescence quenching caused by the short distance between molecules. Nevertheless, the glass materials can still be used not only as dopants for OLED applications, but also for lasing applications. Good thermal properties present a possibility of using them also for nonlinear optical (NLO) property studies.

Author details

Elmars Zarins and Valdis Kokars
Institute of Applied Chemistry, Riga Technical University, Riga, Latvia

Aivars Vembris and Inta Muzikante
Institute of Solid State Physics, University of Latvia, Riga, Latvia

Acknowledgement

This work has been supported by the European Social Fund within the project «Support for the implementation of doctoral studies at Riga Technical University» and «Support for Doctoral Studies at University of Latvia» and by Latvian State Research Programm No.2 in Materials Sciences and Information Technologies. Authors are grateful to Janis Jubles for assistance on reactant synthesis, Kristine Lazdovica for absorption and emission measurements in solutions and conducting thermogravimetric analysis, Dr. chem. Baiba Turofska for electrochemical measurements, Kaspars Pudzs for activation energy measurements, Raitis Grzibovskis for photoelectrical measurements, Dr. phys. Saulius Jursenas for photoluminescence quantum yield measurements, Dr. phys. Vidmantas Gulbinas for useful discussion and Dr. phys. Mikelis Svilans for providing language help.

5. References

[1] L.S. Hung, C.H. Chen, Recent progress on molecular organic electroluminescent materials and devices. Mater. Sci. Eng:R, 39, 143-222, (2002).

[2] P. Strohriegel, J.V. Grazulevicius, Charge-transporting molecular glasses. Adv. Mater., 14, 1439-1452, (2002).

[3] D.M. Bassani, L. Jonusauskaite, A. Lavie-Cambot, N.D. McClenaghan, J.L. Pozzo, D. Ray, G. Vives, Harnessing supramolecular interactions in organic solid-state devices: Current status and future potential. Coord. Chem. Rev., 254, 2429-2445, (2010).

[4] Y. Shirota, Y. Kuwabara, D. Okuda, R. Okuda, H. Ogawa, H. Inada, T. Wakimoto, H. Nakada, S. Yonemoto Y, Kawami, Starburst molecules based on π-electron systems as materials for organic electroluminescent devices. J. Lumin., 72-74, 985-991, (1997).

[5] L. Yang, M. Guan, D. Nie, B. Lou, Z. Liu, Z. Bian, J. Bian, C. Huang, Efficient, saturated red electroluminescent devices with modified pyran-containing emitters. Opt. Mater., 29, 1672-1679, (2007).

[6] G. Kwak, S. Wang, M.S. Choi, H. Kim, K.H. Choi, Y.S. Han, Y. Hur, S.H. Kim, 2D-π-A type Pyran based Dye Derivatives: Photophysical Properties Related to Intermolecular Charge Transfer and their Electroluminiscence Application. Dyes and Pigments. 78, 25-33, (2008).

[7] Y.S. Yao, J. Xiao, X.S. Wang, Z.B. Deng, B.W. Zhang, Starburst DCM-Type Red-Light-Emitting Materials for Electroluminescence Applications. Adv. Funct. Mater., 16, 709-718, (2006).

[8] C.Q. Ma, Z. Liang, X.S. Wang, B.W. Zhang, Y. Cao, L.D. Wang, Y. Qiu, A novel family of red fluorescent materials for organic light-emitting diodes. Synth. Met., 138, 537-542, (2003).

[9] F.G. Webster, W.C. McColgin, US patent 3 852 683, (1974).

[10] C.W. Tang, S.A. VanSlyke, C.H. Chen, Electroluminescent of doped organic film. J. Appl. Phys., 65, 3610-3616, (1989).

[11] C.H. Chen, C.W. Tang, J. Shi, K.P. Klubek, Recent developments in the synthesis of red dopants for Alq3 hosted electroluminescence. Thin Solid Films, 363, 327-331, (2000).

[12] Y.J. Chang, T.J. Chow, Highly efficient red fluoroscent dyes for organic light-emitting diodes. J. Mater. Chem., 21, 3091-3099, (2011).

[13] S. Wang, S.H. Kim, New solvatohromic merocyanine dyes based on Barbituric acid Meldrum's acid. Dyes and Pigments, 80, 314-320, (2009).

[14] D.U. Kim, S.H. Paik, S.H. Kim, Y.H. Tak, Y.S. Han, S.D. Kim, K.B. Kim, H.J. Ju, T.J. Kim, Electro-optical characteristics of indandione-pyran derivatives as red emission dopants in electroluminescent device. *Materials Science and Engineering: C,* 24, 147-149, (2004).

[15] C. Jianzhong, H.J. Suh, S.H. Kim, Synthesis and properties of conjugated copolymers with 2-pyran-4-ylidene malononitrile. Dyes and Pigments. 68, 75-77, (2006).

[16] D.H. Hwang, J.D. Lee, M.J. Lee, C. Lee, Organic light-emitting diode using a new DCM derivative as an efficient orange-red doping molecule. Curr. Appl. Phys., 5, 244-248, (2005).

[17] R. Andreu, L. Carrasquer, J. Garin, M.J. Modrego, J. Orduna, R. Alicante, B. Vilcampa, M. Allian, New one- and two-dimensional *4H*-pyranylidene NLO-phores. Tetrahedron Lett., 50, 2920-2924, (2009).

[18] R. Andreu, S. Franco, E. Galin, J. Garin, M.N. De Baroja, C. Momblona, J. Orduna, R. Alicante, B. Villacampa, Isophorone- and pyran-containing NLO-chromophores: a comparative study. Tetrahedron Lett., 51, 3662–3665, (2010).

[19] S. Wang, S.H. Kim, Photophysical and electrochemical properties of D-π-A type solvatofluorchromic isophorone dye for pH molecular switch. Curr. Appl. Phys., 9, 783-787, (2009).

[20] D.Y. Do, S.K. Park, J.J. Ju, S. Park, M.H. Lee, Nonlinear optical polyimides with various substituents on chromophores: synthesis and glass transition temperature. Opt. Mater., 26, 223-229, (2004).

[21] P.J. Kim, O.P. Kwon, M. Jazbinsek, H. Yun, P. Gunter, The influence of pyrrole linked to the π-conjugated polyene on crystal characteristics and polymorphism. Dyes and Pigments, 86, 149-154, (2010).

[22] P.Y. Chen, Y.U. Herng, M. Yokoyama, New double graded structure for enhanced performance in white organic light emitting diode. J. Lumin., 130, 1764–1767, (2010).

[23] X.H. Zhang, B.J. Chen, X.Q. Lin, O.Y. Wong, C.S. Lee, H.L. Kwong, S.T. Lee, S.K. Wuu, A New Family of Red Dopants Based on Chromene-Containing Compounds for Organic Electroluminescent Devices. Chem. Mater., 13, 1565-1569, (2001).

[24] Z. Guo, W. Zhu, H. Tian, Dicyanomehtylene-*4H*-pyran chromphores for OLED emitters, logic gates and optical chemosensors. J. Mater. Chem., DOI: 10.1039/c2cc31581e, (2012).

[25] P. Zhao, H. Tang, Q. Zhang, Y. Pi, M. Xu, R. Sun, W. Zhu, The facile synthesis and high efficiency of the red electroluminescent dopant DCINB: A promising alternative to DCJTB. Dyes and Pigments, 82, 316–321, (2009).

[26] C.J. Huang, C.C. Kang, T.C. Lee, W.R. Chen, T.H. Meen, Improving the color purity and efficiency of blue organic light-emitting diodes (BOLED) by adding hole-blocking layer. J. Lumin., 129, 1292–1297, (2009).

[27] H. Fukagawa, K. Watanabe, S. Tokito, Efficient white organic light emitting diodes with solution processed and vacuum deposited emitting layers. Org. Electron., 10, 798–802, (2009).

[28] A. Vembris, M. Porozovs, I. Muzikante, J. Latvels, A. Sarakovskis, V. Kokars, E. Zarins, Novel amourphous red electroluminescence material based on pyranylidene indene-1,3-dione. Latvian J. Phys. Tech. Sci., 47(3), 23-30, (2010).

[29] A. Vembris, M. Porozovs, I. Muzikante, V. Kokars, E. Zarins, Pyranylidene indene-1,3-dione derivatives as an amorphous red electroluminescence material. J. Photon. Energy, 1, 011001-1-011001-8, (2011).

[30] E. Zarins, J. Jubels, V. Kokars, Synthesis of red luminescent non symmetric styryl-4H-pyran-4-ylidene fragment containing derivatives for organic light-emitting diodes. Adv. Mater. Res., 222, 271-274, (2011).

[31] E. Zarins, V. Kokars, M. Utinans, Synthesis and properties of red luminescent 2-(3-(4-(bis(2-(trityloxy)ethyl)amino)styryl)-5,5-dimethylcyclohex-2-enylidene) malononitrile for organic light-emitting diodes. IOP Conf. Ser.: Mater. Sci. Eng. 2, 012-019, (2011).

[32] A. Vembris, E. Zarins, J. Jubels, V. Kokars, I. Muzikante, A. Miasjedovas, J. Saulius, Thermal and optical properties of red luminescent glass forming symmetric and non symmetric styryl-4H-pyran-4- ylidene fragment containing derivatives. Opt. Mater., 34, 1501-1506, (2012).

[33] R.H. Wiley, C.H. Jarboe, H.G. Ellert, Substituted 3-cinnamoyl-4-hydroxy-6-methyl-2-pyrones from dehydroacetic acid. J. Am. Chem. Soc., 77, 5102-5105, (1955).

[34] I.P. Lokot, F.S. Pashkovsky, F.A. Lakhvich, A New Approach to the Synthesis of 3.6- and 5,6-DIalyl Derivatives of 4-Hydroxy-2-pyrone. Synthesis or *rac*-Germicidin. Tetrahedron, 55, 4783-4792, (1999).

[35] M.L. Miles, C.R. Hauser, 2-(p-METOXYPHENYL)-6-PHENYL-4-PYRONE. Org. Synth. Coll., 5, 721 and 46, 60. (1973 and 1966). http://www.orgsyn.org/orgsyn/pdfs/CV5P0721.pdf.

[36] N.S. Vulfson, E.V. Sevenkova, L.B. Senyavina, Condensation of dehydroacetic acid with aromatic aldehydes. Rus. Chem. Bull., 15(9), 1541-1546, (1964).

[37] E. Zarins, V. Kokars, A. Ozols, P. Augustovs, Synthesis and properties of 1,3-dioxo-1*H*-inden-2(3*H*)- ylidene fragment and (3-(dicyanomethylene)-5,5 dimethylcyclohex-1-enyl)vinyl fragment containing derivatives of azobenzene for holographic recording materials. Proc. of SPIE, 8074, 80740E-1-80740E- 6, (2011).

[38] C. Jianzhong, H.J. Suh, S.H. Kim, Synthesis and properties of conjugated copolymers with 2- pyran-4-ylidene malononitrile. Dyes and Pigments, 68, 75-77, (2006).

[39] Z.R. Grabowski, K. Rotkiewicz, W. Rettig, Structural Changes Accompanying Intramolecular Electron Transfer: Focus on Twisted Intramolecular Charge-Transfer States and Structures. Chem. Rev., 103, 3899-4031, (2003).

[40] V.A. Pomogaev, V.A. Svetlichnyi, A.V. Pomogaev, N.N. Svetlichnaya, T.N. Kopylova, Theoretic and Experimental Study of Photoprocesses in Substituted 4-Dicyanomethylene-4H-pyrans. High Energy Chemistry, 39, 403-407, (2005).

[41] J.C. Mello, H.F. Wittmann, R.H. Friend, An improved experimental determination of external photolumiescence quantum efficiency. Adv. Mater. 9, 230-232, (1997).

[42] P.R. Hammond, Laser dye DCM, its spectral properties, synthesis and comparsion with other dyes in the red. Opt. Commun., 29, 331-33, (1979).

[43] S.I. Bondarev, V.N. Knyukshto, V.I. Stepuro, A.P. Stupak, A.A. Turbanov, Fluorescence and Electronic Structure of the Laser Dye DCM in Solutions and in Polymethylmetacrylate. J. Appl. Spectrosc., 71, 194-201, (2004).

[44] R. Kapricz, V. Getautis, K. Kazlauskas, S. Juršenas, V. Gulbinas, Multicoordinational excited state twisting of indan-1,3-dione derivatives. Chem. Phys., 351, 147-153, (2008).

[45] V.G. Kozlov, S.R. Forrest, Lasing action in organic semiconductor thin films. Current opinion in Solid State and Materials Science 4(2), 203-208, (1999).

[46] A. Vembris, I. Muzikante, R. Karpicz, G. Sliauzys, A. Misajedovas, S. Jursenas, V. Gulbinas, Fluorescence and amplified spontaneous emission of glass forming

compounds containing styryl-4H- pyran-4-ylidene fragment. J. Lumin., 132, 2421-2426, (2012).

[47] E.M. Calzado, J.M. Villalvilla, P.G. Boj, J.A. Quintana, R. Gomez, J.L. Segura M.A. Diaz Garcia, Amplified spontaneous emission in polymer films doped with a perylediimide derivative. Appl. Optic. 46, 3836-3842, (2007).

[48] J.Y. Li, F. Laquai, G. Wegner, Amplified spontaneous emission in optically pumped pure films of a polyfluorene derivative. Chem. Phys. Lett., 478, 37-41, (2009).

[49] E.A. Silinsh, V.V. Capek, Organic Melecular Crystals. Interaction, localization and transport phenomena (AIP Press, New York), (1994).

[50] E.A. Silinsh, M. Bouvet, J. Simon, Molecular Materials 51, (1995).

[51] E.A. Silinsh, Organic Molecular Crystals Their Electronic States. Springer Series in Solid-State Sciences 16, Berlin, (1980).

[52] A. Rose, Space-Charge-Limited Currents in Solids. Phys. Rev. 97, 1538-1655, (1955).

[53] A. Rose, Concepts in Photoconductivity and Allied Problems. Interscience, New York, (1967).

[54] M.A. Lampert, P. Mark, Current Injectio in Solids. Academic Press, New York, (1970).

[55] S. Nešpůrek, O. Zmeškal, J. Sworakowski, Space-charge-limited currents in organic films: Some open problems. Thin Solid Films, 516, 8949, (2008).